A Retrospective Technology Assessment: Submarine Telegraphy

The Transatlantic Cable of 1866

Prepared by:
The George Washington University
Program of Policy Studies in Science and Technology

Vary T. Coates, Ph.D.,
and Bernard Finn, Ph.D.
(Curator, Division of Electricity,
National Museum of History and Technology)

Co-principal Investigators:
Thomas Jaras, Ph.D., Research Associate
Henry Hitchcock, Research Associate
Robert Anthony, Ph.D., Consultant

Est. 1959

San Francisco Press, Inc.

547 Howard Street, San Francisco, California 94105

547 Howard Street, San Francisco, CA 94105

Printed in the U.S.A.

Library of Congress Catalog Card No. 78-65157
ISBN 0-911302-39-5

Preface

The Retrospective Technology Assessment of Submarine Telegraphy was one of four retrospective technology assessments sponsored by the National Science Foundation (NSF) in 1975 and 1976. These studies are intellectual experiments; they were to test the proposition that studies of the past can guide and improve studies of the future.

The Introduction to this report explains for the reader who may be unfamiliar with these terms what is meant by ''anticipatory technology assessment'' and ''retrospective technology assessment,'' and the purpose of these studies.

Chapter 6 summarizes our findings and conclusions, and supplements the brief summary introduction. *The reader who is seeking a short version of our report, therefore, is advised to read the Introduction and the final Chap. 6.*

The Research Team

The Principal Investigators for this study were Dr. Vary T. Coates, Associate Director of the Program of Policy Studies in Science and Technology at The George Washington University, and Dr. Bernard Finn, Curator for Electricity at the National Museum of History and Technology of the Smithsonian Institution.

Other members of the research team were part of the Technology Assessment Group of the Program of Policy Studies:

Dr. Robert Anthony, physicist and systems analyst
Dr. Thomas Jaras, historian of science
Mr. Henry Hitchcock, political scientist
Mr. Phillip Vernon, attorney
Mr. James Stice, summer intern

They were assisted by several consultants and by a distinguished advisory committee.

The interdisciplinary team worked closely together on all aspects of the study. Mr. Hitchcock took the lead in preparing the material and the initial writing of Chap. 1, portions of Chap. 2, and Chap. 4, the social history surrounding the advent of submarine telegraphy. Dr. Anthony was responsible for the scientific and technical analysis, especially in Chaps. 2 and 5, and for analysis of the Delphi survey. Dr. Jaras took the lead in the search for impacts of submarine telegraphy and the preparation of Chap. 4 summarizing those impacts. Dr. Finn, as co-principal investigator, provided overall guidance for the historical research. Dr. Coates planned and managed the assessment, prepared the introduction and final chapter, and was responsible for integrating and editing this report.

The assessment team wishes to acknowledge with gratitude the contributions made by Dr. G. Patrick Johnson, NSF program manager; by Mr. Joseph Coates, who as his predecessor at NSF conceived the idea of retrospective assessments, and later reviewed and commented on several drafts of our report; by Dr. Howard Moon, consultant on military history; by all who participated in our workshops and Delphi; and particularly by our advisory committee, listed below.

Mrs. Gertrude Delman was our strong support during most of this study and prepared much of the final manuscript; her contribution was invaluable. To the deep sorrow of all who knew her, Mrs. Delman died before the completion of the report. To Ms. Terry Parsons and Ms. Alexandria Minetto, who came competently to our rescue, we also express our appreciation.

Vary T. Coates

Bernard Finn

Washington, D.C., 1979

Advisory Committee

Dr. Raymond Bowers, Director
Program on Science, Technology, and Society
Cornell University
Ithaca, New York

Dr. James E. Brittain
Social Science Department
Georgia Institute of Technology
Atlanta, Georgia

Mr. Robert C. Cooper
Deputy Director—Europe
AT&T Long Lines
New York, New York

Professor Robert W. Kenny
Department of History
The George Washington University
Washington, D.C.

Professor Richard Sylla
Department of Economics
North Carolina State University
Raleigh, North Carolina

Dr. Ward Stoneman
Quarry Hill Associates
Arlington, Virginia

Dr. Ronald Overmann
History and Philosophy of Science Program
National Science Foundation
Washington, D.C.

Table of Contents

SUMMARY

Introduction

Retrospective technology assessments are historical case studies of the development of a new technology and the consequences of that development for society. Retrospective assessments are undertaken in the hope of providing new insights into the relationship between technological change and social change, and thus improving the anticipatory assessment of technologies emerging today and in the future.

Submarine telegraphy, beginning with the transatlantic cable of 1866, provided society with a significant new capability: rapid communication between continents. For fifty years, no other technology could provide this capability—one that had dramatic and irreversible consequences and went a long way toward creating today's world.

The Story of the Cable

Cyrus Field, American entrepreneur, began the effort to lay a transatlantic cable in 1854, a decade after development of landline telegraphy. After an initial failure in 1857 a cable was successfully laid in 1858 but failed after a month. The American Civil War and the necessity of raising capital and developing political support delayed further attempts until 1865 (another failure). In 1866, when the cable had become a predominantly British enterprise, a successful transatlantic cable was laid. During the next two decades submarine telegraphy created a global communications network tying together all parts of the world.

Challenge and Response

The initial phases of creating a new technology can have important impacts on society long before the technology becomes fully operational and widely disseminated, as new legal and social institutions are created to embody the technology and technical, entrepreneurial, and policy-related challenges are met and overcome.

The solution of technical problems in submarine telegraphy contributed greatly to the nascent trend toward application of basic science to industry.

The challenge of raising capital for an enterprise that was unprecedented and highly risky and the successive competitive challenges to the endeavor drove toward the creation of multinational monopolistic organizations. The greater availability of risk capital in Britain and the greater depth of scientific, managerial, and financial expertise in that country caused the control of the enterprise to shift from American to British entrepreneurs. This shift prefigures the problems faced by developing countries today vis-à-vis nations with more mature scientific and industrial establishments.

In the early development of submarine telegraphy, potential public benefits and private-sector motivation were highly congruent. However, requests for governmental support in both America and Britain met with strong existing policies of limited intervention in the market and preservation of free enterprise. Conservative public policies of limited technical assistance and conditional guarantees, with no risk of public funds, were appropriate and successful in this case, but leave unanswered the question of whether similar policies are adequate today for emerging technologies that promise significant societal benefits.

Potential social costs, such as the creation of new vulnerabilities in time of war and the necessity of control over information, were dimly perceived but left for future corrective action.

How Cable Communication Changed the World

Communication is so basic to the functioning of organized society that the contribution of submarine telegraphy to creating a global economy and a global polity can hardly be overestimated. However, in bringing about these consequences submarine telegraphy interacted symbiotically with the development of steam shipping and the slightly older technologies of landline telegraphy and rail transportation, as well as with continuing industrialization and urbanization.

The significant and long-lasting effects of submarine telegraphy derived not from the physical characteristics of the technology itself, but from the new capability it provided.

Cable communications—combined with the swifter, more reliable, and more extensive transportation provided by steamships—made the efficient coordination and direction of the purchase, transport, routing, insuring, receipt, and distribution of goods possible and thus created the vast and complex network through which modern international commerce is carried on.

The venture system of commerce was replaced by futures trading of commodities and stocks and led to an organized system of international markets. Fluctuation of prices was greatly reduced. A system of grading and standardizing commodities developed. Commodity markets moved from centralized consumer areas to the world's producing areas. Middleman costs were significantly reduced; so were merchants' inventories. Arbitrage developed as an important mechanism in leveling prices across world markets. New York became, along with London, a focus of international finance. New mechanisms for commercial credit were quickly created, fluctuation in sterling exchange rates was reduced, and the national currencies became highly interdependent. The natural monopolies enjoyed by large merchant houses, especially the British, eroded, and competition in world markets was enormously increased.

Cable communications stimulated both the development and the standardization of equipment and techniques for weather forecasting and international cooperation in this important area.

Organizations for the efficient collection and distribution of international news were created very quickly, a development that contributed greatly to the expansion of the public press that was already under way. The enormous growth in volume, scope, regularity, and speed of news transmission fed on itself and created an irresistible public hunger for information, especially evident in national policy and the conduct of war. Public attitudes now had an immediate and powerful impact on official actions; information reaching the domestic public also reached the enemy; and modern techniques of propaganda and of censorship developed.

Changes in diplomatic institutions and practices were somewhat slower than changes in economic institutions, perhaps because diplomacy is an area where judgment and experience are the stock in trade, objectives are conserving and protecting, and the substance of communication is often more important than its speed. The most immediate consequence was the creation of constant readiness, or 24-hour-duty operations, since the ability to respond quickly to crisis creates the necessity of doing so. Somewhat more slowly, authority and responsibility became centralized, as the constant flow of information from around the globe had to be evaluated and integrated into decisions. National headquarters for foreign affairs and colonial affairs expanded and—after such systems for a time appeared on the verge of breakdown from an overload of information—specialization and expertise of staffs greatly increased, and decisionmaking became much more complex and bureaucratized.

International organizations were created for continuing cooperation in problem areas such as world health, criminal justice, and disaster relief. Geographically remote areas of the world became involved in international affairs. It became necessary to deal with local problems and crises in the context of international interests and problems. New interdependencies between nations created new political vulnerabilities.

The status of submarine cables in peacetime was embodied in international law in 1884; despite continuing efforts no international agreements to protect cables in time of war could be reached. The right to lay cables with the agreement of both countries to be connected and their safety from interference in time of peace are firmly established as a part of the law of freedom of the seas. (In time of war, prevailing policy is that even cables belonging to a neutral but connected to the territory of a belligerent will be cut if "military necessity dictates," that is, if the neutral's presumed right to communicate with a belligerent is "misused" to provide the belligerent with assistance or support.) The opportunity to create a new international right to free communication was rejected, even as applied to peacetime, by the international recognition in 1884 of the right of any country to intercept or censor messages crossing its land mass.

In military strategy, submarine telegraphy reinforced the importance of naval power. Together with landline telegraphy, it made possible the central control and direction of far-flung, highly mobile armed forces. Britain, for example, could now reduce and concentrate at strategic locations the forces required for protection of its colonies. More important, rapid communications created both the possibility of and the necessity for a state of continual readiness and the ability to mobilize a nation's military and industrial resources at short notice. By tremendously expanding the range and increasing the speed and dissemination of news of foreign and military events, rapid communications involved the general public in such affairs and created strong pressures and strong constraints on governments—creating modern techniques of military intelligence, spying, propaganda, and censorship. Although the impacts of improved transportation, industrialization, urbanization, weaponry, and bureaucratization converged in all these trends, communications was itself central to all these technologies. The seeds of Total Warfare may lie in the Napoleonic Wars, but it was electric communications that brought them to fruition.

Aside from minor and shortlived damage to local ecology at cable ends, the only significant impact derived from the physical characteristics of submarine telegraphy (as contrasted with the capabilities it provides) lies in the contributions made by its development to physics—for example, invention of the mirror galvanometer—to chemistry—development, much later, of synthetic substitutes for natural plastic insulation (gutta percha)—and the significant contributions that cable laying made to oceanography.

The most subtle but perhaps also one of the most important impacts of rapid communications was the general change in the perception of time. For the first time, it became significant that dawn in one country was midday, or midnight, in another. The ability to intervene in events occurring at a great distance, while they were still unfolding, created both opportunity and anxiety, and the pace of society—and of personal life itself—seemed to speed up. The simplest institutional response was the creation of international standards of time measurement and time zones. More generally, institutional response was a state of constant readiness to respond and react to distant events.

Hopes and Expectations

An interest in scientific prediction and forecasting was just beginning in the mid-19th century, growing out of the development of a half-century's systematic statistics collecting and early efforts to compile time-series data. This interest was primarily in the developing disciplines of economics and sociology. But most information was personal, proprietary, and not aggregated or systematically developed. A "technology assessment"—in the modern meaning of the word—was not performed in the mid-19th century, and would have been handicapped in the attempt to make systematic forecasts of social change by this lack of data

But there was a very large volume of social observation and comment concerning submarine telegraphy. In general, Victorians expected science and technology to produce Progress—advancement not only of material wellbeing but also of moral and religious principles, of democracy and equality, and of international brotherhood. There was, to be sure, a minor theme of regret for lost pastoral simplicity and natural goodness, but the dominant mood was one of optimism and exhilaration. Victorians particularly admired daring feats of entrepreneurship and challenge to the forces of nature. They applied all these general hopes and expectations to the transatlantic cable enterprise and expected that it would bring about great Progress—and the benefits of British and American civilization—to the entire world.

Most of the major impacts summarized above were anticipated by mid-19th century commentators. Impacts most often predicted were great increases in volume of commerce (both between Britain and America, and worldwide), an end to war (between Britain and America, and universally), better colonial management and closer ties between colonies and mother country, proliferation of news services and newspapers, the spread of religion (of the Christian Protestant variety), stimulation of scientific advancement, and general improvement of mankind through an increase of education and knowledge. There were occasional extreme prophecies, for example obsolescence of the postal service and of the diplomatic service. The one forecast most often and most fervently made was that communication would bring about understanding and hence love, brotherhood, and peace among nations.

There were occasionally dire warnings: Britain would control the cable to the commercial and military disadvantage of the USA; British colonies would never gain independence; only speculators and Wall Street financiers would benefit from the cable; news would degenerate into trivial gossip. But these forecasts were far outweighed in volume by the more optimistic forecasts.

But the clearest failure of foresight was the inattention to the consequences of submarine telegraphy for warfare. The social costs of continual threat and continual readiness, of enormously increased effectiveness and efficiency in social aggression, and of the dangers posed by public dissemination of information were at best only dimly perceived by a very few commentators. Whether the lag between the cumulative detrimental impacts of technological advance and the development of societal controls is inevitable, or whether such subtle detrimental impacts can be collectively anticipated and forestalled, remains a question today. But in the late 19th and 20th century only immediate and concrete risks and costs were assessed, and the fact that information creates vulnerabilities as well as opportunities was largely overlooked. For example, democratic nations, when suddenly faced with realization of danger, having prepared no alternative strategies, slid easily and without real protest into domestic and international censorship; and in more recent decades the interdependence of national currencies and markets cascaded local panics into worldwide depression.

Since the Victorian forecasts were unofficial, informal, nonaggregated, and almost always very general and abstract, and since they were never integrated or presented systematically to decisionmakers, we cannot tell whether decisionmakers (either investors, or members of Parliament or Congress) relied on such social commentary or simply shared the common hopes and expectations about the cable.

In the course of research presented in the report, we tried two experiments: role playing by groups of experts representing a "technology assessment team" of 1860, and a Delphi questionnaire that also used role-playing techniques. Participants generated a list of "potential" impacts that corresponded well with those identified in Victorian literature. They tended to assign highest "probabilities" to impacts that had the clearest and most immediate causal links to submarine telegraphy, especially economic impacts, and to assign lowest "probabilities" to impacts that could be attributed to the cable only through intervening events, trends, and technologies.

Cables in the Modern Era

Submarine cables proliferated around the world in a very few years. Not until the decade prior to World War I was there a competing technology (radio) and since World War II two additional competitors have appeared (submarine telephony and satellite communications). In the meantime the cable industry had grown into a multinational, multimillion-dollar, highly integrated giant. National policies toward the competing industries (for example, private vs public ownership of common carriers) were conditioned by considerations of invested capital and national security (maintaining economically viable alternative communication technologies).

Lessons Learned from Retrospective Technology Assessment

Despite profound changes in the nature of decisionmaking, the structure of national and world economies, and the nature and speed of technical innovations and their dissemination, there are illuminating analogies and lessons to be gleaned from historical case studies of technological change and social change. Much can be learned from analysis of the differences, as well as the similarities, between the political and social environments of past and present.

The difficulties encountered in performing retrospective technology assessment are of three kinds: those common to all historical research (incomplete and biased historical records, nonexistent or inaccessible data, difficulties of interpretation of motives, etc.); special problems arising from the anachronistic nature of the questions; and problems that are analogous to those of futures research. (The difficulty of constructing credible "alternative pasts," for example, is parallel to the problem of constructing credible futures scenarios.) None of these problems appears to be so critical as to defeat the purpose of doing retrospective assessment, although they may frustrate the attempt to use some of the more sophisticated analytical tools used in present-day analysis.

One historical case study can only be suggestive, not definitive. The advantage of retrospective assessment is the opportunity for comparative analysis of many technological developments. We therefore believe that retrospective technology assessment is a fertile field for research for technology assessors, an extremely useful structure for historians of science and technology, and an area well worth continued and expanded support from funding agencies.

On the basis of our assessment of one critically important technology, and extensive consideration of related research projects and related literature, we are prepared to advance the following conclusions, which we believe will be supported by further studies:

1. In many technological developments the most socially significant and long-lasting impacts will derive not from the physical characteristics of the technology but from the capabilities, services, and products it delivers. This feature should be considered both in choosing subjects for assessment and in allocating efforts within an assessment project.

2. In assigning priorities for allocating limited resources for anticipatory technology assessments, highest priority should be given to technologies that

(a) will provide new societal capabilities rather than incremental increases in existing capabilities or in the economy and efficiency of technologies providing those capabilities;

(b) are systems technologies, and therefore create new interdependencies between markets, institutions, social functions, communities, regions, or nations; and

(c) promise large public benefits while potential private sector benefits are balanced or outweighed by threats to existing, heavily capitalized technologies and industries.

3. It is important to assess the capacity of social institutions relevant to a technological development to adjust and accommodate to the new technology and new societal capability—including adjustments aimed at utilization of, support of, or defense against, the technology and resulting social change.

4. These institutional adjustments appear to follow a pattern: first, simple substitution of one technology for another with little change in activities and functions; second, adaptations in organizational behavior and operations to use the new technology more efficiently; and third, generation of new functions and activities made possible by the new technology.

5. Institutions differ in the speed and effectiveness with which they make such adjustments. In the case of submarine telegraphy our study suggests that the most rapid utilization of the technology, and the swiftest adjustments to it, were in institutions where

(a) organizational objectives were profit and/or growth;

(b) the organizational mode of operation was transactional and aimed at mutual benefit (exchange of cash, goods, ideas); and

(c) the organizational resources traditionally utilized to achieve objectives were economy, efficiency, and risk taking.

The slowest to achieve utilization and adjustment were institutions where

(a) organizational objectives were protection and preservation of existing advantages;

(b) the mode of operation was strategic, aimed at maximizing the benefits to one side; and

(c) organizational resources traditionally mobilized to achieve objectives were experience, judgment, and minimization of risk.

In the first category were merchants, financiers, and newsgatherers; in the second, diplomats and national armed forces.

We conclude that building a body of retrospective technology assessments holds significant promise for improving anticipatory assessments, by providing improved insight into the ways in which technological change may stimulate societal changes of many kinds. In this case, we have demonstrated that the provision of a hitherto nonexistent capability for society caused profound changes in the organization of human activity and in human perception of the world. Other technologies (the steam engine, electricity, the airplane) similarly created dramatically new capabilities and had profound impacts on society. We suggest that certain emerging technologies such as genetic engineering have the possibility of creating totally new human interventions into nature, and hence should have high priority for anticipatory assessment.

WHAT IS RETROSPECTIVE TECHNOLOGY ASSESSMENT?

A technological development such as the automobile, the airplane, or the mechanization of agriculture often changes the world in ways that were not fully anticipated. For example, farm tractors, reapers, and cultivators made it possible to feed the world's multitudes, and also transformed farming from family-owned into capital-intense businesses and sent farm workers streaming into the cities.

Technology Assessment is a type of policy analysis developed over the past decade to assist public decisionmakers in formulating policy about new technologies. It is a systematic attempt to anticipate the range of potential impacts of technology on the economy, the environment, social institutions, and behavior; and to lay out for the decisionmaker the implications of these possibilities in terms of public policies that may be necessary to assure that potential benefits are realized and equitably distributed, and that potential disbenefits or detrimental impacts are minimized, controlled, or avoided. Technology assessment is particularly concerned with the unintended and unplanned consequences of technological development, which are most likely to necessitate social interventions to help people and institutions adjust to dislocations and discontinuities that result from such changes.

Technology assessors make no claim to "predicting the future." They try rather better to inform the public, and the public's representatives, of the range of possible impacts and to illuminate as much as possible the uncertainties (both opportunities and hazards) that lie ahead.

Most new technologies are born in, and disseminated by, the private sector. Traditionally technological change is governed by the economy and the marketplace. When a technological innovation is conceived, the private-sector decisionmaker (an entrepreneur or a corporation) asks: can we make it? will it sell and produce a good return on investment? and (one hopes) is it safe?

But in modern industrial nations, governments increasingly intervene in this private-sector process. They act both to regulate private enterprise in the public interest and to stimulate technological development and innovation. As regulator, a government intervenes in the operations of the market (e.g., in antitrust legislation and in labor relations), and in the production and utilization of products to assure safety for consumers, workers, the general public, and the physical environment. Governments stimulate innovation through procurement for their internal operations or for national security or civilian public works. They can also stimulate development of products or services thought to be in the public interest (pharmaceuticals, new forms of energy, computers) through research and development, demonstration programs, tax incentives, and other kinds of subsidies.

Thus public policy and technological development constantly interact. Technology assessment is a way to help decisionmakers and the public (the ultimate decisionmaker) make wiser decisions about science and technology.

As a form of applied futures research, technology assessment suffers from a difficulty common to all futures studies: the long delay in self-correcting criticism of its conclusions. The impacts forecast as possible consequences of new technology are in the future; in the meantime we cannot know how accurately we forecast. This makes it difficult to work toward improving forecasts. Indeed, if technology assessment helps produce better decisions, the impacts of the technology will be affected accordingly—and we can then only conjecture how things would have turned out without such interventions.

We do know that social forecasting is handicapped because our understanding of the relationship between technological change and social change is still rudimentary. We need much more information and analysis about this reciprocal relationship before we can hope to develop a "general theory" or at least useful models that will help us forecast the consequences of new technology.

Development of general theories, or useful models, requires a reiterative and complex process of generating hypotheses, testing them in the real world, rejecting or revising them, postulating new explanations, examining them in the light of additional data, and so on. The time scale for social change makes this process almost impossible in terms of social forecasting. But case histories of technological change in the past may give us material with which to work. Careful analysis of the technological and social changes that have already occurred can, one hopes, help us to understand their interactions. Examination of the expectations that people had about common technologies when they were introduced, and identification of the assumptions on which those expectations were based, can give us some insight into the hazards of present-day forecasting. Such at least is the hope that led the National Science Foundation in 1975 to fund four "retrospective technology assessments," of which this study of submarine telegraphy is one.*

In choosing a subject for our retrospective assessment, we had several criteria in mind. We wanted to study an important technological development, one that is basic to the world as we know it, but recent enough so that we can understand what society was like before it developed. We wanted to study a technology that brought about significant changes in a wide range of institutions and behaviors. Therefore we looked for a technology that gave society a capability it did not possess before, a capability related to one of the basic functions of organized human society.

Communication is such a function. Before submarine telegraphy, and specifically before the transatlantic cable, "real-time" communication across oceans was impossible.

We also wanted to study a technology that originated in a unique and describable event in time, and that remained "unique" for a significant length of time so that we could hope to distinguish its consequences more clearly—a study of "before and after." It was nearly impossible to satisfy this criterion, since technological change has been so swift and all-pervasive over the last century or more. Submarine telegraphy came as close to meeting this requirement as any technology we could think of. The ability to talk back and forth between Europe and North America sprang into existence (after years of struggle and several failures) in August 1866, when the cable was laid. For about fifty years, until the development of radio, no other technology could provide this capability.

Other cables were soon laid and within a few years a network of cables spanned all continents, so our study necessarily deals with submarine telegraphy in general rather than with the transatlantic cable alone. The cables fed into landline telegraph systems constructed over the preceding quarter-century, which themselves had produced major impacts on society. Moreover, the transition from sail to steam shipping was made in the two decades after the cable was laid; the interaction of these two technologies was extremely important. Despite all these qualifications, submarine telegraphy offered an advantage for retrospective assessment over technologies which represented an improvement over existing technologies (such as steamships over sailing vessels) or those which (like automobiles and telephones) are adopted household by household, so that their impacts build gradually in proportion to their geographical dissemination.

Our study focused on three main objectives: to extract from historical records and from the welter of technological and social changes occurring in the late 19th century those which could be directly related to submarine telegraphy; to identify and understand the public and private decisions made about the cable and the way in which those decisions affected the technology and its social impacts; and to understand the expectations of those who observed the early development of submarine telegraphy, and the assumptions about the future and about social change that produced those expectations. (We speak here of expectations, rather than social forecasts, because it seems anachronistic to use the latter term to describe something that was informal, unsystematic, unintegrated, and only partially intended to influence decisionmaking.)

---

*The other three: telephone (MIT), management practices (Forecasting International), and wastewater technology (Carnegie-Mellon University); for a summary of the last, see J. A. Tarr, Ed., *Retrospective Technology Assessment—1976* (San Francisco Press, 1977, 165-190).

The basic method for this study was simply the use of traditional techniques of historical research and analysis. It differed from them only in that the questions we sought to answer were those posed by technology assessment: what were the primary, secondary (indirect and unplanned), and higher-order impacts of the cables on the way society is organized, uses its resources, and allocates benefits and disbenefits?

We ran into the difficulties that beset all historical research. Data we wanted were not collected and preserved because the questions in which observers are interested change with time. The bias represented by recorded data can only be estimated; for example, were the expectations expressed in speeches, private diaries, and published accounts really representative of the expectations (and fears) held by observers? How safe are we in postulating causal connections between historical trends, and how far can they be generalized? Such methodological and epistemological problems are common to historical analysis. Others closely paralleled the problems of anticipatory assessment. For example, we found that it was fully as difficult to postulate an alternative past (one, for example, in which submarine telegraphy was unknown in 1885 or 1900) as it is to generate credible scenarios of alternative futures. (Herman Kahn has observed that the only "believable" future is the one which is most unlikely to come about—continuation of things as they are.)

In the course of this study we also tried a few experimental techniques, or *Gedanken* experiments. In addition to the basic retrospective assessment (what were the impacts of submarine telegraphy?) and the "historical assessment" (what were the expectations about impacts of submarine telegraphy?) we also attempted an "hypothetical assessment." Here we asked the question, If an interdisciplinary research team had existed in 1864, asking the questions technology assessors ask today, but limited to the data available in 1864, what would their forecasts have been? To do so, we assembled in two workshops teams of experts and asked them to play the roles of technology assessors in 1864; and followed that up with a Delphi questionnaire. The results of these two experiments were more suggestive than informative, but (if our serious readers will forgive us the comment) they added to the pleasure and excitement of the research team in carrying out this study.

Perhaps we—the readers as well as the authors of this report—all are historians at heart. We are intrigued by a look back into the past to try to discern where we have come from, and into the future to try to anticipate where we are going. We invite the reader to wonder, along with us, at the ways in which submarine telegraphy changed our perception of the world we live in and the ways in which we try to control that world; and to question along with us what insights this history can offer about the changes that emerging technologies may set in motion.

Chapter 1

THE STORY OF THE CABLE

COMMUNICATION at a distance has been recognized as a need since early times. The development of written language only partly answered that need, since messages between distant points could be transmitted only as fast as a human being could travel. For most of history transmission of messages by successive relay of visual or audible signals was the only solution.* Many cultures developed systems using relay of drum signals. Smoke signals were used by the American Indians. In 1588 news of the defeat of the Spanish Armada was sent to London from the coast, 200 miles away, by a series of bonfires lit sequentially on hilltops and towers about eight miles apart.

These methods had obvious drawbacks. Only very limited messages could be sent by the simple codes appropriate to the medium. Visual signals are fast, 186 000 miles per second, but work only by line of sight, which means frequent relays; sound travels 1100 miles per second but has limited range and diffuses, obscuring the message. The speed of both is considerably reduced by the necessary human interventions to receive and relay the message. Neither is useful for long distances over water, where relays are difficult to arrange.

Chappe's Semaphore System, developed in the late 18th century, was the most sophisticated of these systems; flags on movable arms were set on towers 6–10 miles apart. A Paris-to-Toulon system covered 475 miles with 120 stations and could relay one signal per minute, the elapsed time from one end to the other being only about 12 minutes. Napoleon established a longer line, from the Prussian frontier via Warsaw to St. Petersburg, about 1200 miles. The system used 220 stations and 300 operators.

In the 18th century research on the nature of electricity had led many to think of the possibility of using electric signals to convey long-distance messages. Salva, Soemmering, and others experimented with electrostatic generators and later wires and batteries.

A practical electrical telegraph system became possible in the 1820s with the discovery of electromagnetism which was followed by the perfection of the electromagnet. One might almost say the telegraph was inevitable, since viable systems were developed independently in Germany (Steinheil), in England (Cooke and Wheatstone), and in the United States (Morse) in the 1830s. Successful telegraph systems were demonstrated in England in 1842 and in the United States in 1844.

---

*An early example of communication over distance by visual signals is Agamemnon's use of bonfires along the Aegean coast to notify his wife, Clytemnestra, of the fall of Troy. The opening lines of Aeschylus's tragedy *Agamemnon* were those of the watchman waiting for the signal:

> I wait; to read the meaning in that beacon light,
> A blaze of fire to carry out of Troy the rumor
> And outcry of its capture; . . .
>
> *A light shows in the distance.*
>
> Oh hail, blaze of darkness, harbinger of day's
> Shining, and of processions and dance and choirs
> Of multitudes in Argos for this day of grace.

By coincidence, one of the ships used in the first laying of the Atlantic cable was *H.M.S. Agamemnon*.

Samuel F. B. Morse* was granted a U.S. patent in 1840 and the U.S. Congress appropriated $30 000 for an underground telegraph cable from Washington to Boston. This arrangement failed and was replaced by an overhead wire. Morse tried but was unable to sell this line to the Federal government for $100 000; in 1845 it was turned over to the U.S. Post Office. By then a private line had been established between New York and Philadelphia. In 1847 telegraphy was a prosperous and expanding enterprise.

The first underwater telegraph cable was laid in 1838, only five years after Gauss and Weber first devised a telegraph and six years before Morse initiated the first land line operating between Washington and New York.

The early cables were laid across rivers or harbors: O'Shaughnessey's Hugli River cable (1838), Morse's New York Harbor cable (1842), Ezra Cornell's Hudson River cables (1845), West's Portsmouth Harbor cable (1846), and the Siemens Brothers' Kiel Harbor cable (1849). The first nonriver/harbor cable was John and Jacob Brett's† Dover-Calais cable in 1850–51. Their success was soon followed by the laying of an England-Ireland cable in 1853.

The idea of a transatlantic submarine cable was advanced very soon after the first harbor cable was laid. In a 10 August 1843 letter to John C. Spenser, Secretary of the Treasury, Samuel Morse expressed the belief, based on scientific principles, that "a telegraphic communication might be established across the Atlantic Ocean."[1] John and Jacob Brett in 1845 registered The General Oceanic Telegraph Co. for the purpose of uniting the British colonies (including Canada) by means of submarine telegraph. They sought the assistance of Prime Minister Robert Peel. Several government departments reviewed the scheme. Since none of these departments felt it was in its jurisdiction, the necessary active government support never developed.

The press was already calling for uniting America and Britain by telegraph. The New York *Tribune* suggested that a cable be suspended across the Atlantic at that point under water where the density of the cable and the density of the water were equivalent.[2]‡

A. A Dream is Born

In 1854 Cyrus Field¶—a recently retired prosperous New York paper merchant in his mid-

---

**Samuel B. Morse* (1791–1872) is best remembered for his work on the telegraph. Although not the first to invent a telegraph, he was among the first to make the telegraph commercially viable. Morse was a professional artist who first became interested in telegraphy in 1832. Using the work of Joseph Henry, he was able to strengthen and simplify the mechanism Wheatstone and Cooke had developed in England. He and his supporters, with an appropriation from Congress, demonstrated the feasibility of the telegraph with a line between Baltimore and Washington. Unable to sell the line to the government, Morse went into the telegraph business. He was not a good businessman; he was continually being victimized by promoters desirous of using his patents. According to Amos Kendall, his partners Cyrus Field and Peter Cooper were two such promoters.

†*John W. Brett* (1805–1863), a prosperous antique dealer, first became involved in submarine telegraphy through his brother Jacob. After successfully spanning the English Channel in 1850–51, Brett became a leading figure in submarine telegraphy. Included in his promotional ventures were cable routes to the Continent and in the Mediterranean. Brett's most lucrative enterprise was the Submarine Telegraphy Co., which had a monopoly on all telegraphy communications between France and Britain.

‡The proposal to suspend the cable under water was based on the popular belief that the deeper the water the greater the density; cables could therefore be held at the point where the density of the water was greater than that of the cable. Several other means were proposed for establishing a transatlantic telegraph. One suggestion was to festoon the cable across to America by a series of balloons. Others suggested suspending the cable between buoys moored to the ocean floor, which would allow ships at sea to tap into the cable. These suggestions were never taken seriously by cable promoters.

¶*Cyrus Field* (1819–1892) retired from business at 35 after making a small fortune in the paper trade. Restless, Field became interested in several grand projects. The first was an abortive attempt to open the Amazon region to American commerce. It was soon after his Amazon trip that he was attracted to submarine telegraphy. In his later years he became involved with a Pacific cable scheme and with New York's elevated railways. Through his investment in the latter, Field lost his fortune as a result of Jay Gould's manipulation of the stock. Impatient, gregarious, and industrious, Field was well known for his perseverance and his ability to persuade.

thirties—was visited by Frederick Gisborne,* a Canadian telegraph engineer. A year earlier, in the spring of 1853, Gisborne had begun work on a New York-to-Newfoundland telegraph line. After he erected about 40 miles of wire through the wilderness of Newfoundland, Gisborne's New York supporters withdrew their backing, leaving him with a debt of £50 000 to the workers on the line. Gisborne now sought Field's backing. arguing that the patronage of the press and commercial interests of the Eastern Seaboard would make such a line a financial success.[3] Being unfamiliar with the geography of the area and knowing telegraphs to be risky ventures, Field said he would have to consider the idea at length before replying.[4]

The next morning after reviewing the geography of the plan Cyrus asked his brother Matthew†: why not try to lay a submarine cable all the way to England and have something big in which to invest? Matthew had heard of such an idea and thought it could be done.[5]

The idea of an Atlantic cable was not new with Cyrus Field, but it was new *to* Cyrus Field; until his mention of it, the discussion had been limited to whether or not to take over the New York-to-Newfoundland route which, when connected with packet ships, would shorten communications between England and America by two or three days.[6]

Before committing himself to the laying of a transatlantic cable Field felt he needed advice on two matters: the present state of development of telegraphy, and whether the cable could indeed be laid along the ocean floor. On the second question, he turned to his former acquaintance, Lt. Matthew Maury‡ of the U.S. Naval Observatory. Maury replied by sending him a copy of a letter he had just written to Secretary of the Navy J. C. Dobbins. Dobbins had requested the latest research findings on winds and currents; Maury used this request as an excuse to push the idea of a transatlantic cable.[7]

Using the recent deep-sea soundings of Lt. O. H. Berryman, Maury described the ocean bed as "a plateau, which seems to have been placed there especially for the purpose of holding the wires of the submarine telegraph, and of keeping them out of harm's way."[8] He went on to suggest to Secretary Dobbins that a prize be awarded to the first company that would successfully lay a cable and transmit a message through it. He did not suggest that the U.S. Navy should undertake the project.

For advice on the physics of laying a transatlantic cable, Field turned to Samuel F. B. Morse. Morse enthusiastically supported a transatlantic cable and believed, as he had in 1842, that a transatlantic telegraph link could be established.[9]

With the support of Morse and Maury, Field was ready to seek financial support. Before doing so, he consulted with his older brother, David Dudley Field,¶ a leading New York lawyer, who

---

**Frederick Gisborne* (1824–1892), an English telegraph engineer, was involved in building landlines in Eastern Canada. He laid the first submarine cable in North America (between Prince Edward Island and New Brunswick), which was incidental to his main task of establishing landlines in Newfoundland; Gisbourne originally intended to use packet steamers or carrier pigeons to carry messages from Newfoundland. When Field took over the Newfoundland project, Gisborne stayed on as chief engineer for a short time.

†*Matthew Field* (1811–1870), one of Cyrus Field's older brothers, also started in the paper business. When paper manufacturing became too competitive in western Massachusetts, Matthew decided to become an engineer. With his wife he travelled throughout the Midwest working on bridges, dams, and railroads for 17 years.

‡*Matthew F. Maury* (1806–1873), a leading oceanographer of the 19th century, made significant contributions to navigation and oceanography during his career in the Navy. His wind and current charts reduced the voyage to California by almost 50 days; his *Physical Geography of the Sea* is considered the first text of modern oceanography. Maury was involved in international scientific cooperation, the founding of the American Association for the Advancement of Science, and the attempt to open the Amazon River to American trade. With the outbreak of civil war Cdr. Maury resigned from the U.S. Navy and went to England to plead the Confederate cause.

¶*David Dudley Field* (1805–1894), Cyrus Field's eldest and closest brother, lived next door to Cyrus in Gramercy Park, New York; their houses were connected by a small passageway. Dudley argued several cases before the U.S. Supreme Court, including *Ex parte Milligan* (1866) (with James Garfield), *Cummings v.*

agreed to handle all legal matters involved in the project and to lend his name in seeking further support. Soon Field had promises of support from four of New York's wealthiest men: Peter Cooper,* Moses Taylor,†Marshall Roberts,‡ and Chandler White.¶

Cyrus Field, Cooper, Roberts, Taylor, White, and Dudley Field took over the charter of Gisborne's bankrupt Newfoundland Electric Telegraph Co. Cyrus and Dudley Field, Chandler White, and Frederick Gisborne left New York on 14 March 1854 to secure a charter for the new company from the Newfoundland legislature in the name of the New York, Newfoundland, and London Telegraph. The charter, drafted by Dudley on the trip to Newfoundland, asked for:

- 50 years of exclusive landing rights
- a grant of 50 square miles of land
- a grant of 50 square miles more when the transatlantic cable was laid
- guaranteed interest on £50 000 of bonds§
- £5000 in cash for a road to be built alongside the overland telegraph on Newfoundland
- waiver of all duties on materials imported for the construction of the telegraph

Newfoundland's Governor Kerr-Bailey Hamilton consented to the new charter. It was quickly approved by the Governor's Council, passed by the legislature with only one dissenting vote, and duly confirmed by the Colonial office of the British government.[10]

From the other governments along the path of the New York-to-Newfoundland route, Field and his associates received equally generous concessions. The administration of Prince Edward Island granted them 50 years of exclusive landing rights, 1000 acres, and £300 per annum. The colonial government of Canada authorized the remission of duties on all imported materials. Nova Scotia and the State of Maine granted New York, Newfoundland, and London Telegraph Co. 25 years of exclusive landing rights in their respective jurisdictions.[11] These concessions, which were of a limited duration and valid only if the venture succeeded,were based on a belief that the existence of a telegraph route would be a positive factor in the development of the colonies.

With concessions in hand, White, Dudley Field, Cooper, Roberts, Taylor, Morse, and Cyrus Field met to subscribe stock ($1 500 000 or £300 000) and elect officers: Cooper, president; White, vice president; and Taylor, treasurer. Morse was appointed advising electrician and given a one-tenth share in the company for $10 000; in exchange he agreed to use his influence to get Atlantic cable messages carried at half cost by telegraph companies using Morse patents.[12] Cyrus Field accepted no office; he would be too busy traveling and supervising the operation to attend to routine business.

Field's first task was to contract for submarine cables for the distance across the Gulf of St. Lawrence between Cape Ray, Newfoundland, and Cape Breton Island, Nova Scotia, and the

---

Missouri (1867), and *Ex parte McCardle* (1869). His clients included such notables as "Boss" Tweed and Jay Gould. As a result of his participation in the cable venture, Dudley became interested in the codification of international law.

**Peter Cooper* (1791–1883), a New York manufacturer and philanthropist, made his initial capital through a glue factory. Cornering the glue market, Cooper invested in a Baltimore iron works. Through the iron works he became involved in the Baltimore and Ohio Railroad and the design of the *Tom Thumb* locomotive (so named for its small size). He later invested in foundries, iron mines, wire manufacturing, and telegraphy. Cooper Union, a center for lectures on art and science in New York City, was his creation.

†*Moses Taylor* (1806–1882) had by 1850 made a fortune in the exporting business, sufficient to be appointed president of City-Bank of New York in 1855. Five years later, at the outbreak of the Civil War, he chaired the committee that gave the Union its first loan.

‡*Marshall Roberts* (1814–1880) made his fortune in shipping. His major investment was the U.S. Mail Steam Ship Company. In 1846 the company obtained exclusive government rights to connect with the company in charge of mail on the Pacific Coast; in 1849 gold was discovered in California.

¶*Chandler White*, a wealthy retired paper merchant, died before the cable was completed. His place was taken by *Wilson G. Hunt*, a New York merchant and former mayoral candidate, who joined as a favor to Peter Cooper even though he was not interested in the project. Hunt eventually developed an interest, joined the American Telegraph Co., and supported Cyrus Field during the difficult years 1858–1864.

§Throughout this period the pound sterling was equivalent to $5 U.S.

A "Cable Shop."

distance from Cape Breton Island to the Nova Scotia mainland. Failing to secure two steamers from the U.S. government to lay the Gulf cable, Field turned his attention to contracting for a cable. On the advice of Gisborne, who was now working on the overland telegraph with Cyrus's brother Matthew, Field wrote to Kuper and Co.,* the British wire-rope and cable manufacturers, and asked them to prepare samples for his inspection.

In January 1855 Field sailed for England, with no acquaintances there except people in the paper trade, to contract for a cable and ships to lay it. Relying on a letter of introduction from Frederick Gisborne, Field contacted John Brett, who took him on an instructive tour of the cable manufacturers, including the newly formed Glass, Elliot and Co.† Brett also secured the services of Samuel Canning (an electrician who had been involved with Mediterranean cables) for supervision of the laying of the Gulf of St. Lawrence cable. Moreover, Brett subscribed some money to the project and became Field's advocate, contact, and champion in Britain. Through Brett, Field met many of the technical and commercial people who were to support the laying of an Atlantic cable.

When Field returned to New York, John Brett saw to the building of paying-out machinery and its installation on the sailing ship *Sarah K. Bryant*. The plan was to have a steamer tow the *Sarah K. Bryant* across the Gulf. The captain of the steamer proved to be inexperienced and the first attempt failed.‡

Field, who had brought his family and friends to watch the cable laying, was not defeated. After going to the Board for more financing, Field this time was able to contract with Glass, Elliot and Co. to construct *and* lay the cable at its own risk. The second attempt across the Gulf of St. Lawrence was a success. When combined with the completed Newfoundland landline, telegraphic communication was established between St. John's in Newfoundland and New York City.

With the completion of the New York-to-Newfoundland portion of the route, Field and his supporters faced three problems:

**Kuper and Co.* (1847–1854), a British firm, was one of the first companies to manufacture the outer core for submarine cables. Most of Kuper and Co.'s involvement in submarine telegraphy came as subcontractor to R. S. Newall and Co. In 1854 the company became Glass, Elliot and Co.

†*Glass, Elliot and Co.* (1854–1864) was founded by Richard Glass, a London accountant and inventor, and George Elliot of Kuper and Co. For ten years the firm operated successfully in the new field of submarine cable manufacturing. In 1864 it merged with the Gutta Percha Co. to form the Telegraph Construction and Maintenance Co.; Richard Glass became managing director of the new firm.

‡This 1855 cable-laying venture was a series of disasters. Cyrus Field hired the luxury steamer *James Adger* to tow the sailing ship *Sarah K. Bryant*. During the first attempt the *Bryant*'s anchor stuck and had to be cut loose; a collision with the *Adger* resulted. On a second attempt the weather proved to be a problem. The captain of the *Adger* failed to coordinate his speed with that of the paying-out machinery on the *Bryant* and the latter had to cut the cable to avoid being dragged under.

Left to right: Peter Cooper, David Dudley Field, Chandler White, Marshall O. Roberts, Samuel Morse, the artist Huntington, Moses Taylor, Cyrus Field, and Wilson G. Hunt. *From an original painting by Daniel Huntington in New York Chamber of Commerce.*

a. How to insure effective linking of the Canadian lines with the myriad American telegraph companies, many of which used competing patents

b. How to use this New York-to-Newfoundland link before the laying of an Atlantic cable

c. How to use this link to get support for an Atlantic cable

Field solved the first of these problems by forming, with nine of his associates, the American Telegraph Co. (1855), which within a year controlled telegraph lines from New York to the Canadian border. The interlocking directorates of American Telegraph Co. and the New York, Newfoundland, and London meant that Field had an unobstructed route to New York City for information coming over the transatlantic cable he meant to build.

The other U.S. telegraph companies, especially Western Union, were understandably alarmed by American's growing power. The best defense, they judged, was cooperation. Therefore, the six most powerful companies, including American, signed an agreement that came to be called the "Treaty of the Six Nations," by which each company gained a monopoly in its own territory. American Telegraph Co.'s territory covered Newfoundland, New Brunswick, and the states of the Eastern seaboard from New England to Florida—and thus the major business centers of America.

The answer to the other problems—how best to use the New York-Newfoundland link and at the same time build substantial support for the cable—at first appeared to be the packet-drop system,*which was however never extensively used.

*The idea behind the packet-drop system was to insure that when a packet steamer left America or Europe it carried the very latest news. News was telegraphed from New York to Newfoundland for 2–3 days after the steamer left New York. At Halifax or off the coast of Newfoundland, the latest news was placed on board. When the steamer reached Ireland, the news was telegraphed to London. The same process was used to speed up transmission of news from London to New York. Although this packet-drop system was used several times by people looking for an advantage in the highly competitive business of foreign news, it never became fully established.

Despite the advantages of quicker communication than demonstrated by the packet-drop system, Field was unable to interest capitalists in New York or Boston in supporting the laying of an Atlantic cable. It was only with great difficulty that he received the routine support of the government.

B. The First Attempt: 1856-1857

With the cable project becoming more certain Field suggested to Maury that, given recent improvements in sounding techniques, it would be a good idea to survey the ocean floor again. Maury agreed and requested the necessary ships and equipment from Secretary of the Navy Dobbin. Dobbin refused; he had been persuaded by Alexander Bache,* head of the U.S. Coast Survey, that deep-sea soundings were Bache's responsibility, not Maury's. Professional jealousy between Bache and Maury would cause problems later; now Maury had to accept the decision to send Bache's man, Lt. O.H. Berryman, to do the soundings.

Having had no success in America at raising capital for the cable, Field went to England with Samuel Morse to try to raise capital there. He was empowered to raise money either by selling stock in the New York, Newfoundland, and London Telegraph Co. or by establishing an English company.

Isambard K. Brunel,† the great civil engineer of 19th century England, was interested in the cable from its inception. He became a consultant to the project and was soon writing Field valuable memoranda on cable laying.[13] Brunel cautioned against the hurried, underbudgeted, haphazard way in which Field and his associates were approaching an engineering feat of such magnitude. He suggested that Field ought to wait until work on the new ship, the *Great Eastern,*‡ was completed. This, the largest ship in the world, would be best suited to laying the cable. Field rejected the proposal, reasoning that the use of two untried mechanisms (the cable and the *Great Eastern*) would multiply the chances of failure.

Over and over again the unquenchable enthusiasm, drive, and determination of Field—the qualities that finally brought the cable into being over tremendous odds—would lead him to ignore the warnings and cautions of men much more technically competent than himself. The speed with which he pursued his goals left him impatient with the methodical planning and testing approach of the scientists and engineers whose expertise he sought; the exigencies of raising capital impelled him to strike when the iron was hot, and his impatience sometimes precipitated mistakes and problems that could have been avoided. Yet the cable's ultimate success may well have come because of these "shortcomings" rather than in spite of them.

Unable to sell shares in his own company, Field found it necessary to form a special English company. It was to be the first step in the process through which control of the cable enterprise passed from the American to the British side of the Atlantic. Brett and others from his Magnetic Telegraph

---

**Alexander Bache* (1806–1867), a great-grandsonn of Benjamin Franklin, entered the U.S. Military Academy at West Point at age 15 (unheard of in those times) and graduated with honors at 19. In 1842, after teaching at the University of Pennsylvania and serving as the chief of research at the new Franklin Institute, he was appointed Superintendent of the U.S. Coast Survey, a post he held until his death in 1867. Bache was the first president of the National Academy of Sciences.

†*Isambard Kingdom Brunel* (1806–1859), a civil engineer, was widely known (because of his short stature) as the Little Giant. At age 20 he was supervising work on the Thames tunnel. Prominent in many engineering works, he is best known for his design of ships. At 32 he launched the *Great Western*, the first ship powered solely by propeller. Brunel's greatest work was the *Great Eastern*, which was launched in January 1858, shortly before his death.

‡The *Great Eastern* (1858–1889) was, until the 20th century, the largest ship in the world; it weighed 25 000 tons and measured 692 ft in length and 83 ft across the beam. Unfortunately, her size proved to be a major drawback for her intended operation as a passenger ship. An economic disaster, the ship was sold at great loss and refitted for the 1865 Atlantic cable attempt. After laying the first Atlantic cables, the *Great Eastern* was used for other cable expeditions. With the introduction of specialized cable ships, it was converted into a show boat and finally scrapped in 1889.

Co.—such as his solicitor, J.A.M. Pinniger, and Samuel Stratham of London Gutta Percha Co.*—formed the Atlantic Telegraph Co., Ltd. (under the newly enacted Limited Liability Act) and appointed themselves the provisional committee.

Several tasks fell to this provisional committee. First was selection of the appropriate cable. After many experiments at the works of Glass, Elliot and Co. (successors to Kuper and Co.), they selected a cable of nine interwoven copper wires, surrounded by a sheath of gutta percha, itself encased in 18 woven strands of iron. Brunel and others suggested a cable with a thicker, more flexible outer covering.

Next, a contract was let for the construction of the cable. The 2500-mile inner core consisting of wire and gutta percha was to be manufactured by Samuel Stratham's London Gutta Percha Co. The outer covering was handled differently: according to George Saward,† as a result of intrigues ". . . they were unfortunately driven, as an act of policy rather than prudence, to the necessity of dividing the work into two contracts. . . ."[14] Half was given to Glass, Elliot and Co., who were originally to have made the whole line; the other 1250 miles went to R.S. Newall and Co.‡

Having selected and conditionally contracted for a cable, the provisional committee set about raising money for the enterprise. The estimated capital (£350 000) was divided into shares of £1000 each. It was hoped that the £1000 shares would mean that a "much more comprehensive and national character might be given to the undertaking . . . and that all who were inclined to participate in the honor of lending a hand to the grand project should be invited to subscribe for these shares."[15]

The investment effort was informal: there were no brokers, no formal prospectus, no advertisements, and no public announcement of the project. Instead Field and Brett concentrated on the pre-sold market of investors previously involved in other submarine cable projects, most connected with the Magnetic Telegraph Co.¶[16]

Using the facilities of Magnetic Co., Field travelled throughout Scotland and England selling shares in Atlantic Telegraph Co. Potential investors were offered the opportunity to help in bringing about the unity of the "Anglo-Saxon race," furthering the brotherhood of mankind, stimulating commerce, and reducing the possibility of warfare. More immediately, they were offered an opportunity to make a handsome profit. The British government had given a conditional guarantee of a return of 8% on the investment regardless of future competitors using parallel routes. The backing of preeminent scientific and engineering men, an unprecedented monopoly on landing rights in a British territory, and the promise of an acceptable dividend return if the cable was successful also stimulated investment.

**London Gutta Percha Co.* (1845–1864) was formed only two years after gutta percha's first appearance in England. A rubber-like substance obtained from the gum of gutta trees indigenous to the Malay Peninsula, gutta percha can be molded easily when placed in hot water. Soon after its introduction in England it was used for a wide variety of purposes including toys, whips, coffin linings, golf balls, and shoe soles. Insulating cables became the primary use of gutta percha and between 1851 and 1864 the London Gutta Percha Co. monopolized the manufacture of gutta percha-covered cable core. In 1864 the company merged with Glass, Elliot and Co. to form Telegraph Construction and Maintenance Co.

†*George Saward* (d. 1876) was one of Field's chief contacts in Britain. He had been secretary and manager of the British Telegraph Co. before he was appointed secretary of the Atlantic Telegraph Co. In his later years, he wrote a short history of the Atlantic Telegraph, giving the particulars of the financial arrangements.

‡*R. S. Newall and Co.*, like Kuper and Co., became involved in cables through experience as a wire-rope manufacturer. The director, R. S. Newall, was a highly competitive and inventive man. He personally directed all aspects of his cable ventures. Newall laid several important cables in the pioneering era 1851–1854—the Black Sea cable during the Crimean War, the Irish Sea Cables, and numerous cables in the Mediterranean for the French, Sardinian, Greek, and Ottoman governments. His 1859 Red Sea cable proved such a complete failure that it contributed to the decline in Newall's importance as a cable-laying company.

¶*Magnetic Co.* was formed by John Brett and several Liverpool merchants and shipowners to facilitate communications between England and Ireland. In 1858 Magnetic merged with the European and American Telegraph Co. to form the British and Irish Magnetic Telegraph Co., one of the major domestic telegraph companies in Britain until nationalization of the telegraph service in 1868–1870.

Field was able to raise the necessary £350 000 in two weeks, selling 250 shares to businessmen in London, Liverpool, Glasgow, Manchester, and elsewhere.[17] The remaining stock was divided between Brett, who took 12 shares, and Field, who took 88. Most of the selling was done through the offices of Brett's British and Irish Magnetic Co. to its shareholders, who had a great deal to gain as "the Magnetic Company's routes to Ireland would benefit greatly from an Atlantic cable to Ireland."[18]

Field felt very strongly that the cable should be a joint project in more than name; not all the capital should come from England.[19] Unfortunately, he was able to sell few of the 88 shares he had retained for American investors. Only a few businessmen and social leaders contributed, leaving him with an investment of $335 000 in the cable. Some years later, after more failures to raise money in America, he defended the American role in the project when he stated, "Our little company [the New York, Newfoundland, and London] raised and expended over a million and a quarter dollars before an Englishman paid a single pound sterling."[20]

While working on these problems, the provisional committee decided the constitution conferred by the Limited Liability Act* was insufficient; they therefore chose to seek incorporation through an act of Parliament. Such an act would confer "various important and requisite powers and privileges," including the power to enter into contracts with the British and American governments, the right to land cables in the United Kingdom, a confirmation of the agreement with Newfoundland Co., and sanction for the alternate priority policy outlined in the government guarantee.[21]

In November 1856 the provisional committee filed the necessary notice of their intention to seek an act of incorporation from Parliament in the session of 1857. The committee granted the promoters (Field, Brett, Whitehouse,† and Bright‡) certain stock rights and completed an agreement with New York, Newfoundland, and London Telegraph Co. giving it one-third of the cable's receipts. The provisional committee then resigned, after appointing permanent directors.

With the establishment of the permanent board of directors of Atlantic Telegraph Co., Cyrus Field returned to America. In six months he had formed a company, selected a cable design, and contracted for a cable; and he had secured the support of a group of influential business, professional, and technical people. Perhaps his greatest feat was securing initial agreement on the first contractual involvement of the British government in a private cable venture.[22] Until he received the recognition of the British government in the form of a guarantee, Field had dealt with national governments as one of many private citizens seeking assistance in a private enterprise. It was for a limited purpose (making soundings and borrowing ships) that he first sought the assistance of the British government. After writing a letter to the Admiralty requesting soundings and ships to lay a cable, Field and Morse visited Lord Clarendon, the Foreign Secretary. They explained that state aid was vital to the success of this venture because of the magnitude of the undertaking. Specifically, Atlantic

---

**Limited Liability Acts* of 1856 and 1862 limited a person's liability to the amount he invested in a corporate venture. This arrangement allowed the creation of joint-stock companies, run by directors and responsible to shareholders. Besides limiting personal liability, the Acts continued a trend toward making it easy to establish commercial organizations. Though business continued to be dominated by personal enterprises until the late 19th century, a number of large enterprises such as railroads benefitted from the passage of these Acts.

†*E.O.W. Whitehouse* made some of the earliest experiments on the technical feasibility of an Atlantic cable. Originally a Brighton physician, he left his practice to devote full attention of the Atlantic cable. Too ill to go on the first cable-laying expedition, Whitehouse stayed in Ireland while Professor William Thomson (later Lord Kelvin) sailed with the expedition. Whitehouse's affiliation with the Atlantic cable project came to an end after he put an enormous charge through the failing 1858 cable and thereby worsened the faults in the gutta percha insulation. Whitehouse later served as consulting electrician on the 1861 Malta-Alexandria cable.

‡*Sir Charles Bright* (1832–1888) had two dozen patents for electrical inventions by age 23. As engineer for Brett's Magnetic Telegraph Co., he supervised the laying of the cables to Ireland. In 1861 he resigned from that company and went into business with Latimer Clark. Between 1861 and 1872 Bright laid cables in the Mediterranean, the Persian Gulf, and the West Indies; he also served in Parliament (1865–1868). He was consulting engineer on the 1865 and 1866 Atlantic cable layings. Bright developed many of the basic techniques of cable laying and wrote on them extensively. He was knighted for his role in the 1858 Atlantic cable venture.

Telegraph Co. would require government ships and crews to lay the cable. Lord Clarendon assured Field that the British government was interested in the project and asked him for a request in writing.[23]

The request was referred to James Wilson,* Financial Secretary of the Treasury. Wilson, who had previously been exposed to the cable scheme by John Brett, was in the midst of formulating a method for state support of private telegraph schemes to India.

The problem facing Wilson was a difficult one. The British government had an interest in seeing a cable laid to its North American possessions. As a free-trader and supporter of private enterprise, Wilson considered his government's primary interest in supporting this particular private venture to be the stimulation of other private efforts in the new field. If the Atlantic project were to be shown to be technically feasible and commercially viable, private enterprise would establish the strategic cable routes to India and the Far East. Wilson formulated a solution that called for just enough state support to encourage the venture. Above all he wanted to avoid establishing a precedent of state-financed cable projects, a proposal he had encountered several times with regard to cable projects in the Mediterranean.

The agreement he prepared and Field gladly accepted assured the Atlantic Telegraph Co. of a minimum annual payment for services rendered (once the cable was working) of £14 000, equal to a 4% return on the assumed capital of £350 000, until such time as the net profits of the company would reach a 6% annual return on investment. At that time the minimum annual payment would be reduced to £10 000 for a period of 25 years. The capital requirement of £350 000 was that established by the company itself. By establishing the minimum payment guarantee Wilson identified the company's success with the national interest but relieved the nation of any responsibility if the cable should fail to work. The agreement also stated that the British government would provide ships for necessary ocean soundings and "consider favorably" any request for aid in laying the cable. The company in turn agreed to giving British government messages priority over all others (with parity for the U.S. government should it offer a similar agreement) and to a provision that the British government must approve its tariff rates.

Wilson preferred to see the state's actions as the purchase of a "service" similar to a mail contract. The government only promised to pass a minimum volume of traffic over the system or pay the difference necessary to insure a dividend to the stockholders. The Treasury handled the entire matter and it passed through Parliament with little effort.[24] It did not pass the House of Lords without some comment; Lord Redesdale felt the whole project should be underwritten by the British government, to avoid any need to share responsibility and control with a foreign government. But the cable project was really never in jeopardy from Parliamentary actions.[25]

In his dealing with the British government, Field had several advantages. Important were the support and position of James Wilson, who was aware of the obstacles facing cable promoters. Wilson's interest was in stimulating transoceanic communications for reasons of empire, as far as possible without resort to government revenues. Also, Field had the support of some of the most influential commercial and technical personages in the nation. Further, the British were familiar and experienced in dealing with large engineering projects, such as railways and other, similar ventures. Finally, an excess of investment capital was available in the United Kingdom.

Similar factors either were not present or did not work to Field's advantage in the United States. Though most of the capital had been raised in Britain, Field was unable to secure the support of

James Wilson (1805–1860), the son of a Quaker wool manufacturer, was a major figure in financial affairs of 19th century Britain. He made his initial fortune as a hat manufacturer and lost it in indigo speculation in 1836. After recouping his lost wealth, he retired from business to devote full attention to his unique journal, *The Economist*. Supporting free trade and repeal of the Corn Laws, Wilson wrote the entire periodical during its initial years. Elected to Parliament in 1842, he gained a reputation as a financial expert with a business imagination and rapidly acquired a position in government. After serving on the India Board in the 1840s, Wilson was appointed financial secretary to the Treasury in the 1853 Palmerston government. In 1859 Wilson accepted the task of organizing the India government's financial structure. Soon after his arrival in India, Wilson died of dysentery.

prominent American businessmen. Samuel Morse was the only prominent scientist who supported the project. Moreover, given the nature of American government, Field could not get the necessary support through the informal tactics he used in England. Instead, he had to submit a bill to Congress. Obtaining support from Congress for such a venture was a difficult task, especially since the cable joined two British territories. Also, there was little precedent for U.S. government support for large engineering projects.

The desire for U.S. government support for the cable project came largely from Field's determination that the cable should be an international project and from the expectation of the British government that the U.S. would provide a guarantee similar to the one Britain had granted.[26] While in England, Field wrote Peter Cooper asking him to write President Franklin Pierce requesting support similar to that given by Great Britain. No official documents passed between the governments; all the details were arranged through Field and his associates.[27]

The letter to Pierce noted the generous support of Britain and asked for the same terms. The President and Secretary of State William Marcy individually favored the idea but were fearful of exacerbating the sectional conflicts looming over the Congress and the nation. But other political figures were already interested. On 23 December 1856 the Senate passed a resolution asking the President to transmit all information in regard to "present condition and prospects of a proposed plan for connecting by submarine wires the magnetic telegraph lines on this continent and Europe."[28] Pierce responded by sending a copy of the letter Cooper had written at Field's request.

Field, meanwhile, had appealed to his friend, Senator William Seward* of New York, who introduced S. 493, "A Bill to Expedite Telegraphic Communication for the Use of the Government in Its Foreign Intercourse." The bill asked for the same conditions Field received in England—ships to lay the cable, message priority, tariff setting, and an annual subsidy of $70 000. It was presented as essentially a mail agreement. The bill was introduced on 9 January 1857 and was immediately referred to the Committee on Post Offices and Post Roads, which returned it to the floor of the Senate without amendment on 12 January 1857; debate commenced nine days later.

Throughout January Field remained in Washington to lobby for passage of the bill. He visited every Senator and Representative he could, to try to persuade them of the worth of the project. This persistence almost defeated his cause, for "so vehement were the prejudices against Mr. Field, for what was then regarded as presumption and officiousness on his part, that the great bill was only saved by his withdrawing . . . and remaining unobtrusive and unseen in his own lodging."[29]

The debate over the bill was lengthy and bitter. Proponents of the cable agreement spoke of the benefits of closer ties with "the other English speaking races," of the great promise for international peace and brotherhood through better communication, of its uses for American diplomacy. They appealed to national pride and to the ideal of progress. While insisting that the cable agreement should be no more controversial than "another contract for delivery of mail," the advocates promised great benefits for American commerce and trade, for farmers, and in particular for cotton planters, who would "know what their crop was worth in the markets of Liverpool," and thus have a defense against "the speculators of Europe" who usurped their profits.

The last was an appeal to the large group of Southerners in opposition to the cable agreement. Sectional rivalry was indeed the chief source of hostility to the cable. Southerners and Westerners opposed government support of the cable, which they saw as a potential benefit chiefly to the industrial and commercial power of the Northern states—the merchants, traders, and speculators. But suspicion of Great Britain—a rival in the northwest, and a traditional enemy (it was, after all,

---

**William H. Seward* (1801–1872) is best remembered for Seward's Folly, the purchase of Alaska from Russia in 1867. His first major political office was Governor of New York from 1838 to 1842, followed by his election to the U.S. Senate as a Whig. Strongly opposed to slavery, Seward joined the Republican Party in 1855 and attempted to get the Party's nomination for President in 1860. When Lincoln was nominated and elected, Seward accepted the position of Secretary of State, serving in that post until 1869. As Secretary, Seward favored expansion. He pressed for the acquisition of Alaska, Hawaii, and Santo Domingo. Part of his expansionist policy was the support of any American cable venture—Atlantic, Pacific, or Caribbean.

only 45 years since the British had burned the White House) was also a strong factor. The British, with their predominance in international trade and their colonial empire to govern, would, some Americans believed, reap the most benefits from a cable connecting two points on British territory. The motives of the British were also suspect to many in both North and South, since it was not clear which way their sympathies lay in their sectional struggle.

The bill embodying the cable agreement was passed by a narrow margin in the House (19 votes) and by only one vote in the Senate (20-19, on 25 February 1857), and was signed by President Pierce as his last act before leaving office.

As soon as the bill was signed, at the suggestion of Maury, the steam frigate *Niagara** (the largest in the world) was placed at Field's disposal.[30] The brand-new *Niagara* had not yet been outfitted in heavy armor. Navy Secretary Dobbins and Lt. Maury felt this assignment would be a good test of the ship's capabilities.[31] Maury also charted the exact path of the cable and made suggestions on the procedures for its laying.

Later in March 1857 Field sailed for England to supervise the manufacture of the cable. The manufacturing process was proceeding at a very rapid pace. Professor Thomson examined the cable during manufacture and found that the quality of the copper core varied widely; he could do little about it since the specifications had been pushed through "by the dynamic energy of Mr. Field."[32]

The reason for the haste was the short period of time allotted by Field and the provisional committee for manufacturing and testing the cable. Newall and Co. and Glass, Elliott and Co. were told to cover part of the core and send it to Birkenhead near Liverpool to be loaded onto the cable-laying ships; no provision was made for testing the cable in one length.[33] Again Field's impatience carried the day. He knew that the month between July 10 and August 10 would be the best time to attempt to lay the cable as the sea would then be relatively calm.[34] Anticipating a short duration for loan of the ships, pressures from investors, and pressures from the interested governments, Field and his associates would not delay another year.

The cable was ready by the first week in July. *Niagara* and her escort *Susquehanna* had already arrived from America, with the addition of the Royal Navy's steamer *Agamemnon*.† The crews soon began loading the cable into their holds.

Now the directors and officers of Atlantic Telegraph Co. met to settle the question over how to proceed. The chief engineer, Charles Bright, appointed to that post at the urging of the Liverpool directors,[35] thought that *Niagara* and *Agamemnon* should meet in mid-ocean and lay the cable as they steamed away from each other. The directors, believing it important to keep in touch with Ireland, decided that one of the ships would start laying the cable from Ireland; when it reached the end of its part of the cable, it would splice that with the part on the other ship.

After several celebrations in England and Ireland,‡ *Niagara* and *Agamemnon* left Valentia Bay on 5 August 1857. The former laid the first portion of the cable. With 380 miles of the 2500-mile cable laid, an untimely application of the brake on the machinery for paying out the cable caused the

---

**U.S.S. Niagara* was the finest vessel in the U.S. Navy at the time it was sent to England to help lay the cable. The largest steam frigate in the world (5000 tons), it was modelled by the great shipbuilder George Steers along the lines of his famous yacht *America*. The Navy felt that *Niagara's* participation in the cable expedition would be a good test of the new ship's capabilities as well as a gesture of international goodwill. When the vessel appeared in the Thames in 1857, it was the first American warship seen in England for many years. To add to its uniqueness, two officers of the Russian Navy were on board to witness the cable-laying experiment.

†*H.M.S. Agamemnon* was *Niagara's* counterpart in the cable-laying expedition. One of the British Navy's best ships of the line, the 3200-ton vessel was, like *Niagara*, driven by propellers. *Agamemnon* was particularly renowned in Britain because, during the siege of Sebastopol in the Crimean war, it had steamed within a hundred yards of the enemy's guns.

‡At the celebration at Valentia, the Earl of Carlisle, Lord Lieutenant or Ireland, spoke of the impact of the cable on existing ties: "We are about, either by this sundown or to-morrow's dawn, to establish a new material link between the Old World and the New. Moral links there have been—links of race, links of commerce, links of friendship, links of literature, links of glory, but this, our new link, instead of superseding and supplanting the old ones, is to give a life and an intensity which they never had before."

After splicing the cable in mid-ocean, *Agamemnon* and *Niagara* part for Valentia and Trinity Bay in the second temporarily successful expedition of 1858. *From Harper's Weekly, 14 August 1858.*

cable to break. Several attempts were made to recover the lost cable. Since no one had ever recovered a cable in deep sea previously, it was not surprising that these attempts failed.

C. The Second Attempt: 1858

The two ships returned to England, where a meeting of the Board of Directors of Atlantic Telegraph Co. was held (17 August 1857) and the following decisions were made:

1. There would be no more attempts to lay a cable until
   (a) the paying-out machinery was refined;
   (b) 1858, when the next favorable season would occur.
2. The cable designed for the first attempt would be suitable for subsequent attempts.
3. More cable would be carried in the next attempt (3000 rather than 2500 miles).
4. The American engineer, William Everett, would, at the pleasure of the United States government, return to England and help design new paying-out machinery.
5. In the next attempt, paying out would begin in mid-ocean; and more money would have to be raised to construct the necessary 600 miles of new cable and outfit the ships that would lay the cable in 1858.

First Atlantic cable (1857–58).

The request for money (£100 000) was met by the previous backers of the enterprises. Indeed, the prevailing attitude was that with the lessons of the first attempt the second would surely succeed. Field, in speaking to the board of directors, saw no reason to stop; the only question was when to try again. In fact, both he and Bright, the chief engineer, minimized their disappointment.* Redesign of the paying-out machinery began and Everett was granted leave by the U.S. Navy to help with the improvements. Meanwhile *Niagara* and *Agamemnon* had to discharge the remaining cable because of other commitments. The cable was stored in a wooden tank on the Keyham Docks in Plymouth.

Between August 1857 and May 1858 progress continued on all fronts. Several new inventions were forthcoming, including Smith's new adhesive and insulating compound, Thomson's† mirror galvanometer, and J.G. Appold's self-releasing brake. William Everett applied Appold's principle to the paying-out machinery. By March he had shown a model to Field and some of the foremost engineers and machinists in Britain; they were impressed with the machine because it seemed to know when to hold on and when to let go.[36]

Atlantic Telegraph Co. also benefited from the prompt cooperation of both governments. U.S. Secretary of the Navy Toucey gave Field everything he asked for: Everett, *Niagara*, *Susquehanna*, and their crews. The British Admiralty was equally helpful, continuing its agreement with the company and thereby facilitating the appropriation of the necessary capital. The Admiralty allowed *Valorous* to take the place of *Susquehanna*, which was detained in the West Indies as a result of a yellow-fever epidemic.

Many of the men involved in the project refused to accept pay for their work. Field, who was appointed general manager of American Telegraph Co., turned down offers of remuneration. So did the engineers and machinists advising Everett on the paying-out machinery. Thomson never received payment for his efforts on the cable. Most had other incomes, but their willingness to devote extensive time and energy to the project at no charge reinforces the view that this was a great national undertaking rather than a purely commercial venture.

The four ships, *Niagara*, *Agamemnon*, *Valorous*, and *Gorgon*, left Valentia Bay on 10 June 1858 after some preliminary testing of the paying-out machinery. The plan was to meet in mid-ocean, splice the cable, and lay it as the ships steamed apart. Despite a gale, the ships met in the mid-Atlantic on June 25. After several mishaps it was agreed that if the cable broke before each ship had paid out 100 miles they would rendezvous at the same spot; otherwise they would return to Ireland. On the next attempt the cable broke after 160 miles had been laid; both ships returned to Ireland independently.

The failure of this second attempt was most discouraging. Several board members, including the chairman, recommended abandonment of the project. Some even resigned. Others felt more time was needed to investigate the suitability of the cable. The strong supporters of the project, including Field, argued that delay would be fatal; the governments would surely force the ships to unload the already damaged cable. The capital for refitting a new expedition and reloading the cable would be very difficult to raise.[37] This reasoning prevailed; the ships got under way immediately. After rendezvous on July 29, they spliced the cable halves and steamed to their respective home countries.

**Field*: Although this unfortunate accident will postpone the completion of this undertaking for a short time, the result of the experiment has been to convince all that took part in it of the entire practicality of it, for, with some slight alteration in the paying-out machinery, there appears to be no great difficulty in laying down the cable . . . .

Bright: It has been proved beyond a doubt that no obstacle exists to prevent our ultimate success; and I see clearly how every difficulty which has presented itself in this voyage can be effectually dealt with in the next.

†*William Thomson (Lord Kelvin)* (1824–1907), who for fifty-three years held the chair of natural philosophy at the University of Glasgow, became interested in the practical applications of scientific knowledge and eventually went into the business of making measuring devices and electrical regulators. A prolific writer on a wide range of subjects, Thomson worked with Faraday and many other prominent pioneers of electricity. After his work on the Atlantic cable (for which he was elevated to the peerage, with the title Lord Kelvin), Thomson worked on making more reliable mariners' compasses, perfecting deep-sea sounding devices, and improving electrical lighting.

The second telegraph fleet assembles in the Bay of Biscay in late May 1858 for a rehearsal designed to avoid the errors of the previous year's expedition.

Fireworks set New York's city hall ablaze in the two-week celebration of Field's fleeting triumph in late August 1858.

Niagara reached Newfoundland on the same day *Agamemnon* reached Ireland: 5 August 1858.

The news of the success of the project caught the American public by surprise, generating considerable excitement. Banquets, speeches, parades, and sermons erupted throughout America. The British also celebrated but were less enthusiastic in their reaction to the success of the cable. After all, Britain was merely linked to another overseas area; America, on the other hand, was finally linked to the outside world—to Europe.

The cable was used immediately. George Saward[38] presented an excellent description of its uses:

> Among the services accomplished by this cable during the short time it was in operation may be mentioned a message of peace and congratulation between Her Majesty and the President of the United States, and similar messages between the Corporations of London and New York.
>
> Intelligence was also conveyed of the collision of two steamers of the Cunard line, the *Europa* and the *Arabia*, and the same message conveyed to the relatives of the passengers the assurance of the safety of all on board.
>
> Directions were also despatched to the American Colonies, on behalf of her Majesty's Government, which saved large sums of money to the country by countermanding orders which could not have been arrested in time in any other way.
>
> In all some two hundred and seventy-one communications passed through this cable. Messages, and portions of messages; half-formed words; letters, and fragments of letters, with unintelligible signals, make up the curious and interesting contents of the Newfoundland and Valentia diaries.

The cable operated for only a month. From the beginning, transmission of messages was difficult. The situation became gradually worse, until the cable was presumed dead on 18 September 1858*, only a few days prior to a major scheduled celebration of its success. Again Saward aptly described the situation:

> The despatch of ninety-eight words from the Queen to President Buchanan occupied sixteen hours in transmission. The same message could now be sent in less than eight minutes, while the President's reply of 149 words, which through the present cable would take about ten minutes, required more than ten hours to transmit.
>
> All these utterances, however, were but the result of stimulating doses of battery power upon a cable in a state of dissolution. They became more and more indistinct as the fault enlarged, and at length, on the 18th September, 1858, this great historical and experimental telegraph subsided into entire silence, effacing all those hopes of commercial success in which directors and shareholders had begun to indulge. . . .[39]

After proclaiming the cable one of the wonders of the age, the public was astonished and dismayed at its failure. Many declared the whole thing a prearranged hoax desiged to swindle the public. Several assumed that the stock Field had been unable to sell before completion was sold very quickly after the first news of success—although in fact only one share was sold.

There was widespread speculation about the cause of the failure. Morse cited the poor handling and excessive exposure of the cable to the elements on the Keyham Docks.† Others pointed to the haste in the construction of the cable and the lack of testing. Some held the cable to be "mechanically

---

**George Saward* to London *Times*, 6 September 1858:

Sir,—I am instructed by the directors to inform you that owing to some cause not at present ascertained, but believed to arise from a fault existing in the cable at a point hitherto undiscovered, there have been no intelligible signals from Newfoundland since one o'clock on Friday, the 3d inst. The directors are now at Valentia, and, aided by various scientific and practical electricians, are investigating the cause of the stoppage, with a view to remedying the existing difficulty. Under these circumstances, no time can be named at present for opening the wire to the public.

To C. W. Field, New York, 25 September 1958:

We have received nothing intelligible from Valentia since the 1st of September, excepting feeling a few signals yesterday. I cannot send anything to Valentia. There has been very little variation in the electrical manifestations. De Sauty.

†One of the properties of gutta percha is deterioration owing to oxidation. When left in the sun and air, gutta percha loses its insulating properties. The only way to prevent such deterioration is to keep the material under water, as was confirmed in experiments done by William Fairbairn in connection with the Board of Trade inquiry.

sound but electrically bad.''[40] The electrical problems were numerous; there was a break in the cable 300 miles from the Irish shore. The insulating material had deteriorated as a result of the enormous electrical charge Dr. Whitehouse (over the objections of Professor Thomson) put through the cable. In retrospect, it can be said that all these factors, especially the exposure of the cable and the resulting deterioration of the gutta percha, contributed to the cable's demise.

D. Despair and Reassessment

The reaction to this failure was markedly different from the reaction to the failure of the first cable-laying event. Many felt the cable, though an interesting scientific experiment, was not commercially viable.[41] William Brown, chairman of the board of Atlantic Telegraph Co., resigned and advised others to do so; several followed his advice. Many of the original supporters refused to put up any more money for the scheme. Even the directors of New York, Newfoundland, and London Telegraph Co. withdrew their support.

But Field and his close associates in America and in Britain remained convinced that with new funding a commercially viable cable could be laid. Returning to England, Field called a meeting of Atlantic Telegraph Co.'s board of directors. At the request of the board, Parliament passed a bill raising the total capital ceiling of the company to £2 million. The bill, passed in July 1859, gave the company the power to issue £600 000 of new stock in £5 shares, each dividend bearing at 8 per cent. In anticipation of passage of this bill, the company authorized the issuance of £600 000 of preference stock and issued a prospectus. Their intention was to use £20 000 to recover the old cable and the remainder of the £600 000 to build and lay a new cable.[42] But despite their vigorous efforts the Board was unable to raise any significant capital.

Investors were reluctant to participate. Recently, there had been several other deep-water cable failures; the most monumental was that of the Red Sea cable scheme to India.* Moreover, the political situation in America in 1859 was so uncertain that American cooperation was precluded.

The directors of Atlantic Telegraph Co. realized that the only way they could obtain commercial backing for a new cable would be with an absolute government guarantee, i.e., the government would pay the stockholders whether the cable worked or not. The directors urged the goverment to grant such a request, because (a) the project was national in character, (b) heavy losses had already been sustained proving its feasibility, (c) there was a precedent for such a guarantee in the Red Sea venture, and (d) a cable would benefit science and the administration of government.[43]

Their initial contacts were with the new Whig government headed by Lord Palmerston† and Chancellor of the Exchequer William Gladstone,‡ a firm believer in laissez-faire and non interven-

---

**The 1859–60 Red Sea cable* attempt was one of the greatest failures in cable-laying history. When the India Mutiny (1857–58) dramatized the need for instant communications between Britain and India, the British Government chose to support the only scheme readily available—a cable via the Red Sea. As support, the Government guaranteed 4.5% interest on the company's £800 000 capitalization for fifty years. This guarantee was represented to the public by Red Sea and India Telegraph Co. as an absolute guarantee—whether or not the cable worked. The cable never operated successfully over the entire distance. Parliament was forced to accept full financial responsibility.

†*Henry John Temple*, third Viscount Palmerston (1784–1865) was Prime Minister during the late 1850s. Brought to power as a result of failures in the Crimean War, a war he helped start as Foreign Minister, Palmerston was known as "the people's darling" for allegedly turning defeat into victory in the Crimea. Originally a Tory, he usually acted with the Whigs. His first post was as Secretary of War, in which he served for twenty years (1809–1828). More interested in sport than politics, Palmerston lacked any great political ambition. Only after intense pressure by Canning did he become Foreign Secretary. As such he ran the office in his own manner, often making policy on his own. His first ministry began in 1855; after a brief Tory government in 1885–89, he served as Prime Minister to 1865.

‡*William Gladstone* (1809–1898) held his first cabinet post under Sir Robert Peel as Junior Lord of the Treasury. A financial expert famous for his thrift and budgetary skills, Gladstone became Chancellor of the Exchequer under Palmerston in 1859. A Tory turned Whig over the issue of Parliamentary reform, Gladstone believed in free trade, minimal government interference, peace, and retrenchment. In 1868 he formed the first of his three ministries (the other two began in 1880 and 1892). He retired from public life in 1894.

tion in private schemes. In concert with their personal lobbying, the directors of the company urged prominent men of commerce and science to appeal for government support of an absolute guarantee. Memorials were written by some of the more prestigious groups in the Empire, including the Royal Society, the British Association for the Advancement of Science, and the Royal Geographical Society. The government rejected these petitions on the ground that it would be supporting a monopoly. Moreover, the loss of £800 000 through an absolute guarantee of the Red Sea cable was fresh in their memories.

The directors took a new approach: They offered the government a share of the company with representation on the board.[44] This arrangement would allow the government to control the management. Again the government declined. Finally, as an alternative to an absolute guarantee, the company asked for a bonus in return for the services rendered to the War Office in preventing the unnecessary movement of Canadian troops to India.[45] This too was rejected. The government did, however, agree to increase its conditional guarantee from the original £14 000 to £20 000, but only if the cable worked. A condition for this guarantee was the surrender of New York, Newfoundland, and London Telegraph Co.'s monopoly on cable landing rights in Newfoundland. In return the government agreed not to grant landing privileges to any other companies until Atlantic either succeeded or abandoned the project.[46] The directors of the company, including Field, agreed to this condition in a meeting with Prime Minister Palmerston;[47] they had to agree, since without some form of government support there was no hope of laying a new cable.

The combined U.S. and British guarantees of £34 000 or an 8-percent return on invested capital were not sufficient to overcome the resistance of potential investors. Again the directors issued a prospectus for £600 000 in 8% preference shares. Again, they canvassed the major industrial and commercial centers of Britain for support. In the end the subscriptions totalled £72 000.[48]

Spurred by the Atlantic and Red Sea cable failures, the recognition of the strategic importance of cable communication, and the desire to promote cable development without unduly risking public and private money, the British government created a special commission of the Board of Trade to investigate the question of deep-water submarine cables. The commission was given £3000 to conduct experiments on the construction, testing, insulating, and laying of cables. Chaired by Capt. Douglas Galton* of the Royal Engineers, the commission included Professor Wheatstone,† the famous authority on electricity; William Fairbairn,‡ President of the British Association; the mathematical prodigy George Parker Bidder‡; Cromwell Varley,§ a famous telegraph mechanician; the secretary of Atlantic Telegraph Co. George Saward; James Stuart-Wortley,⋆ Chairman of Atlantic Telegraph Co.; and Latimer and Edwin Clark, both telegraph engineers with the Electric and International Telegraph Co.

For a year (1859-1860) this distinguished group witnessed a parade of equally distinguished

**Captain Douglas Galton* (1822–1899), an officer in the Royal Engineers, was best known for his work in sanitary engineering. He participated in several government investigating commissions.

†*Professor Charles Wheatstone* (1802–1875) began as a musical instrument maker. In the 1820s he became interested in telegraphy and with Cooke devised one of the first telegraph instruments. In the late 1830s Wheatstone became interested in submarine telegraphy.

‡*William Fairbairn* (1789–1874), whose first interest was in building iron boats for canals, later became involved in preventing boiler explosions on steamships. In 1860 he was elected the first president of the British Association for the Advancement of Science.

¶*Cromwell Varley* (1828–1883) studied telegraphy under Cooke and joined Electric and International Telegraph Co. During the Board of Trade hearings he had proposed simulating the environment of the cable in order to test its capabilities.

§*George Parker Bidder* (1806–1875), an engineer, designed the first railway swing bridge. He also worked in hydraulic engineering; he is well known as the designer of Victoria Docks. Bidder became interested in telegraphy very early; he was one of the founders of Electric Telegraph Co., Britain's first telegraph company.

⋆*James Stuart-Wortley* (1805–1881) graduated from Oxford in 1826 and was called to the bar in 1841. He soon became a prominent London barrister, serving as counsel to such clients as the Bank of England. He was also active in politics and served in Parliament from 1842 to 1859.

engineers, promoters, electricians, and scientists, each testifying on experiences with submarine telegraphy. Though many of these witnesses used the opportunity to promote their personal causes, the commission was most thorough in its investigation. The results of the commission, published in September 1860 and formally received in April 1861, are summarized in the following letter:

London, 13th July, 1863

We, the undersigned, members of the Committee, who were appointed by the Board of Trade, in 1859, to investigate the question of submarine telegraphy, and whose investigation continued from that time to April, 1861, do hereby state, as the result of our deliberations, that a well-insulated cable, properly protected, of suitable specific gravity, made with care, and tested under water throughout its progress with the best known apparatus, and paid into the ocean with the most improved machinery, possesses every prospect of not only being successfully laid in the first instance, but may reasonably be relied upon to continue for many years in an efficient state for the transmission of signals.

Douglas Galton	Cromwell F. Varley
C. Wheatstone	Latimer Clark
Wm. Fairbairn	Edwin Clark
Geo. P. Bidder	Geo. Saward

At the time of the British Board of Trade inquiry, a less active inquiry in the United States was in process. Selected by Atlantic Telegraph Co. in 1859,[49] the commission met rarely and never published a final report.*

Field continued to be unsuccessful. In Britain, the government was more interested in the Gibraltar cable, which was eventually laid in 1861–1862 between Malta and Alexandria. Also, there was alarm over first the threat, and then the outbreak, of the American Civil War. A cable would suggest an alignment with the industrial North. The most prudent British policy was to postpone any new action on the cable until the war ended.

In America, Field was himself caught up in a fight within American Telegraph Co., and then in the effect of the secession of Southern States on the telegraph industry as a whole.* American Telegraph Co. lines ran from Maine to New Orleans; the southern portions were immediately confiscated by the Confederacy. American's chief rival, Western Union, on the other hand, had most of its lines running east and west, north of the Confederate States; during this period Western Union built the transcontinental telegraph.‡ When the transcontinental line was completed, Field received a telegram announcing the event: "The Pacific Telegraph calls on the Atlantic cable. . . ."

---

*This commission was appointed by Atlantic Telegraph Co. A major reason for its ineffectiveness was the competition between factions on the commission. Joseph Henry and Lt. O. H. Berryman, USN, supported Alexander Bache, head of the U.S. Coast Survey. Matthew Maury was opposed by Bache and his supporters. William Everett and Samuel Morse did not take sides. Henry did some experimentation with the cable but his results were not published as a report of the commission.

†At one time during the Civil War Cyrus Field was accused of treason on the grounds of conspiring with a public enemy—Great Britain. The source of the allegation is unknown. It was brought before a grand jury; the charges were dropped for lack of proof—all while Field was in England. News of the allegations was nonetheless leaked to the London papers. When Field demanded that something be done, Secretary of State Seward and Attorney General Bates officially exonerated him.

‡The first message across the transcontinental telegraph was sent by Stephen Field, Cyrus Field's older brother and Chief Justice of the California Supreme Court. He was later appointed by President Lincoln associate justice of the U.S. Supreme Court, where he served the longest term of any justice in the history of the Court. The *Des Moines Register* wrote of the transcontinental telegraph:

> Ever since Adam was an infant, the City of Des Moines, or the site where it is located, has been cut off from the exterior world. We have had no Railroads. We have had no telegraph. We have been excluded from the activities of commerce. Situated midway between the two great rivers of the continent, without anything but coaches and stage roads to connect us with the rest of mankind, our condition has not been the most pleasant in the world.
>
> To-day our situation is immensely improved. We have the privilege of reading the latest dispatches in our own paper. The lightning and telegraph company have at length made us even with the Mississippi cities. . . .

and replied, "Your message received. The Atlantic Cable is not dead but sleepeth. In due time it will answer the call of the Pacific Telegraph."

Cyrus Field remained active in telegraphy throughout the Civil War. He was one of the first advocates of a unified system of military telegraphs, and one of the early supporters of the development of codes.[50] His favorite project, which he discussed often and at length with President Lincoln and Secretary of State Seward, was the building of a network of submarine cables along the East and Gulf Coasts that would help the Union monitor Confederate efforts to break the Union blockade.[51] This project, although it received the enthusiastic support of General George McClellan, was never carried out.

Field was in close contact with Seward and Lincoln at the time of the *Trent* affair and was therefore in a position to point out the utility of cables during such international crises. Two Confederate emissaries were seized from the British steamer *Trent* by a Union ship. Indignant at this violation of freeedom of the seas, Britain demanded an apology and release of the two emissaries, and sent 7000 troops to fortify the Canadian border. Lincoln and Seward, on hearing of the event, wrote Lord Russell, the Foreign Secretary, that the seizure was not done under orders and the prisoners had been released, but for a time, while the apology made its way across the Atlantic, there was real danger of hostilities between the Union and Great Britain.[52]

The London *Times* later observed, "We nearly went to war with America because we had not a telegraph across the Atlantic."[53] Field wrote to Seward on 1 January 1862:

> The importance of the early completion of the Atlantic telegraph can hardly be estimated. What would have been its value to the English and United States governments if it had been in operation on the 30th of November last, on which day Earl Russell was writing to Lord Lyons, and you at the same time to Mr. Adams, our minister in London?
>
> A few short messages between the two governments and all would have been satisfactorily explained. I have no doubt that the English government has expended more money during the last thirty days in preparation for war with this country than the whole cost of manufacturing and laying a good cable between Newfoundland and Ireland.[54]

Secretary Seward wrote to the U.S. minister to Britain, Charles Francis Adams, that President Lincoln "entertains most favorably the enterprise in question, and would be happy to cooperate with the British government in securing its successful completion. . . ."[55] Field followed this letter with a trip to England in January 1862 and visited Lord Russell, the Foreign Secretary, to urge his support. Lord Russell informed Field that the government thought it imprudent to become involved at that time.

Unwilling to accept defeat, Field and many eminent supporters tried in vain to persuade Prime Minister Palmerston to change the government policy.[56] Palmerston expressed interest in the project but reminded the delegation that he believed in minimal governmental interference in private schemes.

The most encouraging development during this trip was Glass, Elliot and Co.'s offer to make and lay the cable and to subscribe £25 000 to the project. In return the company would be reimbursed for costs of material and labor in building the cable and would receive an additional 20 per cent of the cost of the line.

Armed with this offer, Field, Lampson,* and Saward canvassed the major industrial centers of the United States and Great Britain seeking financial support for the project. At the urging of Lampson (though it was hardly necessary), Field attempted to persuade the American government to pressure the British to join in the project. Their efforts met with little success.

Field did not despair at the lack of government support. In a letter to Lampson he wrote, "We

---

**Curtis Lampson* (1806–1885) was born in Vermont but left for England at the age of 24. After making his fortune as a London merchant, Lampson became a naturalized British subject. When Field came to England with the cable venture, Lampson signed on as one of the initial directors and, when several of the directors quit, became Vice Chairman of Atlantic Telegraph Co. He received a baronetcy for his work with the Atlantic cable, and later became deputy governor of the Hudson's Bay Company.

can succeed without further assistance from either government, as I believe an appeal to the public will now get us all the money that we want. . . .''[57] Arriving in London in October 1862, he began to work to fulfill his prophecy. Firming up the offer from Glass, Elliot and Co. to make and lay the cable for £675 000, he got a better idea of just how much capital was required.

Field then began calling on his more influential contacts in England. George Peabody* and Junius P. Morgan,† two American financiers in London, expressed an interest in the project; Peabody had been among the original directors of Atlantic Telegraph Co. Field also visited the homes of many distinguished persons, including the Liberal statesman John Bright; Lady Franklin, the widow of the Arctic explorer; Capt. Galton of the Board of Trade commission; and Charles Francis Adams, the U.S. minister. Field was also introduced to Chancellor of the Exchequer William Gladstone.

The culmination of the efforts of Field, Saward, and Lampson was a special meeting of the board of directors of Atlantic Telegraph Co. on 2 December 1862. From this meeting came a prospectus calling for subscriptions to total £600 000 at 8-percent interest; the prospectus also noted the Glass, Elliot and Co. offer to make *and* lay the cable. Finally, a scientific advisory board including several members of the Board of Trade commission (Galton, Wheatstone, and Fairbairn) was appointed to advise on technical questions.

E. The Third Attempt: 1864-1865

Field left for America to try once again to raise American capital for the cable project. Again he made a tour of the major cities of the United States—New York, Boston, Philadelphia, Buffalo, Albany, and Washington. For the most part he addressed meetings of leading merchants such as the Corn Exchange, the Board of Brokers, and the Chamber of Commerce in New York City. Although the Civil War was in progress, Field and his associates were looking ahead to a burgeoning of American manufacture and trade after the war.‡

Field incessantly preached the benefits of the cable to commerce: ''I do not believe that *ten* cables would begin to do the work which would in a short time be given to it. The great commerce of our ports demands prompt communication with Europe.''[58]

> *But from our own interest*, here in the United States, it is greatly to be desired that a larger proportion of the stock should be held this side of the Atlantic. . . . This is a work which concerns us *far more* than it concerns England. England it connects with the United States and the British Provinces, but America it connects, not only with England, but with the continent of Europe and the East, in fact with the whole civilized world.[59]

Field was able to raise £66 000. Although this was only one-tenth of the capital needed for the project, it was much more than he had received in America previously—and the money was subscribed while civil war raged in Pennsylvania and Maryland. Secretary of the Treasury Salmon P.

**George Peabody* (1795–1869) was an influential American financier in London at the time Field was trying to finance an Atlantic cable. He began as a merchant in Georgetown, Maryland. By 1843 he had settled in London and opened up a banking house specializing in foreign exchange and American securities. Peabody was soon competing with the great banking houses such as the Rothschilds and the Barings. As an American financier in Britain, he worked hard to maintain British commercial confidence in America; his Fourth of July dinners were a London social institution. When Congress stalled on appropriating money for an American exhibit at the 1851 Great Exhibition, Peabody supplied the capital. In addition, he gave most of his fortune to educational institutions in America.

†*Junius Morgan* (1813–1890) joined Peabody & Co. in 1854. When Peabody retired the firm was renamed J. S. Morgan and Co. Morgan's most famous loan was $70 million to the French in 1870 despite Bismarck's statement that the loan would not be recognized. Junius remained head of the company until 1890, when he died in a carriage accident in Monte Carlo. After 1880 the business was run by his son John Pierpont Morgan.

‡*Peter Cooper* said of the cable in 1862: ''If we were in immediate communication with Europe we should avoid sending our goods to unprofitable markets. Hence advantageous results would follow to the great agricultural interests of the country. The produce merchants of America are ready and willing and able to supply all parts of Europe with the necessary provisions. That is one great item in itself, and amply sufficient, if there were no other, to pay the cost of putting down and completing such a work.

Chase* wrote Field:

> The war moves too slowly and costs too much; but it moves steadily, and rebellion falls before it. . . . I trust you are succeeding well in your great scheme of the inter-continental telegraph. It is an enterprise worthy of this day of great things. If I had the wealth of an Astor you should not lack the means of construction.[60]

Chairman of the board of Atlantic Telegraph Co. James Stuart-Wortley and secretary George Saward had better luck in England. After canvassing the major cities and some of the smaller towns of the British Isles, they had by May 1863 raised subscriptions of £240 000, including the Glass, Elliot and Co. subscription, which had increased from £25 000 to £40 000.

With this amount in subscriptions, the board decided to advertise for tenders (bids) on the cable construction. Although arrangements had been previously made with Glass, Elliot and Co., the Atlantic company wished to insure that nothing escaped the notice of the board.[61] Eleven firms, including Glass, Elliot and Co. submitted cable samples. These specimens were tested by the scientific consulting committee formed at the December 1862 meeting. As expected, a Glass, Elliot and Co. cable was selected.

The cable chosen would contain three times the copper and one-third more gutta percha than the previous one. In addition, the scientific consulting committee stipulated that elaborate measures be taken to insure the quality of the cable. An important lesson had been learned from the mistakes of 1858. The improvement in materials and quality control substantially increased the cost and thus necessitated the raising of even more capital.

Appeals to the public and to the government for additional financial support fell on deaf ears. All who would subscribe had done so. Such was the situation in January 1864 when Field returned to England; a cable had been selected but less than half the capital had been subscribed. In his efforts to revive the project, Field was introduced (possibly by John Bright) to Thomas Brassey,† 32-year-old London financier and railroad entrepreneur. After questioning Field thoroughly on the nature of the project and its expected return, he agreed to be one of ten to subscribe the necessary capital.

As had been the case with Peter Cooper's support at the beginning of the project, Brassey's support was conditioned on that of others. But also, as in Cooper's case, Brassey's endorsement stimulated that support. The next person Field approached was John Pender,‡ a Manchester industrialist who had been in Brett's Magnetic Telegraph Co. and on the board of Atlantic Telegraph Co. With Brassey's endorsement, Pender agreed to be one of the ten.

A few days after agreeing to support Field, Pender formed a new company, Telegraph

---

**Salmon P. Chase* (1808–1873) started as head of a boys' school in Washington, D.C. Elected to the U.S. Senate as a Democrat in 1849, he soon became an ardent, outspoken opponent of slavery. By 1855 he was Governor of Ohio. After two unsuccessful attempts to gain a presidential nomination, Chase became Lincoln's Secretary of the Treasury. Under his guidance the national banking system was instituted, providing reliable notes backed by government securities. Chase, consistently opposed to Lincoln's policies, resigned in 1864 to pursue his own presidential ambitions with the support of Horace Greeley and Roscoe Conkling. After another unsuccessful bid, Chase was appointed Chief Justice by Lincoln, and later presided over the impeachment trial of Andrew Johnson.

†*Thomas Brassey* (1805–1870) was one of Britain's foremost railroad promoters of the 19th century. Beginning as a surveyor, he soon became a railway contractor—the man who coordinated the multitude of activities that comprise building a railway. Brassey became best known for his railway projects in France, India, Australia, Argentina, Canada, and the Crimea during the Crimean War. Like other successful contractors, he had to become part financier and it was in this capacity that Field first approached him.

‡*John Pender* (1815–1896) made his initial money as a textile merchant in Manchester and Glasgow. However, he is best remembered as the primary figure in British cable expansion. After the success of the Atlantic cable, Pender and his associates financed a French transatlantic cable. Subsequently, the Pender-controlled Eastern and Associated Telegraph companies laid cables in the Mediterranean, Atlantic, West Indies, Persian Gulf, Indian Ocean, and the Pacific. By the late 19th century his companies controlled so many of the world's cables that Pender became known as Lord Cable. He was also active in areas other than cables. In 1869 he published *Statistics of Trade of the United Kingdom from 1840,* and from 1872 to 1885 he served in Parliament. In later life Pender became interested in electrical lighting in London.

The *Great Eastern* dwarfs her escorts as she heads out from Valentia, Ireland, on her first cable-laying expedition to America in 1865.

Construction and Maintenance (TC&M). TC&M was created by a merger of Glass, Elliot and Co. and Gutta Percha Co.; for the first time all phases of the construction of cables would be done within one organization. Pender did not accomplish this goal alone; he was aided by Brassey, Field, Daniel Gooch,* Richard Glass, and C. Elliot. Pender became chairman and Glass, managing director. Among the company's directors were Thomas Brassey and Daniel Gooch.

The creation of TC&M in April 1864 ended Field's search for capital. The company agreed to subscribe the remaining £315 000 and to purchase an additional £100 000 in bonds. This single subscription thus solved the problem that had plagued the cable for almost six years—lack of capital.

With the necessary capital and the right cable, the only problem remaining was to find a ship to lay the cable. Brunel had once urged Field to wait for completion of the *Great Eastern*, the largest ship afloat. The *Great Eastern*, having failed as a liner in the cost-cutting wars among the steamer fleet, was now available at a bargain price. Daniel Gooch, a director of both the company that owned the ship and Atlantic Telegraph Co., bought the ship outright and made it available for laying the cable. If it failed, there would be no charge; if the cable was successfully laid Gooch was to receive £25 000 in Atlantic Telegraph stock.†

**Daniel Gooch* (1816–1889) like Brassey pioneered in the areas of telegraphy and railways. With Brunel, Gooch was involved with the construction of the Great Western Railroad. A locomotive superintendent at 21, he went on to develop locomotive engines faster and safer than had previously been thought possible. In 1864 he resigned from the Great Western and devoted his time to the Atlantic cable. When the cable was successfully laid, Gooch, as chairman of TC&M, sent the first message from England to America. Aside from his duties as Chairman of TC&M he served in Parliament from 1865 to 1885 and helped save the Great Western Railroad from bankruptcy.

†Gooch did not buy the ship alone. Cyrus Field and Thomas Brassey put up the necessary capital to enable him to purchase it on the auction block. In January 1864, the ship was put up at auction. No satisfactory bid was received and a second auction was scheduled. Gooch, a director of the company trying to sell the *Great Eastern*, communicated with Field and Brassey between the first and second auctions. At the second, Gooch, using Brassey's, Field's, and his own funds, bought the ship for £25 000; the vessel had cost £1 million to build.

The final contracts for the construction and laying of the cable were signed on 5 May, 1864. When construction commenced, Field returned to America to secure U.S. ships to escort the *Great Eastern*. Given the resolve of the Board of the Atlantic company to take stringent quality control measures throughout the process, Field had little influence on the cable project's day-to-day operations. Before he left England, he gave a banquet in honor of the renewal of the project. James Stuart-Wortley described the scene:

> Collected around his table last night was a company of distinguished men/members of Parliament, great capitalists, distinguished merchants and manufacturers, engineers and men of science, such as is rarely found together even in the highest houses in this great metropolis. It was very agreeable to see an American citizen so surrounded. It was still more gratifying, inasmuch as we were there to celebrate the approaching accomplishment of the Atlantic Telegraph.[62]

During the banquet, U.S. minister Charles Francis Adams described the recent events as ". . . the reneval of a scheme which, if successful, will be among the most memorable events in the history of the world."[63] Others cited the increase in understanding that would result from the cable and the decreasing likelihood of crises owing to misunderstandings.

Before returning to New York, Field travelled around the Maritime Provinces of Canada to inspect the telegraph system that would carry the Atlantic cable's messages to New York and to select a new site for landing the cable. He selected Hearts Content— closer to the open sea, deeper, and more sheltered than the previous landing place, Bull's Arm.

Visiting Washington in January 1865, Field presented President Lincoln with a petition from Peter Cooper and himself requesting government supply ships to aid the expedition. Lincoln left the question to the discretion of Secretary of the Navy Gideon Welles. Welles rejected the project because of British "partiality" to the Confederacy, such as the British 24-hour rule limiting how long Union ships could remain in English ports.

By then 1000 miles of cable had been manufactured and loaded in the water-filled tanks of the *Great Eastern*. As the cable venture began to develop, it received increasing attention. Newspapers followed the progress of construction; the Prince of Wales visited the *Great Eastern* and sent the message, "I wish success to the Atlantic cable," through the completed cable. Cyrus Field, again a celebrity, was invited to speak at the opening of the Suez Canal.[64]

On 23 July 1865 the *Great Eastern*, accompanied by the British warships *Terrible* and *Sphinx*, began laying the cable from Ireland to Newfoundland. All went well until August 2 when, during an attempt to repair a faulty portion of the cable, the ship drifted and the cable snapped, only 600 miles from North America. As in 1858 there were attempts to recover the cable with grappling hooks, without success; this time the failure to lift the cable was not the failure of the grappling or lifting methods but of the lifting machinery. After unsuccessful attempts the ships returned to Ireland, arriving on 17 August 1865. Though the voyage had failed, it had successfully demonstrated the value of the paying-out methods employed. "From that time the task of laying deep-sea cables became a comparatively easy business. . . ."[65]

F. Success: 1866

The engineers and electricians on board the *Great Eastern* expressed this optimism to the board of Atlantic Telegraph Co. The board shared their confidence. The decision was made to raise money to build a new cable, as well as to lift and splice the 1865 cable.

The capital was raised by forming a new company, the Anglo-American Telegraph Co. The Anglo-American, as agent of Atlantic Telegraph Co., would construct, lay, and operate the cables. For these services it would receive £125 000 annually from Atlantic Telegraph revenues and another £25 000 annually from revenues of New York, Newfoundland, and London Co.—a 25-percent return. TC&M put up a large part of the necessary capital. The rest was easily raised from a few British capitalists.

With the money raised, work began anew on the cable. The *Great Eastern* was secured once again. The British government sent *Terrible* and *Albany* to aid the expedition.

The cable laying began at Valentia Bay on 13 July, 1866, with three British ships—*Terrible*,

Albany, and *Medway*—escorting the *Great Eastern*. On July 27 they arrived in Newfoundland, and landed the cable in good working condition. Immediately the *Great Eastern* left Newfoundland, returned to where the 1865 cable had been lost, spliced on a new section, and laid the remaining 600 miles. By 8 September 1866 there were two operating Atlantic cables.

Queen Victoria and U.S. President Andrew Johnson exchanged messages. But there was little popular enthusiasm in America compared with the 1858 celebration. The public had lost interest in the project during the Civil War. Besides, this attempt was even more the product of British promoters, scientists, financiers, and capital than the 1858 venture.

In Britain, on the other hand, there was a good deal of pride in this, the latest of Great Britain's triumphs. At the suggestion of Prime Minister Lord Derby, Anderson, Thomson, Glass, and Canning were knighted by Queen Victoria; Daniel Gooch and Curtis Lampson received baronetcies. (Cyrus Field, since he was not a British subject, received honorable mention from the Queen.) A grand banquet was held in Liverpool to honor the cable's success.

The event was not wholly ignored in America. Field was honored at a New York Chamber of Commerce banquet attended by such notables as Admiral Farragut and General George Meade. John Greenleaf Whittier commemorated the event by composing a cable hymn. Congress passed a resolution giving thanks to Mr. Field for working to establish the transatlantic communications link. A medal was to be struck for the occasion, but it was lost; Field was given a facsimile. Seven years later he purchased the original for $600 just before it was to be melted down at the Philadelphia mint.[66]

Chapter 2

CHALLENGE AND RESPONSE

IN HUMAN enterprises, there is almost always an element of uncertainty. When this uncertainty is large—as in an enterprise or endeavor for which there are many unknowns—a challenge is created, a test of human ingenuity and human capability. Whether or not the response to such a challenge is successful, the people who attempt it and their institutions and parent society are affected, because the lessons learned and the experience gained have altered the uncertainties and affected future patterns of response to similar challenges. Therefore a technological enterprise can and does produce significant impacts even before the technology itself is successfully implemented.

The historical record provides our only real source of information on the consequences flowing from the introduction of a new technology into a complex social, economic, and cultural environment. This analytical insight, gained through retrospective technology assessment, can give us significant clues as to how future technologies may affect society. Thus retrospective technology assessment promises to be a useful tool in guiding prospective or anticipatory technology assessments.

Along with the advantage of having an historical record from which to glean insight about the consequences and impacts of innovations comes the disadvantage of historical research: the inaccessibility of the real decision process that led to establishing that innovation. Alternative courses of action that were considered, unsuccessful attempts and false assumptions, the perceptions of the uncertainties that led the entrepreneurs—many aspects of the original challenge and response—were only selectively recorded and preserved. In contrast, in the anticipatory assessors' prospective view the potential impacts remain uncertain and ambiguous; yet the alternative courses of action and the varying perceptions of the challenge are accessible. This dichotomy must be recognized in the examination of the challenge and response to the opportunity of creating a transatlantic telegraph link.

Different tasks in that enterprise posed different uncertainties and the response strategies revealed varying degrees of confidence in coping with these uncertainties. Three distinct types of challenge can be discerned: technological, entrepreneurial, and political.

A. The Technical Challenge

In 1854, when Cyrus Field first conceived of the possibility of creating a submarine telegraphy link between the North American Continent and Europe, "practical" men clearly dominated the developing telegraph industry. These practical men were technological entrepreneurs just as Field was a financial entrepreneur: their desires were to make things happen, both for profit and for glory. These "practical" men included inventors and tinkerers such as Samuel F. B. Morse, and engineers such as Charles Bright and Frederick Gisborne, who had first approached Field with a proposition to build a link to Newfoundland. Field turned to Matthew Maury of the Naval Observatory and Samuel Morse for advice on the feasibility and practicality of such a venture. Both enthusiastically supported the idea.

At the same time, William Thomson and Professor George Gabriel Stokes were exchanging letters on the theory of submarine telegraphy.[1] These were leading men of science. Thomson was later to become Lord Kelvin, renowned for his work in thermodynamics and electricity; the Kelvin scale of absolute temperature is named after him. Stokes was an eminent mathematician and hydrodynamicist. Neither man questioned the feasibility of building submarine telegraph lines across the Atlantic, but their investigations cast considerable doubt on the practicality of obtaining a rapid rate of signal transmissions.

It may have been fortunate for the early entrepreneurial efforts that these theoretical investigations did not dampen the investors' enthusiasm. However, the failure to appreciate the significance of these investigations certainly contributed to, if they did not directly cause, the demise of the 1858 transatlantic cable.

By the end of the decade several major cable project failures had completely destroyed the confidence of potential investors. The entrepreneurial drive was stalled until the Joint Committe of the British Board of Trade and of Atlantic Telegraph Co. had completed their systematic study of the advisability of such ventures in 1861. This well-organized study shows how a retrospective view of the experience gained by all parties engaged in completing cable projects uncovered a failure to disseminate experience and learning among the competitors in the cable industry. But most of all it demonstrated that a systematic decomposition of a complex technological venture into its component requirements could produce a high degree of assurance of future success. Thus technology assessment has a firm basis on which to assess *prospectively* the likelihood of *technical success*.

Understanding Long Submarine Telegraph Lines

In 1854, with data from the 356-mile Varna-to-Balaklava Black Sea cable and similar operations, Thomson was attempting to anticipate the behavior of much longer cables. It had already been noticed that submarine cables could not transmit as many message pulses per unit time as could landlines. This anomaly worsened with increased length.

For all practical purposes, the rate of signal propagation along landlines was instantaneous, even though scientists did define and measure this speed. In submarine telegraph lines the speed of transmission, and, more important, the sharp rise and fall of voltage as the batteries were connected and disconnected to encode signal pulses, were delayed, attenuated, and smeared out. The cause and extent of this signal deterioration could greatly reduce the practicality of transatlantic telegraphy.

Thomson derived an equation to describe the behavior of submarine telegraph cables. He also calculated the time sequence "arrival curve" of the received pulse when the signal initiation is a discrete and sharp connection to batteries and its end is neutralization of residual charge at the cable-sending terminal.

The basic differential equation obtained by Thomson was known as the "heat equation"—first studied in detail by Baron Jean Baptist Joseph Fourier in his analysis of the penetration of heat into the earth through seasonal changes.* From a theoretical point of view it is fortunate if one can represent the behavior of a novel device by a well-understood equation. But from an engineering point of view,

---

*The heat equation is a first derivative in time but a second derivative in space. For a submarine telegraph of electrical capacity c per unit length and electrical resistance k per unit length, the heat equation is

$$ck\, dv/dt = d^2v/dx^2 \qquad (1)$$

where v is the voltage along the cable with distance coordinate x. (The wave equation would involve a second derivative of time as well as distance.)

The solution to Eq. (1) derived by Kelvin was

$$v = \tfrac{1}{2}TV\,[kc/\pi t^3]^{1/2}\,\exp[-(x^2kc)/4t]$$

where TV is the duration and potential of the initial pulse sent from $x = 0$ at $t = 0$. Note that the maximum of this pulse travels as $t = kcx^2/6$; that is, pulses are delayed on the receiving end of cables in proportion to the square of the distance, rather than in proportion to distance as with wave propagation.

The corresponding signaling rate over such a cable would vary as $S = A/kcL^2$, where S is the number of discernible pulses per second, A is a constant, and L is the length of the cable.

Real telegraph cables tend to operate somewhat more favorably for reasons explained by neither Kelvin nor Bright. An empirically derived formula gave:

$$S = (1000/kc)[(826/L) + (27\,690/L^2)]$$

the heat equation has many disappointing properties. As one would expect, the propagation of "pulses of heat" is strongly attenuated by conducting media. It would have been much more encouraging if one had instead obtained the "wave equation" that describes the propagation of sound and similar transmissions of weak pulses over long distances.

As was the common practice of the time, Thomson used a mechanical analog to improve his intuitive grasp of a long cable's electrical responses.[2] Imagine a long pipe of India rubber containing water and stuffed with a porous spongy matter filled with minute apertures. The water is analogous to electrical charge, the elasticity of the walls of the pipe is analogous to the electrical capacitance of each section of cable, and the porous material creates the analog of electrical resistance. The pressure of the water in the pipe is analogous to electrical voltage; the flow of water represents electrical current.

Transmission of an electrical signal would be equivalent to injecting a spurt of water into the sending end of the pipe and consequently raising the pressure. If the pressure were maintained by continuing addition of more water, eventually the entire pipe would become distended as the water worked its way through the porous material. Removing the pressure at the sending end would permit the cable to return to its original size and pressure, but only slowly. At the receiving end one could either observe the flow of water out of an open end or detect the increased pressure of a closed end. In either case one can easily imagine there would be a retardation between sending and receiving of signals.

Operation of submarine telegraphs consisted of abruptly attaching the sending end to a battery voltage and very quickly returning the connection to a neutral condition. In the analog, one would suddenly pressurize the sending end of the rubber pipe and quickly release that same pressure. *Obviously, most of the water introduced into the pipe would reverse itself and flow back out of the sending end*! This is exactly what happens with the telegraph line; most of the charge introduced by the sender is discharged at the sender's end. It is hard to imagine that any flow is induced at the receiving end.

A quantitative analysis is necessary to understand the magnitude and duration of the received pulse. Electrical phenomena stretch the human capacity for heuristic reasoning to its limits even with the assistance of extensive mathematical training. Here the inventors and majority of engineers were at a considerable disadvantage in grasping the phenomena associated with submarine telegraphy. There were five principal quantitative implications of Thomson's theory:

a. The time duration of the smearing out of a cable pulse from two cables of similar construction but different lengths increased as the *square* of the lengths. Since the ability to distinguish successive pulses determines the rate of sending messages, any method applicable to a shorter cable would be slowed down in proportion to the smearing of the signals. That means that a transatlantic cable of 2200 miles would be $(2200/356)^2 = 38.4$ times slower than the Varna-to-Balaklava cable if both were of similar construction.

b. Amplitudes of signal pulses decreases as the inverse *cube* of the ratio of cable lengths. Therefore, from the previous example, the transatlantic cable signal would be only one part in 238 as strong as the Varna cable for comparable sending apparatus.

c. Enlarging the size of the cable cross section while holding constant the ratio of the central conductor diameter to the total diameter of insulation would increase the signaling rate only in direct proportion to the added weight of cable core. Doubling or tripling a cable's weight might be practical, but increasing it by 200 or even 40 times would be out of the question. Moreover, the manufacture and laying of a heavier cable would be a formidable task. No ship available at that time could carry 2200 miles of cable—even of smaller size.

d. If leakage of currents through the insulation were significant, then the resistance per unit length of cable to such currents must scale as the inverse *square* of the length. This condition is quite unlikely to be met. One could only hope that the leakage currents would be negligible in all cables.

e. In a paper on the "peristaltic induction of electric currents"[3] Thomson showed that signal pulses sent over one conductor of a multiple conductor cable of significant length would produce pulses on the neighboring conductors even if they were electrically insulated. The induction resulted from the capacitive couplings between the separate conductors, which acted to "induce" charges in the secondary conductors. Fortunately, Thomson's analysis and advice were heeded in the design of the transatlantic cable; only a single-stranded conductor was used and the promoters gave up the intention of sending several messages simultaneously.

These five findings would suggest caution in the design, manufacture, and operation of very long submarine cables, but most of the warnings implicit in these findings were ignored by the promoters. Fortunately, the actual behavior of submarine cables deviates favorably from the theory; but that could have been known only *retrospectively*. Unfortunately, poor understanding of the cable's electrical properties by a chief engineer of the 1858 cable project may have caused its demise.

Thomson, continuing his heuristic line of reasoning, based upon the water analogy, was looking for evidence of inertial effects in the transmission of pulses over cables.[4] He knew that magnetic induction introduced inertia-like behavior to the movement of electrical charges. But inertial effects were not to be found in the observed cable responses. A controversy arose over "true and false discharge of a coiled electrical cable" that was resolved by Thomson's recognizing the mutual magnetic induction of the coil. The phenomenon consisted of a persistence of current flow in a coiled cable even after the battery had been disconnected. No such phenomenon occurred once the cable was uncoiled as it was paid out behind a ship.

It was Oliver Heaviside who first theorized (1885) that increasing the magnetic induction of a submarine cable would greatly enhance transmission.[5] If we return to the water analog, magnetic induction would correspond to the inertia of the water. Suppose the rubber pipe were much larger in diameter and that the porous resistive material were negligible; then a pulse at the sending end would be transmitted along the pipe as a *wave* of stretching and contracting pipe. With sufficiently small resistance, the output pulse of water would be undiminished from that of the input. Moreover, adjacent pulses would not smear into one another, so that the rate at which they could be sent and received would be greatly increased. This theoretical insight had to wait more than thirty years for the development of magnetic materials sufficient to produce the necessary induction.

Thomson's 1854–1873 papers contain schemes to encode different letters with different amplitudes of pulses. If this proposal had been carried out, it would have been a very early example of pulse code modulation, in which digital information is represented by discrete amplitudes of a continuous analog signal. He also proposed to shape the input signal to avoid exciting the longest wavelength modes of the cable; since these modes were the slowest to discharge, they inhibited faster signaling. Again, it was over twenty years before signal shaping was introduced into transatlantic telegraphy in the form of "curbing." Since curbing is accomplished by the transmission of pulses canceling counter pulses, automatic sending equipment was required to generate the proper pulse sequence for each signal.

Thomson's conceptual model is to a large extent vindicated by the 1879 success of duplex telegraphy—simultaneous transmission and reception of messages. In duplex telegraphy the powerful outgoing signals are applied in a balanced fashion to a Wheatstone bridge circuit. Weak incoming signals upset this balance and are consequently detected when they cause a current to flow in the very sensitive receiver spanning the bridge. The ability to balance the bridge circuit depended on the degree to which an electrical model, or "artificial line," could be made equivalent of the real cable. These artificial lines, first proposed by Varley, were constructed of resistors and capacitors assembled in exactly the arrangement described by Thomson as the basis for his theory.

In retrospect we see that these theoretical investigations had a mixed impact on the first transatlantic cable venture. In a positive sense, the design of the cable was improved when all suggestions of using multiple conductors to send simultaneous messages were dropped.

Fortunately for the confidence of investors, Thomson's results did not lead engineers to question whether a practical rate of signaling could be achieved or whether leakage currents would be increased by the conditions at the bottom of the Atlantic Ocean. Nature was good to the enterprise since signaling rates are higher for real cables than Thomson's theory predicts and cables perform better in the ocean's depths than they do on the factory's wharf.

Good fortune will go only so far. The near-total disregard of theoretical investigations possibly resulted in the failure of the 1858 cable. Thomson realized that the incoming signals would be exceptionally weak if normal battery power was applied to the sending terminals. He devised the "mirror galvanometer" to detect these weak currents. This instrument had been used aboard ship during the laying of the cable. Charles Bright, the son of Sir Charles Bright who laid the cable, gave this account of the cable's failure:

> When the cable passed out of the hands of Charles Bright and his assistant engineers, and from Professor Thomson and the electricians on board, to Mr. Whitehouse on shore, it was in good condition—far better (on account of the low temperature and pressure at the bottom) than before laying. It had only been subjected to battery currents, derived from about seventy ordinary Daniell's cells, by which all the signals were interchanged between ships.
>
> Unfortunately for the life of the cable, Mr. Whitehouse, the chief electrician—a man highly respected for his scientific attainments—was imbued with a belief that currents of very high intensity (i.e., a high potential) were the best for signalling; and he had enormous induction coils constructed, *five feet long*, excited by a series of very large cells of the potent Smee type, and yielding currents estimated at about 2,000 volts potential. Instead of being satisfied with the comparatively mild and innocuous currents from the cells used on board ships, he brought these induction coils into operation, with the result that faults were developed. The insulation was, in fact, unable to bear the electrical strain, and thus the signals began to fail gradually. Ordinary battery power was resumed, but too late, for then the number of Daniell's cells had to be continually increased, till they actually numbered 480! On this, the cable soon ceased to speak, even when using Professor Thomson's mirror galvanometer.
>
> As a proof of the intense and destructive power of the induction coil currents employed by Mr. Whitehouse, the following experiment may be mentioned. It was carried out by Mr. E. B. Bright and Mr. C. F. Varley, at Valencia, shortly after the cable broke down. A very fine prick was made with a needle in a spare length of the core of the cable, sufficiently deep just to touch the conductor, which was then placed in a large earthenware jar filled with water and connected to earth, as was one end of the great induction coil. On the other end of the secondary coil being joined to the conductor of the piece of cable, and the current applied, the interior of the jar was immediately lit up as if it were a lantern, and on withdrawing the specimen of core, the gutta-percha was found to be melted away to about an inch around the tiny needle puncture.[6]

The dialog between Whitehouse and Thomson on the signaling rate of cables revealed that the differences in their opinions predated the application of these induction coils.

Whitehouse's use of large voltages did not destroy a well-working cable; poor insulation was a major cause of the failure. Between the 1857 and 1858 expeditions, Thomson had made tests on the cable and found numerous faults. One cannot say *for certain* that the fate of the cable would have been greatly altered had Whitehouse not used high-voltage signals, but surely its chances would have been improved.*

The Rush to Construct a Transatlantic Submarine Telegraph

There was ample reason for the financial promoters, engineers, and innovators engaged in creating a telegraphic link across the Atlantic to take an optimistic prospective view of their endeavor. They lived in a century of unprecedented technological innovation. For the first time in human history machines driven by energy derived from fossil fuel were substituting for human and animal muscle power. Table 2-1 places the expanding telegraphic communications technology in the context of related developments in transportation and related communications systems.

The Brett brothers, Jacob and John, were the first to complete a significant submarine telegraph connection—that between Dover and Calais spanning the English Channel. This undertaking was not

---

*Dr. Mervin J. Kelly and Sir Gordon Radley, prominent engineers and planners of the first transatlantic telephone cable (1956), held this impression of the cause of failure of the 1858 submarine telegraph cable:

> These failures are today seen to be the result of insufficient appreciation of the relation between the mechanical design of the cable and the stresses to which it is subjected as it is laid in the deep waters of the Atlantic. The making and laying of deep sea cables was a new art and designers had few experiments to guide them. (*Bell Syst. Tech. J.* 36:1-5, 1975.)

without its setbacks,* but the second attempt in 1851 was a lasting success. By 1861, more than ten cables operated between Great Britain and the Continent. These shallow-water cables were a marked financial success. The earlier cables drew the attention of engineers and financial backers to sponsor more ambitious projects in deeper water and spanning greater distances.

Entrepreneurial brashness and optimism continued on into the early approaches to deep-water long-distance submarine telegraphy. When Field and the other promoters launched their ambitious plan to construct a 2200-mile cable (the longest prior to that being 171 miles) and to lay the cable in depths exceeding 2000 fathoms, the reaction of the practical men in charge to the unprecedented problems was to continue with the established practices. When these proved inadequate, then improvisations were made.

Cyrus Field, anticipating a short duration of the loan of the warships, along with pressures from investors and governments, and desiring to lay the cable during the August calm in the North Atlantic, left only four months for the cable's manufacture. This was an incredible feat for the time as it required drawing 20 500 miles of copper wire, 300 tons of gutta percha, and 367 500 miles of protecting wire from 1687 tons of charcoal iron.[7]

In meeting the four-month schedule, Field overrode the advice of his engineers, Messrs. Bright and Whitehouse. Advice that a larger diameter cable was needed was also ignored, dismissed without adequate consideration.**

Competition among manufacturers also complicated matters. To insure quick manufacture and avoid the necessity of choosing one over the other, the manufacture of the outer covering of the cable core was divided between the two major manufacturers of the day: Glass, Elliot and Co. and R. S. Newall and Co.

Engineering judgment was consistently overruled by the entrepreneurial promoters. Engineers suggested splicing the two halves in mid-ocean and laying the cable as the ships steamed towards opposite shores. This scheme would cut laying time in half and allow the choice of the most appropriate time and weather for the difficult task of splicing a cable in the middle of the North Atlantic. The engineers were overruled by the board of directors of Atlantic Telegraph Co., who wanted to maintain constant contact with the expedition. Thus, *Niagara* laid the first half of the cable; *Agamemnon* was to take over in mid-ocean, but the cable broke and the attempt was abandoned for that year.

In 1858 the engineers' method was used, although the failure of 1857 had not resulted from the procedure for laying the cable. (In 1866 a ship capable of laying the cable in one piece from shore to shore had become available, and the simpler means of laying a cable could again be used.)

---

*Charles Bright gives the following account of the demise of the first cable:

It subsequently transpired that a Boulogne fisherman had—accidentally or otherwise—raised it to the surface with his trawl. Imagining that he had discovered a new kind of brown kelp seaweed, snake, or coral 'with gold at its center' he cut out a considerable length. As has since been abundantly proved, there was, indeed, gold in its heart, though not in a literal sense. This fisherman has been aptly referred to by Dr. W. H. Russell as a *piscatore ignoble*, though probably only anticipating by a few hours the fate to which such a like was surely doomed. (*Submarine Telegraphs*, 9.)

[*Note*: The obvious fate was predicted, yet not considered or ignored by the promoters.]

**Bright comments:

This race against time was the outcome of an unfortunate engagement (insisted upon by Mr. Field in connection with his American arrangements) on the part of the company towards its shareholders and the public. Messrs. Bright and Whitehouse urged that more time should be given, to ensure greater care in manufacture, and the former advocated a different type of cable, with a conductor more than three times as large, and a much greater thickness of insulation....Unfortunately no experience in these matters was at hand, or no doubt the extreme importance of such questions would have been better appreciated.[8]

Comments from the Joint Committee hearings:

Mr. Whitehouse (Question 1838) states that Mr. Cyrus Field objected to further experiments because it would have put off the laying of the cable for twelve months, and other persons admit that additional experiments should have been made.[9]

TABLE 2–1.—The development of transportation and communications, 1880–1878.

Year	Telegraph	Steam Shipping	Railroads	Other T/C Systems
1800	Voltaic battery described			
1802			Trevithwick's first steam locomotive, Gt. Britain	
1807		Fulton's paddle steamer		
1809	von Sommering's water voltameter telegraph			
1811		Steamboat service on the Mississippi		
1816		Regular steam service on rivers, Gt. Britain		
1818				Erie Canal started
1820	Oersted discovers electromagnetism	First iron-hulled ship		
1825	Sturgeon electromagnet		First passenger-carrying railroad	Erie Canal completed
1828	Henry's improved electromagnet			
1829			Stevenson's *Rocket* wins Rainhill trials	
1830	Schilling demonstrates electric telegraph	Steam on Ohio, Mississippi, Missouri rivers	*Rocket* used London to Manchester	
1831	Henry's electromagnetic signal		Pennsylvania RR opens	Goodyear—vulcanized rubber
1832	Morse's telegraph			
1833	Gauss & Weber telegraph operates over 9000 ft			Canadian clipper crosses Atlantic in 25 days
1834	"Wheatstone bridge" devised by S. H. Cristie			Pittsburgh-Philadelphia canal opened
1835	Morse's first electromagnetic instrument (short distance)			
1836	Cooke's first electromagnetic instrument (short distance)			
1837	Morse can transmit over 10 miles			First daguerreotype First penny newspaper, U.S.
1838	Cooke's instrument (long distance)	First scheduled steam lines cross Atlantic *Great Western* crosses in 15 days		10 000 miles post roads, U.S.
1840			3000 miles of railroads in U.S.	3300 miles canal in U.S. Postage stamp introduced in Great Britain
1841		U.S. Gov't orders its first iron ship		
1842	Cooke-Wheatsone line: London to W. Drayton, 13 miles			
1843	Congress grants Morse $30K for telegraph line Gutta-percha introduced to England	Launching of first propeller-driven ship to cross Atlantic		

Table 2–1 (Cont'd)

Year	Telegraph	Steam Shipping	Railroads	Other T/C Systems
1844	Telegraph line between Baltimore and Washington			
1845	English Channel cable proposed by Britain	M'Naught's compound steam engine		
1846	Washington-Boston linked House's printing telegraph			First penny daily newspaper in Great Britain
1847	Siemens extrudes gutta-percha sheeth	Founding of Hamburg-Amerika Line	RR Zurich to Baden	NY Associated Press founded Postage stamp introduced in U.S. Double-cylinder rotary press introduced
1848	Telegraph lines reach New Orleans			
1850	Telegraph line to Halifax, 1200 mi in U.S.	5% British fleet steam	First cast-iron RR bridge	
1851	English Channel cable			N.Y. Times founded Wells Fargo founded
1852	2500 mi telegraph line in U.S.			Loads carried by coastal vessels twice those of RR
1853	India telegraph system begun England-Ireland cables			
1854		First ship with compound expansion engine	Turin-Genoa RR	
1855	Hughes printing telegraph Crimean cable laid American Telegraph Co. founded	Iron Cunard steamer crosses Atlantic in 9.5 days		
1856	Western Union founded		Black Forest RR with 40 tunnels	
1857	Atlantic cable attempt fails Treaty of Six Nations	N. German Lloyd steam line founded	European financial crisis coused by speculation in U.S. RR shares	
1858	Atlantic cable works one month	*Great Eastern* launched		
1859	Failure of Red Sea cable			First oil well, Titusville, Pa.
1860		10% British fleet steam		Suez Canal begun First practical internal combustion engine First primitive typewriter
1861	Telegraph to India Parliamentary inquiry on cables Transcontinental (U.S.) line			
1862				
1863			First RR New Zealand	First U.S. free city mail delivery
1864	Persian Gulf cable	Civil War naval battles		
1865	Atlantic cable laid, breaks Telegraph Union founded	1700 die in *Sultana* explosion, Mississippi	35 000 mi RR (U.S.)	San Francisco *Chronicle* and *Examiner* founded First oil pipeline (U.S.)

Table 2–1 (Cont'd)

Year	Telegraph	Steam Shipping	Railroads	Other T/C Systems
1866	TRANSATLANTIC CABLE Western Union takes over American Telegraph		RR through Brenner Pass	Prussia establishes government mail service
1867	Kelvin's siphon recorder patented Russian-Alaskan telegraph abandoned			
1868	Permanent International Telegraph Bureau established			
1869	French transatlantic cable laid		Transcontinental (U.S.) RR completed	Suez Canal completed Brooklyn Bridge begun
1870	British cable line to India completed	20% British fleet steam		Clipper *Cutty Sark* launched Standard Oil established
1871	British cable line to China and Australia completed	First modern luxury liner, *S.S. Oceanic*		
1872	Stearn's duplex telegraph		Automatic RR airbrake	
1873	Transatlantic cables			First production of typewriters
1874	Siphon recorder on Atlantic line Edison's quadruplex telegraph St. Petersburg Conference of International Tel. Union			International Postal Union Eads Bridge across Mississippi river
1875	Madeira-Brazil cable			*Glory of the Seas* (sail) sets record 35 days from San Francisco to Sydney
1876			First Chinese RR	Bell invents telephone Edison invents phonograph
1877		Frozen meat first shipped Argentina to Europe	Whipple truss RR built, Ohio River	First public telephones, U.S.
1878	Duplex introduced on Atlantic line			

In short, Field and the other promoters had prejudiced their ambitious undertaking, and quite possibly delayed its success for nine years, by insisting on unreasonable speed and ignoring the need for quality control in manufacturing the cable. In doing so they overruled the warnings of Bright, Thomson, Whitehouse, and other experts.

If the investment in submarine telegraphy could be examined carefully, surely 1859 would have been near the low point for return on capital. By 1861 only a little over 3000 miles of 11 364 laid to date were still operable.[10] The heaviest losses were on the long and deep lines. Four cables accounted for over half of all that had been laid: 2200 miles in the 1858 Atlantic cable (not working), 3499 miles of the Red Sea cable's six sections (none of which was working), 550 miles of the Singapore-Batavia cable (out of order by 1861), and 700 miles of the Sardinia-Malta-Corfu cable, which had worked for 12 months before it had to be abandoned. Losses on the Atlantic cable were £350 000; those on the Red Sea cable, guaranteed by the British Government, were £800 000.

Aroused by these failures and recognizing the unwillingness of investors to take further risks, the British Government resolved to investigate thoroughly the question of submarine telegraphy. A Joint Committee was appointed, four members by the Board of Trade and four by Atlantic Telegraph Co. Care was taken to obtain impartial and informed views of the subject. This committee's report was to become the first systematic assessment of submarine telegraph ventures.

The Benefits of Technical Planning

By 1860 the technology had developed far enough so that a retrospective view—that is, a careful review of past experience—suggested a systematic framework for evaluating the technology of constructing submarine telegraphs. This framework was based on a systematic decomposition of the functional requirements of rapid long-distance communication.

A systematic decomposition begins with the general purpose for which the technology is intended and extracts the subordinate capabilities that must be provided in order to accomplish this higher purpose. The articulation of subordinate capabilities proceeds until all contingencies are covered. The decomposition *insures a comprehensive examination* of all necessary subordinate capabilities.

A fully articulated decomposition could then be used as the basis for a *prospective examination* of the requirements of future efforts. Because technology can be subdivided into relatively discrete, noninteracting components, each component can be refined once and for all without a complete rethinking of all the other components. Moreover, a technical problem once solved remains solved—future contingencies do not negate previous solutions, although they may demand refinement. This systematic divisibility of complex technical requirements into many tractable component capabilities provides the basis for confident pronouncements of an examining team as to whether the complex requirements could be met or not.

Table 2–2 illustrates the generic nature of a functional decomposition of rapid communications into component technical requirements. Five requirements are common to all eleven communications technologies: a communication medium, a physical signal that can be transmitted, a coding of messages in terms of these signals, a means of initiating the signals, and a means of receiving the signals. Each communication technology encounters its own peculiar limitations; key limitations are listed.

This decompositon applies to post-telegraphic communication as well as pretelegraphic. Very primitive methods such as drums, signal fires, or semaphores relied on human senses as the ''means of receiving signals from the medium.'' One early electrostatic telegraph used human tactile responses to high-voltage pulses as the detector, but for the most part telegraphy was the first communications technology to interpose instruments between the incoming physical signal and human senses. Today long-distance telephone communication is carried on among computers without interception, interpretation, or exercise of human intervention. More ''human'' capabilities are embodied within the technology's repertoire of capabilities. Yet the principles of decomposing complexity remain unaltered.

The Joint Committee adopted a systematic framework for their retrospective view of the accomplishments and failures of submarine telegraphy during the 1850s. The first successful shallow-water submarine telegraph from Dover to Calais had established the requirements of the basic components of the cable itself: (1) the copper conductor as a core, (2) insulation from sea water by gutta-percha, and (3) protection from damage by a sheathing of iron rope. The laying of cables posed new challenges, especially the steady navigation independent of weather offered by steam ships. These topics formed the major categories in the Joint Committee's decomposition of the functional requirements of constructing successful submarine telegraphs (Table 2–3).

All the key technologies for laying shallow cables were available in 1850 as the first attempt was made. Copper could be drawn into wire by the early 1800s. Gutta-percha was relatively new to British manufacturing; it was introduced in 1843, but its insulating properties were determined by 1847. In the same year Werner Siemens developed a machine for extruding gutta-percha in a continuous flow around copper wire. The technique for drawing iron wire and twisting it into rope was relatively new but available; the key British patent was taken out during the 1840s. Steam-powered ships were introduced in 1820 and had been refined to a degree capable of providing the steady navigation necessary for paying out telegraph cables.

The basic components of the cable and laying equipment were only the organizing framework for a description of the more demanding task of constructing long-distance submarine telegraphs in

TABLE 2-2.—Functional requirements of rapid long-distance communications by any means: Retrospective—prospective (ca. 1861).

	Talking Drums[11]	Signal Fires	Semaphores	Electrostatics	Landlines
1. Communications Medium	Atmosphere	Optical portion of electromagnetic spectrum	Optical portion of electromagnetic spectrum	Insulated electrical conductor	Elevated conducting wire
2. Types of Signals That Can Be Transmitted	Sound waves	Bursts of light	Positions of signal flags or light sources	Presence of high voltage or neutrality	Pulses of electrical current at moderate voltage
3. Coding of Messages by Basic Signals	Two-tone phases representing words in sender's tonal language	Pre-arranged signals or codes	Letter codes for phonetic written languages	Letter codes for phonetic written languages	Letter codes for phonetic written languages
4. Means of Entering Signals into the Medium	Man striking drum	Man shadowing light source or interrupting smoke column	Man orients flags or lights or shutters	Man generates voltage with friction device and switches it on or off conductor	Man switches electrochemical batteries on and off the line
5. Means of Receiving Signals from the Medium	Human listener	Human observer	Human observer with aid of telescope	Tactile sensation, gold leaf electrometer, chemical reaction	Electrochemical reaction, Electromagnetic click, needle swing, mirror swing, mechanical decoder
Principal Limitations	Distance of transmission between stations, need for quiet	Weather, time to establish fire, slow rate	Weather, message rate, number of men to operate network	Must have very well-insulated conductors	Signal rate with human operators, protecting wires

TABLE 2–2 (Cont'd)

Submarine Telegraphy		Telephone	Broadcast	Satellite	Fiber Optics
Short-Shallow	Long-Deep				
Insulated and mechanically protected conducting wire (cable)	Insulated cable capable of supporting its weight and of maximum size	Insulated electrical conductor	Electromagnetic spectrum of Earth's atmosphere	Electromagnetic spectrum through Earth's atmosphere	Transparent glass fiber of very small diameter
Sharp pulses as with landlines	Elongated pulses of weak electrical current	Oscillating electric currents at low voltage	Electromagnetic waves for which atmosphere is transparent	Electromagnetic waves that penetrate atmosphere	Interrupted or varying pulses of light
Letter codes as with landlines	Letter codes or cryptographic codes to increase message capacity	Conversion of voice induced sound vibrations into electrical current oscillations	Discrete pulses for codes or vocal oscillations on specially modulated waves	Similar to broadcast but coded to prevent undetected transmission errors	Very high message capacity needs electronic coding
Same as landlines	Same as landlines plus emphasis on increasing signal rate	Microphone: electrical switch activated by sound vibrations	Varying the amplitude or frequency of a high-frequency oscillator driving an antenna	Same as broadcast but at higher frequencies and focused into a beam	Solid-state light-emitting diode or laser, electronically modulated
Electromagnetic click, mirror swing	Same as short lines plus greater sensitivity to weak input signals	Speaker: an acoustic agitating device magnetically driven by currents	Detecting variations in the resonant oscillations in a receiving antenna	Same as broadcast but at higher frequency and using a radio-telescope antenna	Photo-cell or light-sensitive sold-state switch with electronic decoding
Protecting shore ends from anchors and wave chafing	Signal rate, weak signal detection, cable size	Signal decay along wires, cost of total network	Finite capacity of electromagnetic spectrum	Satellite cost and inaccessibility, satellite power from sun, finite spectrum	Signal entry, receiving much less rapid than the theoretical message rate of the optical beam in the fiber

TABLE 2–3.—Outline of the Joint Committee's report.[12]

I. An account of the principal telegraphs which have been laid
II. The construction and laying of submarine cables
III. Summary of principlies which should govern the construction and laying of submarine telegraphs

(Anticipated success for transatlantic cables in the future.) (Disastrous consequences of rapid initial successes.)

1. *The conducting wire.* (Always copper.) Stranded wire. Quality of copper. Effect of admixture of foreign substances with copper. Standard of resistance.

2. *The insulating covering.* (Problems with india rubber.) (Effect of exposure to air and sunlight.) (Purity of gutta percha.) Cause of decay of gutta percha and india rubber. Effect of temperature on insulating substances. Absorption of water by insultating substances. Induction. Experiments on induction and insulation. (A limit to signal rate achievable.) Mr. Varley. To increase speed.

3. *The external protection.* (Bearing its own weight.) (Abrasion on ocean floor.) (Kinking while paying out.) (Stretching under stress; buckling upon release.) (Table of types of cable design.)

4. *Laying and maintenance of submarine cables.* (Ocean floor survey.) (Powerful ship to maintain control in foul weather.) (Trimming ship to compensate for shifting load.) (Paying-out equipment and self-acting brakes.) (Cable drag affects ship's steerage.) (Commentary on contracting practices.) (Guarantees of performance.)

Note. Items in parentheses indicate contents of sections with annotations similar to marginal notes printed in the original report (shown here without parentheses).

---

deep water. The new requirements were principally: (1) paying-out equipment capable of holding back the weight of the suspended cable while it sinks to the sea floor, (2) ships with large holds capable of carrying long cables, and (3) the signal-detection equipment based upon an understanding of long-cable transmission as discussed in the previous section.

Unfortunately, the technological innovations necessary for long-distance, deep-water submarine telegraph cables were not yet perfected at the time the promoters of the first Atlantic cable launched their projects. These technologies were still under development throughout the early attempts of 1857–58. Moreover, there was little prior experience to fall back on. The only experiences with deep-sea cables before the Atlantic cable were on the short cables between Italy and Corsica (1856, 300 fathoms) and Italy and Sardinia (1857, 1800 fathoms).

(1) *Paying-out equipment* was first developed in these two Italian connections. Brett introduced the idea of "holding back" machinery—two flanged drums for regulating the speed of the cable's descent. Werner Siemens suggested applying a dynamometer to the paying-out machinery to measure the strain on the cable and enable the engineers to regulate the cable's speed better. The application of a brake on the paying-out drum was introduced.

In 1857 existing paying-out machinery was inadequate to meet the challenge of laying a transatlantic cable, and chief engineer Bright was forced to devise his own version, using four drums and dynamometer. The machinery worked well in both shallow and deep water, but in a particularly difficult moment it lacked the necessary flexibility and a misapplied brake caused the cable to break.

In the resolutions of the board of Atlantic Telegraph Co. following the 1857 failure, the key area of improvement deemed essential for success in 1858 was better paying-out machinery. Bright and the American engineer Everett devised a simpler, lighter, more flexible apparatus with a self-releasing brake and a dynamometer. These improvements gave the engineers much better control over the speed of the cable's descent.

Before actual use of this machinery in the laying of a cable, the board of directors arranged to test the machinery in the Bay of Biscay (1800 fathoms). Experiments with paying-out cables, splicing cables, and retracting cables were performed to general satisfaction. In both 1858 expeditions the paying-out machinery worked so well that in 1865 the same machinery, with some minor modifications, was used.

Development of paying-out machinery by "cut and try" methods had cost investors the loss of the 1857 cable.

(2) *Ships with large holds* capable of carrying 2200 miles of cable were not available in 1857, so a procedure for laying two sections and joining them at sea had to be devised. The controversy between the engineers and the promoters over how that was to be done has already been discussed. By 1865, when the next major attempts were made, the *Great Eastern* was available—a ship with hold capacity sufficient to carry the entire cable of even larger size and the water tanks to keep the cable immersed.

(3) *Signal-detection equipment* was refined during the short life of the 1858 cable. The controversy between Whitehouse and Thomson was settled—at some cost. And all objection that transatlantic telegraphy was infeasible was quelled. Eminent scientists, notably Professor G. B. Airy, Fellow of the Royal Society and Astronomer-Royal, had claimed that the feat would be impossible.

In laying both the shallow-water cables and the deep-sea cables, many of the functional requirements of the undertakings were anticipated by the promoters and designers. Unfortunately, many functional requirements were not anticipated and led to difficulties and cable failures. These oversights were not so fundamental as to render the endeavor infeasible. They were overcome by technical refinements but only after the great expense of the initial failures. This cyclical process of anticipating needs, encountering difficulty, and refining methods until success is achieved is shown in Fig. 2–1.

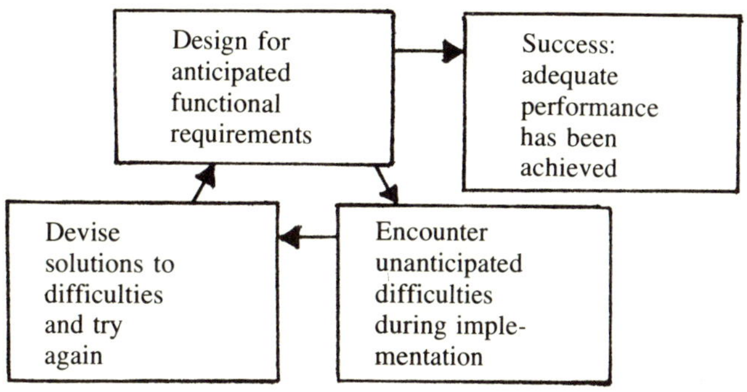

FIG. 2–1.—The cycle of refining submarine telegraph technology.

The Joint Committee systematically enumerated the functional requirements and difficulties encountered in previous submarine telegraph ventures according to the four functional components of cable telegraphy—the conducting wire, the insulating covering, the external protection, and laying and maintenance.

Table 2–4 consolidates these findings under the three steps of development—anticipated functional requirements, unanticipated difficulties, and solutions to the difficulties.

The need for accurate knowledge of the Atlantic Ocean floor illustrates a functional need that was anticipated before the first cable attempt. The choice of location was based on the most recent information. Soundings made by British and American naval expeditions had led Maury to identify the "telegraph plateau" as the ideal spot for laying a submarine cable, and the promoters of the Atlantic cable never seriously considered any other route.

However, many of the difficulties that were not anticipated could have been—damage by fishermen's anchors, rapid corrosion of the protective covering of iron, lightning damage, and the conducting quality and integrity of the cable just following manufacture. Other difficulties would be nearly impossible to anticipate—damage by boring teredo worms, embrittlement of gutta-percha through oxidization and exposure to sunlight, kinking under tension and buckling upon release, and steerage impairment owing to confining of the cable's exit point from the ship. Other failures were due simply to carelessness or the combination of many risks to produce a disaster—paying out the Red Sea cable too tautly over jagged coral, running over the cable with the flange of a drum and not correcting it immediately, permitting water storage tanks to leak, and so forth.

TABLE 2–4.—Detailed functional requirements of early submarine telegraphy.

Functional Requirements Organized by the Basic Elements of the Cable and Its Laying	SHALLOW SUBMARINE CABLES		
	Anticipated Functional Requirements	Unanticipated Difficulties	Solutions to Difficulties
The Conducting Wire Telegraphy requires a physical connection between communicating points capable of conducting electricity	High conduction of drawn copper wire used	Mediterranean Cable: faulty conductor	Stranded conductor offers more independent paths for current and gives greater pliability to copper
The Insulating Covering Submarine telegraphy requires insulation of the conducting path from the "short circuit" of signal currents through the surrounding conducting sea water	India rubber had best insulating properties of materials available Gutta-percha is a practical plastic that could be extruded warm over central conductor	Rubber ages poorly; it becomes gummy and oxidizes even in cable sheath Adding sulfur to harden gutta-percha to reduce core sponginess causes copper core to corrode Dardanelles, Candia, and Athens cable insulation eaten by teredo worm common in coral regions	Don't use India rubber, use gutta-percha Don't add sulfur to gutta-percha; reduce sponginess of wrappings of core and compression from iron wires in protective cover Wrap insulation with copper foil to prevent worm penetration
The External Projection Physical stresses from ship's anchors and handling during laying require greater strength and protection than offered by the conducting wires and insulating covering	First Dover-to-Calais cable was lost to fisherman's curiosity. Iron wire formed into small cables and wound about core used as protection led to success. All subsequent cables used "iron-rope" protective covering	Second Dover-to-Calais line succumbed to corrosion of iron and fisherman's anchor. (Mediterranean cable also.) Anchors yet a probleem: England-to-Holland cable, Singapore-to-Batavia cable Muscat to Karachi; Portland to Guernsey; Bass Straits (1859); all damaged by wave action of deep-sea currents	Steel wires covered by hemp soaked in pitch or tar or other impervious compound to prevent corrosion Bundle more cables together and increase protective covering thickness Increase cable weight and protective covering thickness
The Laying and Maintenance of Submarine Cables Installation and maintenance of submarine cables require new capabilities not previously developed for other purposes	Paying out cable to sea required control of the rate and tension while the ship moved. To accomplish these ends, steam-powered ships were needed and special paying-out brakes were developed	Gulf of Newfoundland experience proved to Cyrus Field that sailing ships did not offer sufficient control for paying-out cable Mediterranean cables: first paid out too slack and broke during retrieval, second ran short due to error in ship's course Toulon and Algiers cable: cable ship caught in a storm, subsequent collision between vessel carrying the cable and that sent to assist Portland-to-Guernsey cable: damaged by lightning Bass Straits (1859) floated by kelp growing on cable	Use steamships Better control of cable tension. Guide ships came into use to avoid compass errors due to paying-out iron (of cable covering) More powerful ships needed Provide better path to ground for lightning Heavier cable

Table 2–4 (Cont'd)

LONG AND DEEP SUBMARINE CABLES

Anticipated Functional Requirements	Unanticipated Difficulties	Solutions to Difficulties
Enhance cables's transmission rate by bundling all copper conductors at center of insulation—don't attempt to obtain multiple independent signal paths within a single insulating covering	1858 Atlantic cable damaged by high voltage used to strengthen signals Gibraltar #4 cable, conductor knuckled through insulation 1858 Atlantic cable's conduction poor owing to impure copper with low conductivity used in some sections 1858 Atlantic cable: faulty splices owing to hurried project	Use more sensitive receiving apparatus (mirror galvanometer) rather than more intense signaling voltage Reduce elasticity of outer covering and sponginess of core insulation and iron rope corrosion protection Determine loss of conductance owing to impurities and require very pure copper Test cable conductance during manufacture
Use gutta-percha and try to keep cable size small so it will fit into ship Gutta-percha was soon shown to withstand (and even improve) with pressure and temperature of the marine atmosphere—1858 cable demonstrated that	1858 Atlantic cable: lay exposed to the sun in Newall's Yard, causing oxidization and embrittlement of the gutta-percha Red Sea cable softened by heat of sun after manufacture and carelessly rolled over by the flange of a drum	Store cables underwater after manufacture until laying Establish quality control procedures to avoid careless damage Wray's compound allows faults in gutta-percha to be detected during manufacture
Cable must be able to support its own weight hanging overboard to the ocean bottom. Repair, retrieval, and even splicing suggest a significant margin of additional strength is necessary	Malta and Alexandria cables: water storage tanks leaked and allowed external iron protection to corrode Gibraltar #4 cable: kinks owing to excessive core elasticity allowed spiral iron wires to unwind under tension	Take care to prevent water tanks from leaking Minimize elasticity of core and use a shallow lead on the spiral windings
Large-scale manufacturing, handling, storage of very long cables unprecedented in 1857–58. Paying-out equipment must give positive control of cable's rate of being paid out without damage to cable Survey of ocean floor necessary to avoid jagged areas, provide slack for traversing mountains or canyons	1857 cable broke when paying-out brake lacked "give" at crucial moment Steerage impaired if cable is confined to a port or hawse-pipe Poor supervision and handling of 1858 cable contributed to its demise Red Sea cable paid out too taut, caused excessive abrasion over coral. Retrieval rendered impossible 1865 cable end lost during laying 1858 Atlantic cable: size limited by ship's hull size; water tank storage capacity limited	Improved dynamometer-controlled brakes proved successful Permit cable to swing across stern after leaving brakes Establish quality-control testing and checks throughout Slow down laying speed and control tension more closely *Perfected deep-water retrieval which then permitted subsequent repair *Steamship *Great Eastern* ideal for cable laying: powerful, with large enough hold to carry entire transatlantic cable in water tanks

*Events that occurred after 1861 Report, yet consistent with the pattern of this chart.

Despite the failure of the 1858 transatlantic cable, the attempt did prove informative. When several miles of cable were recovered near the shore ends it was found that the insulating properties of gutta-percha actually improved under water. Why, then, should the insulation have caused the cable's failure? The Board of Trade committee had several experiments done which showed that sunlight and high temperatures cause gutta-percha to deteriorate, and that the smallest defect (a tear or an air bubble) seriously decreased its insulating properties.

One might question whether these subtle properties of materials and signal-transmission characteristics could have been discovered in some way other than by attempting the project. The following quotation from the summary report of the Joint Committee suggests so:

> We attribute the failure of this enterprise to the original design of the cable having been faulty owing to the absence of experimental data, to the manufacture having been conducted without proper supervision, and to the cable not having been handled, after manufacture, with sufficient care. We have had before us samples of bad joints which existed in the cable before it was laid; and we cannot but observe that practical men ought to have known that the cable was defective, and to have been aware of the locality of the defects, before it was laid.[13]

Strangely enough, no mention is made, here or elsewhere in the Committee's Summary Report, of Whitehouse's application of very high voltages to the troubled 1858 cable.

Comments in the summary report on proposals by responsible, but probably uninformed, parties show that the systematization possible when considering a technology permits the authors to discriminate against unsound advice.

Within the basic framework of essential cable components and functional requirements, at least 26 design variations were examined and compared during the Joint Committee's deliberations (Table 2–5). These variations represent competing views by the principal engineers and manufacturers of the time. The three major categories can be understood in the following way: cables with spiral coverings continued in the tradition of rope and iron cable manufacture—the spiral binds the strands together; cables with longitudinal coverings avoided compressing or excessively elongating the insulating conducting core by taking the tension of laying without deflection or kinking; and the designs by Allen prevented kinking by placing the tension-bearing iron rope inside the compressible gutta-percha core. Paying-out machinery, grappling for broken ends, and abrasion on the ocean floor would act on the jute outer covering of Allen's cables rather than the iron wires of the alternative designs. Whatever the reasons for not using Allen's designs, modern telephone cables are often built in just this fashion.

The Committee's final evaluation, released as a Report to the Lords of the Committee of Privy Council for Trade,[14] was a "technology assessment" only in a narrow sense of the term—it assessed the maturity of the technology as a worthwhile investment. It did not, however, assess the technology in the modern sense of the term "technology assessment," for it did not question the *purpose for which the cables were to be used nor their social or economic impacts*. Neither was the report a technology forecast. Rather, it was a careful review of the state of the art of submarine telegraphy construction and practice, an art deemed fully capable of producing a serviceable connection over long distances and in deep water. The authors stated emphatically:

> We desire, however, in conclusion, to observe that we are clearly of opinion that the failures of existing submarine lines which we have described have been due to causes which might have been guarded against had adequate preliminary investigation been made into the question. And we are convinced that if regard be had to the principles we have enunciated in devising, manufacturing, laying, and maintaining submarine cables, this class of enterprise may prove as successful as it has hitherto been disastrous.[15]

On what basis could the Committee make such an emphatic pronouncement about the *prospective* outlook for submarine telegraphy? On the same basis as they organized the report: technology accommodates a functional decomposition into relatively independent considerations, the whole of which comprehensively covers the requirements sufficient for success. This was not, and is not, true of entrepreneurial or political endeavors.

The Manufacture of Cable

Cable manufacturing became a monopolistic industry almost from the very beginning. Of the 48

TABLE 2–5.— Types of cable design examined for strength and density.

Name	General Description of Arrangements for Strength and Protection	Specific Gravity	Length of Cable in Water in Fathoms Equivalent to		
			0.5% of Elongation	1% of Elongation	Breaking Strain
Cables in which the strength is placed in the outer covering, by means of hemp or wires, laid on spirally					
Clark, L.	Steel wire, with a slight spiral, covered with tape, shellac, and marine glue	1.92	1610 1119	Not reached Not reached	Not broken 3376
Gisborne and Forde (Gibraltar Deep Sea)	Steel wires, coated with hemp, laid spirally	1.9	1176 2440	1947 3924	Not broken Not broken, with a weight = 5683 fathoms in water
Siemens	Strands of hemp, laid spirally, protected by thin copper sheathing	1.5	1342 2046 1694	2552 3672 3284	Not broken Ditto Ditto
Cables in which the strength is placed in the outer covering, by means of hemp or wire, laid on longitudinally					
De Bergue	Tarred hemp lines served round with cord of hemp	1.26	5300 —	8124 5908	Not broken 16982
Godefroy	Steel wires, coated with india-rubber, and covered with india-rubber canvass	1.9	2653	Not reached Not reached	3424 3424
Ditto	Ditto	1.4	2596 2596	3500 Not reached	Not broken 3309
Hall and Wells	Hemp lines kept in place by plaited hemp	1.35	760	1126	4420 4% extension
Ditto	Steel wire and hemp lines kept in space with plaited hemp	1.6	2499	Not reached	2729
Ditto	Ditto	1.9	2437	4525	5597
Silver	Steel wires, covered with plaited hemp	2.8	1642 1832	2260 Not reached	Not broken 3213
Sinnock	Tarred hempen lines, served in a close spiral, with hemp string and iron wire	1.4	622 —	1088 463	Not broken 2855
	Ditto	1.7	997 —	1782 636	— 7132
Cables in which the strength is placed next the copper conductor inside the insulating covering					
Allen	Steel wires, laid spirally on copper conductor, outer covering plaited jute saturated with marine paint	1.6	1519 1283	Not reached 2258	Not broken 2936
Ditto	Ditto, outer covering india-rubber canvass	1.38	3405	5555	7484
Ditto	Ditto, outer covering Godefrey's compound	1.3	3835	6340	6438

cables listed in the Joint Committee's Report, ownership was spread among 20 different firms and national governments, yet the manufacturing was performed primarily by two firms. This early monopoly rested on two essential components of telegraph cables—gutta-percha insulation and iron rope protective coverings.

The first, and unsuccessful, cable over the English Channel did not use iron rope for external protection but did use gutta-percha insulation manufactured at the works of London Gutta Percha Co.

Refined gutta from a variety of gutta-producing trees quickly proved to be an indispensable material for a cable's electrical insulation. Gutta-percha from four varieties of trees found first in Singapore and then in the neighboring Malay Penninsula and Borneo were most sought after; competing guttas such as balata from South America eased the demand on these principal sources. British dominance of the gutta trade acted to restrict the free entry of foreign manufacturers for many years, even though the Germans and Dutch did provide access to other limited sources.

As data and calculation show,* a focused selective pressure was applied to the gutta-producing species of trees for more than three reproductive generations. Has this ravaging destroyed dependent species? What would have happened if gutta had great cultural significance for the peoples of Southern Asia and the Malaysian Islands? Would submarine telegraph cables have had to be looted if a more urgent use for gutta had arisen? If today's environmental concerns were common in the late 19th century, the development of international submarine telegraphy would surely have been inhibited. Hypothetical contingencies such as the above did not arise (to the best of our knowledge); however, from a prospective vantage point, decisionmakers of the time should not have discounted these contingencies out of hand. As is generally the rule, one proceeds until trouble develops, since there are many contingencies, and copes as best as one can as the contingencies emerge.

Credit for the crucial innovation of iron-rope protection and structural reinforcement for cables is shared by several men. William Kuper, the colliery rope manufacturer, suggested it to T. R. Crampton, the engineer on the second English Channel cable (1851). Willoughby Smith, Edward Highton, R. S. Newall, and W. Reid are all credited with having made similar suggestions.[16]

Newall and Co. entered cable manufacturing on a patent-right challenge. They had a patent for inserting a soft core into wire ropes to make them more manageable.[17] Their challenge interrupted the construction of the second English Channel cable then under contract to Wilkins and Weatherly.[18]

Glass, Elliot and Co. took over the business of Kuper and Co., one of the leading wire-rope manufacturers. Therefore, the two contractors of the first Atlantic cable—Glass, Elliot and Co. and Newall and Co.—had established iron-rope patent authority by the early 1850s. In 1864 London Gutta Percha Co. merged with Glass, Elliot and Co. to form TC&M, thereby further consolidating the control of manufacturing.

One wonders whether the near-monopoly control of submarine cable manufacture led to a concentration and hence more rapid accumulation of knowledge, or inhibited rapid development by reducing competition. Possibly the prospect of future business was an incentive to the cable contractors to enlarge facilities and build specialized equipment of a quality that could be used repeatedly. On these points the historical record seems imperfect.

---

*Gutta-percha is a durable natural plastic with insulating properties that are best when the material is refined to greatest purity. According to Bright (*Submarine Telegraphs*, pp. 252–261), of the more than twelve species of gutta producing trees, *Isonandra gutta* proved best; others ranged widely from satisfactory to poor. The Isonandra is a tree fround in the jungles of Singapore, Borneo, Malya, and throughout neighboring regions. These trees grow to heights of 60 to 70 feet; they take 25 years to begin developing commercial quantities of gutta and from 30 to 35 years to flower and bear fruit. A single tree produces 1⅓ pounds of refined gutta-percha. To harvest the gutta, the tree is cut down, the bark slashed, and the sap drained.

In 1881 alone, 3600 tons of gutta-percha were exported from Singapore and 1248 tons from Borneo, which required the felling of over 7 million gutta trees and most likely the destruction of 38 million forest trees in all. That would amount to 294 trees for every square mile of the present Malaysia. So thoroughly had these trees been harvested that Singapore Island was stripped of several species of gutta and natives no longer knew what the flowers of mature Isonandra looked like. The Dutch began to form gutta plantations in 1886. At the height of submarine telegraphy's development, there were 350 000 nautical miles of cable averaging 300 lb of gutta-percha per mile of cable, not including the cable used to replace or repair defective lines or the lines that were abandoned by 1928. The above mileage alone would account for more than 78 million trees; a total of over 100 million must have been harvested in all. There would have been only three generations of gutta growth to support this demand.

All of this harvesting focused on a few species. Gutta from British Guiana, called Balata, diverted the concentration on the principal gutta producing trees, but the refining of Balata was less economical.

Management and Contracting

Prospective assurance derived from a systematic understanding of the functional requirements of the components of a submarine telegraph cable, the accumulated experience of two primary manufacturing firms, and the recognition that each of past technical difficulties could be overcome, were *not* together sufficient to assure future success. The manufacture, laying, and maintenance of submarine telegraphs were a complex technological procedure for its day. There were many ways to go wrong—inadequate quality control, lack of recognition that a never-before-encountered technical problem was arising, and outright carelessness in handling or laying could jeopardize any project. The Joint Committee explicitly mentioned these management deficiencies of previous projects.

The most expensive failure, double that of the transatlantic cable, was the Red Sea cable to connect Suez to Karachi—3499 miles, laid in six sections as a relay. Improper management led to this failure in several ways.*

Contracting procedures were partially responsible for the poor performance of many cables laid. After collecting testimony and evidence, the Joint Committee reflected on much of the current contracting practice:

> All such contracts have been proved to be objectionable, and telegraph companies have acted in respect to the manufacture of telegraphs, in a manner totally different from that in which they would have acted in any other operation. They begin by simply asking the contractor for what amount he would be prepared to lay down a cable in a given position, guaranteeing this performance for a limited time, as though a cable, unike a railway or any other means of communication, were a thing which would necessarily, when once laid down last forever. No greater error could be committed. Nothing can be more precarious than the life of a cable. The company which, without regard to its future maintenance, provides capital for merely one cable must be fortunate indeed if such a course should not prove its ruin. The future repair, maintenance, and restoration of a cable is as important an item as the maintenance of a railway.[20]

The Joint Committee laid out specific recommendations on how to improve the management practices in contracting for construction of a submarine telegraph line:

> If the contractor is made solely responsible for the laying of the line, he cannot be made subject to much interference by other parties; but he will put on a large additional price to cover the risk he incurs, and the extra cost to be paid does not secure the success of the enterprise. If, on the other had, he is not made responsible, he will have little interest in the success of the undertaking, and without such interest it is difficult to secure that close supervision to the details of the manufacture which will alone obtain perfection . . . . Hence we should suggest that a defined, but limited, pecuniary responsibility should be placed on the contractor, who must therefore be allowed a certain liberty of arrangements, but the specifications for the manufacture should be drawn up by the engineer, and include clauses reserving to the engineer the right of approving of all the arrangements to be made . . . We believe that the best plan would be that the contractor for laying the cable should contract for its maintenance for a certain number of years.[21]

These principles seem eminently sound, even today. Cost-plus-fixed-fee development of hardware used by the National Aeronautics and Space Administration for the successful execution of the Apollo manned moon-landing project embodied these principles. They insure that technical decisions are made by those who are best informed, yet the risks are shared among all parties.

Employing a self-regulating adaptive technique in the contracting procedures and the technical approach may have been the pivotal development for submarine telegraphy. The promoters of the Atlantic cable needed the Joint Committee's advice and set up a scientific consulting committee to oversee the selection and manufacture of the 1865 and 1866 cables. Similar adaptivity in the

---

*"It appears that the contract provided that the cable should be maintained in working order for 30 days by the contractor, and that the several separate sections were worked for a longer period, but that the whole was not worked throughout for that period. . . . We consider that the failure of this line is attributable to the cable having been designed without regard to the conditions or climate or the character of the bottom of the sea over which it had to be laid, and to the insufficiency of the agreement with the contractor for securing effectual supervision during the manufacture and control of the manner of laying. It is, moreover, to be regretted that the contract for laying this line was entered into without a full investigation into the question, considering that the success of the Atlantic cable was at the time very doubtful."[19]

technical approach came with the perfecting of deep-sea grappling techniques. Once they were perfected, faulty sections of very long lines could be repaired or replaced. From this time forward, cables were designed to be repaired and most major cables had several sections replaced during their 50- to 80-year period of useful service.

The first successful attempt at repairing a deep-sea cable was made in 1858: 60 miles of the Sardinia-Italy cable were picked up by starting from the shore and proceeding seaward.

The necessity to refine the techniques for finding, hooking, and lifting a broken cable from the bottom of the Atlantic 2000 fathoms below again arose after 1500 miles of the 1865 cable had been laid, when the end being paid out by the *Great Eastern* snapped.

The decision in 1866 to try to grapple and lift the lost 1865 cable was made by Samuel Canning, an engineer who had experience picking up cables in the Mediterranean. Using regular grapnels, wire rope, and buoys, Canning dragged the bottom for the cable. In nine days the cable was hooked and partially lifted three times. The machinery and ropes were strengthened. After several more successful hooks but no successful lifts, a new method was tried: the *Great Eastern* hooked the cable, lifted it to 900 fathoms from the surface, and buoyed it. The *Great Eastern* and the *Medway* then hooked the cable three and five miles west of the buoy. The *Medway*, after raising the cable to within 300 fathoms of the surface, caused the cable to break. This break and the buoy eased the strain on the *Great Eastern*'s remaining grapnel, allowing it easily to raise the 1865 cable, splice it to the onboard portion, and continue laying it to Newfoundland.

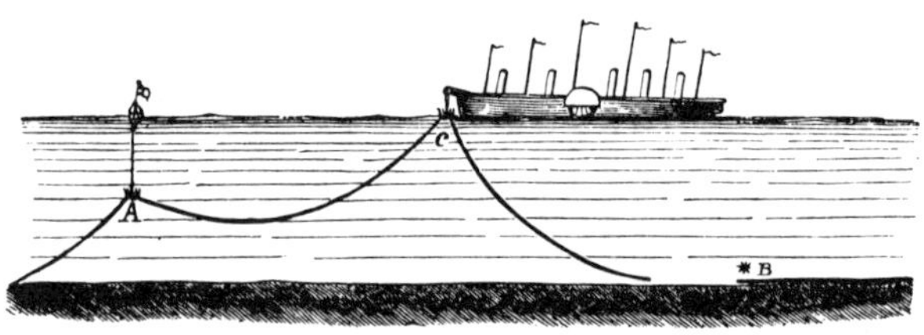

FIG. 2–2.—The final method adopted for picking up 1865 Atlantic Cable:
A, point where cable was buoyed by *Great Eastern*;
B, point where cable was broken by *Medway*;
C, bight of cable ultimately brought to surface by *Great Eastern*.

The Lessons Learned by Meeting the Technical Challenge

The procedures and techniques developed in the numerous attempts to lay the 2200-mile deep-sea cable were modified and adopted in later cable ventures. This was the end of the pioneer stage of oceanic telegraphy.

The technical responses to the problems generated by the Atlantic cable created impacts in another, more general area: the infiltration of science into business. The movement during the nine years of the venture towards stricter quality control had the corollary effect of increasing the importance of the engineer/electrician to the enterprise. Initially the investors' concerns were paramount in selecting cables and setting time schedules. By 1864—much as the result of the Red Sea failure and the Joint Committee report—the concerns of the engineers/electricians were given greater weight. They did not run the enterprise, but they did set the limits within which it could operate.

This infiltration of science into business decisionmaking was not at all common in 19th-century Britain or America. There was a fundamental hostility between the "pure" scientist and the tinkering machinist. Telegraphy was one of the very few industries (chemicals were another) where science and technology were effectively merged. One reason was that telegraphy was not based in a craft, but in recent scientific discoveries. J. D. Bernal wrote in *The Social Function of Science:*

> The most complete integration of industry and science is, however, only reached when the knowledge of the fundamental nature of the processes is so extensive that it is able to lead to the development of entirely new processes unthought of, or indeed, unthinkable, by traditional methods; as for example in the chemical

syntheses of new dye-stuffs or specific drugs. The same result follows even more directly when a purely scientific discovery of a new effect is turned to some industrial use as, for instance, in the telegraph or the electric light. In these cases we have an industry scientific through and through, an industry which owes its inception as well as its development to science. The leading modern examples of these are the electrical industries, both for the production and distribution of power and for the improvement of communications.[22]

The integration of scientists into telegraphy, and the Atlantic cable in particular, was limited. As shown above, scientific understanding and knowledge were most needed in signaling theory and practice. They were to penetrate the enterprise here first. Limited as it was, scientific understanding was recognized to be beneficial. Thomson, Faraday, Wheatstone, and others contributed greatly to the nascent trend toward application of science to industry.

Most technical improvements were still made by machinists or engineers through trial and error. Systematic approaches combining the accumulated experience of the entire industry and pursuing a structured program of experimenting and testing emerged only after numerous failures had undermined investor confidence in the large-scale submarine telegraph projects. However, technology does permit a systematic decomposition of necessary functions to act as a test of confidence for the expected success of future projects. On this basis the Joint Committee of the Board of Trade and Atlantic Telegraph Co. was able to endorse very positively further cable projects in spite of recent failures.

The tendency of investors and entrepreneurial engineering enthusiasts to charge ahead without careful consideration of the *technical* advisability of their actions persists even today. Within the fastest-growing industry of computers and microelectronics little systematic consideration was given to the fundamental characteristics of the systems being created.* Would continuation of present trends place us in a desirable position in the future or a disadvantaged cul-de-sac?

The ability to overcome all technological barriers does not, by itself, guarantee the success of complex projects. Contracting agreements and management processes determine the degree to which sound technical judgment will be applied to decisions and work of good quality will result. Adaptive technological innovations, such as deep-sea grappling that permits repair of faulty work, greatly enhance the prospect of success and long-term profitability for complex projects.

B. The Entrepreneurial Challenge

The challenge faced by the entrepreneurs in laying the Atlantic cable was only partly technical; equally great was the challenge of raising capital for an enterprise that was unprecedented and highly risky and called for far more money than the original small group of promoters could provide from their own resources. The several early failures made it necessary to find additional investors time after time. America was still a small, predominantly agricultural country with only a handful of private fortunes; Britain, with its expanding economy and powerful position in world trade, had greater resources of risk capital. The story of how the cable was laid shows how the project became a predominantly British enterprise despite the determination of Cyrus Field that it should be a joint endeavor.

---

*This point is recognized by thoughtful electronics specialists, as follows. "I am concerned to find that with all our capabilities in complex data-handling problems, we have only very recently ascertained the minimum number of logical levels required for addition. We are functioning in our data-processing environment with limited vision, learning mainly by experience within a particular set of existing conditions. The brilliant intellectual conquests of von Neumann, Turing, and a few others, must be considered as examples to follow. . . . We have methods for examining applications, proceeding from mathematical formulation through numerical analysis and flow charting, and procedures for determining the types and amounts of input and output data and data-processing requirements. Yet, we have not really learned, in general, to characterize the problem formally so that we can predict mathematically the optimum system for a particular problem. We do not have mathematical tools to examine the trade-offs between various approaches and changes in parameters; for example: What should the software be and what do we put into hardware? How large should each memory hierarchy be? How many levels should we have? When do we consider multiprocessing? How should we structure it? Erroneous answers to these questions can have a profound impact on many of our systems. (E. G. Fubini and M. G. Smith in *IEEE Spectrum*, May 1967.)

But even in Britain there were many competing opportunities for risk capital and investors were stubbornly cautious. To make their venture more attractive to reluctant financiers, industrialists, shippers, and merchants, the promoters of the cable had to assure large profits, and to do so they were driven to concentrate great efforts on the critical area over which they had some hope of exercising control—that of avoiding or reducing competition and maintaining the uniqueness of the service they would offer. The efforts of Field, Brett, and others to avoid competition by other cables, and to control connections with domestic lines, created institutions and practices that greatly affected the development of the communications industry and the economic structures of the U.S. and Britain. These early though indirect impacts of the Atlantic cable deserve to be examined.

Reducing Competition

Competition was the central characteristic of the young telegraph industry in Britain and America when Field first envisioned an Atlantic cable. In America the telegraph lines had spread with lightning speed; between 1846 and 1852 miles of telegraph in operation jumped from less than 500 to over 22 000. In 1850, when 12 000 miles had been constructed, 20 separate companies were competing for the public's business: Ohio, in 1852, had 14 companies vying to serve its citizens.[23]

In Britain, the spread of the telegraph was less rapid and more controlled. Beginning in 1846 with the founding of Electric Telegraph Co., the British telegraph network concentrated mostly on the more populous areas of the South and the Midlands. By 1852 the Electric company, the dominant force in British telegraphy, had almost 4000 miles of wire in operation. Its primary competitor was British Electric Telegraph Co., whose lines were located primarily in Northern Britain and Scotland. The English and Irish Magnetic Co. (of which John Brett was a director) had the majority of telegraph business in Ireland and between England and Ireland.

One area of competition among these three British telegraph companies was in establishing telegraph communications with the Continent and Ireland. Soon after English and Irish Magnetic Co. laid submarine cables between England and Ireland, Electric Telegraph Co. laid its own cables. In addition, Electric Telegraph Co. merged with International Telegraph Co. to form Electric and International Telegraph Co., commonly known as the Electric. By merging with the International, the Electric obtained exclusive rights to telegraph communication to Holland. This move brought the Electric into competition with British Electric Telegraph Co. which, through a merger with the European and American Telegraph Co., had a special relationship with Brett's Submarine Telegraph Co., the sole company with submarine cables between England and France.

Field's first efforts were aimed at reducing competition to his venture both from other Atlantic cable ventures and for the business created by an Atlantic cable. From the first, Field envisioned the Atlantic cable as a system of interlocking domestic and cable networks under the control of one group of investors. The core of this group was New York, Newfoundland, and London Telegraph Co., which was formed to provide an unfettered line between Newfoundland and New York for the transmission of foreign news. Field sought monopoly rights in the Canadian provinces and New England states through which the company's lines would pass. Like his British counterparts, he focused on the area most amenable to state control, in this case the right to land submarine cables. The governments of Newfoundland, Nova Scotia, Prince Edward Island, and the state of Maine granted New York, Newfoundland and London Telegraph Co. such rights.

Field's interest in securing the business that would be generated by an Atlantic cable did not end with getting information to New York. His plan, equal in ambition to the Atlantic cable, was to unite the inefficient telegraph companies handling the majority of the business of the Eastern Seaboard into a unified, effective telegraph company. To this end he; Cooper; Hunt of the New York, Newfoundland and London; and seven others formed American Telegraph Co. in 1855. Soon after its incorporation, the American signed several leases that gave it control of extensive telegraph lines in New England.

Field and his associates also developed a special relationship with one of the biggest potential users of the Atlantic cable—the New York Associated Press (NYAP). The relationship between the

NYAP and Field was of long standing. D. H. Craig, NYAP's general agent and a dominant figure in foreign news gathering, had a financial interest in New York, Newfoundland and London Telegraph Co. When Field and Cooper sought to unify the telegraph companies of the Eastern Seaboard under American Telegraph Co., Craig used his and the NYAP's influence to effect leasing agreements. In exchange for his support Craig's dispatches were given absolute priority; although the arrangement was not formalized until 1860, from the first operation of the American no dispatches of any reporter could be transmitted before Craig's were.

The strength of the relationship became especially evident in 1859. Robert Russell, a newly appointed member of the board of directors as a result of the absorption of New York and Washington Telegraph Co., began to push for an end to the NYAP monopoly, the establishment of a first-come-first-served procedure, and higher tariffs for the messages. Craig and his supporters fought, especially in the press, this attempt to destroy their monopoly. After a special committee of American Telegraph Co. had approved Russell's recommendations, Cyrus Field, Wilson Hunt, and Abraham Hewitt joined Craig in his fight against Russell. They felt the company had no right to take arbitrary action against a group lawfully organized to share the costs of telegraph transmission. They also feared that the press would either build their own lines or associate with a rival company.

Despite this opposition, Russell continued his efforts to destroy the NYAP-American Telegraph special relationship. He bought out Cyrus Field and Abraham Hewitt. But the two men Russell appointed to take their place opposed his tactics and finally forced him from the company's management.

The special relationship between American Telegraph Co. and NYAP was resumed, and combined with the American's aggressive consolidation campaign caused concern among the other major telegraph companies. If an Atlantic cable was connected exclusively with the growing network of the American Telegraph Company, the entire telegraph industry would be dominated by the American.[24] In 1857 these companies, especially the recently incorporated but powerful Western Union, initiated negotiations with the American for the purpose of "harmonizing all interests and protecting existing lines."[25]

The result was a six-company contract sometimes called the Treaty of the Six Nations. Each of the six companies, which included the American, was guaranteed "exclusive enjoyment of all telegraphic business" in a specified geographic area; each company would transact business only with the other five signers and refrain from acting unilaterally. Thus the six companies together established an effective monopoly of the U.S. telegraph business. American Telegraph Co. was designated "exclusive owner of the Hughes patent right" for Newfoundland, Nova Scotia, New Brunswick, New England, Long Island and Staten Island, New Jersey, Delaware, North Carolina, South Carolina, Georgia, Florida, and a line through Mississippi and Louisiana to New Orleans (Fig. 2-3).

Thus Field successfully reduced the potential competition for the business of the Atlantic cable in America. He and his associates established a niche in which the cable could operate most effectively and profitably. However, he was not entirely successful. As Fig. 2-3 shows, Western Union was a formidable competitor. Despite the six-party contract, these two giants of the telegraph industry continued to compete for the support and approval of investors. One area of competition was between Western Union's plan for a transcontinental telegraph and the American's Atlantic cable venture.

In Britain, Field was also faced with a competitive domestic telegraph industry. His association with John W. Brett meant that the Atlantic cable project was associated with English and Irish Magnetic Telegraph Co., in opposition to Electric and International Telegraph Co. When Atlantic Telegraph Co. was founded, its directors came from the English and Irish Magnetic, as did the facilities for distributing its stock. When the English and Irish Magnetic merged with British Electric Telegraph Co. (BETC) in 1857 to form British and Irish Magnetic Telegraph Co. (the Magnetic), it meant the Atlantic cable would be well connected to the commercial and industrial centers of Ireland, Scotland, and Northern England—Dublin, Glasgow, Manchester, and Liverpool. Moreover, the

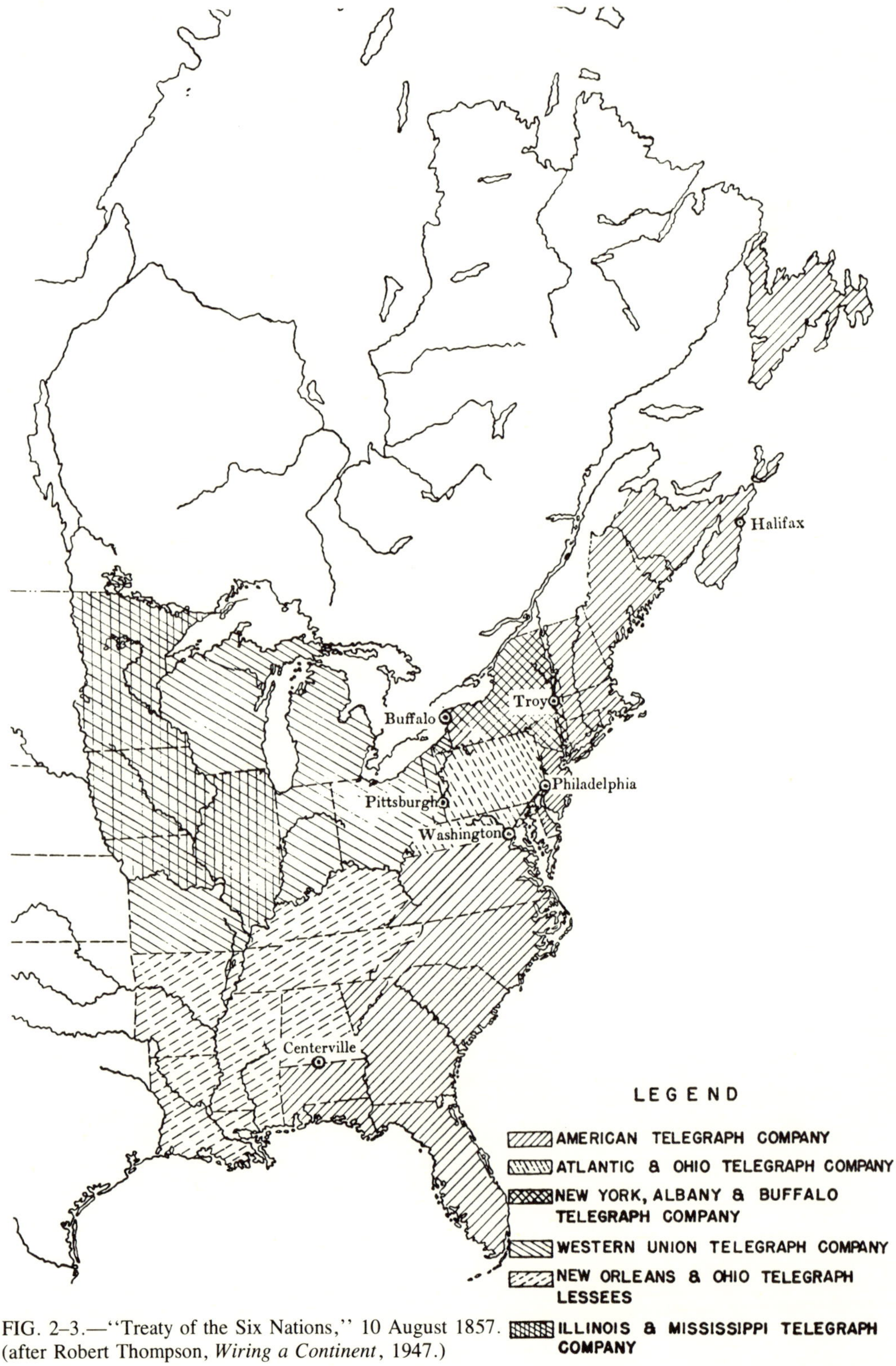

FIG. 2–3.—"Treaty of the Six Nations," 10 August 1857. (after Robert Thompson, *Wiring a Continent*, 1947.)

Magnetic's assumption of BETC's arrangement with Brett's Submarine Telegraph Co. for exclusive transmission of cables from France meant the Atlantic cable had easy access to continental telegraph systems.

Besides fearing diversion of profits through competition over the connection of the cable with existing landline systems, Field and his associates were genuinely concerned about the possibility of a competing Atlantic cable. The demand for monopoly landing rights in the Canadian provinces, the wording of the Congressional bill requesting support, and the conversations with officials of the British government were directed at reducing the threat of competing schemes.

Although this possibility was remote, given the risky and large-scale nature of the project, Field's fears were not entirely unfounded. One serious alternative to Field and the Atlantic company's route was the North Atlantic route proposed by Taliaferro Shaffner and James Wyld. In 1858 Kendall and Smith of Magnetic Telegraph Co. (Morse's company) gave their support to Shaffner's plan to link Britain and America via the North Atlantic islands (Faroe Islands-Iceland-Greenland-Labrador) by short cable spans. Shaffner believed there were too many difficulties with long cables to make them viable. At the same time Wyld, a British geographer, was forming British and Canadian Telegraph Co.; the prospectus of this company claimed that the Canadian government and NYAP had each guaranteed the company £10 000 on completion of the system.

Having attracted the support of several well-known Arctic explorers, Shaffner was able to see Prime Minister Palmerston and ask for British government support. In response the Admiralty selected Sir Alfred McClintock to survey the proposed route in *H.M.S. Bulldog*. When the expedition reported to the Royal Geographic Society in 1861 that it was not clear whether cables would withstand the ice floes of the region, interest quickly died.

A more serious challenge was the Bering Straits route proposed by Perry Collins and supported by Hiram Sibley and the Western Union. Collins had been interested in establishing a telegraph link to the Far East since the mid-1850s. In 1856 he had been appointed commercial agent of the United States to the Amur River, most likely as a cover for his investigations into the possibility of the Siberian telegraph.[26] With the completion of the transcontinental telegraph in 1861, Collins began to promote the idea of a Bering Straits telegraph in earnest. His plan was to construct a telegraph line north through the northwest United States, British Columbia, the Yukon, Russian America [Alaska], across the Bering Straits, and down the coast of Russia to meet the Russian landlines at the Amur River (Fig. 2–4.)

When Congress took no action on Collins's plan, he turned to Hiram Sibley of Western Union, who became an enthusiastic backer of the scheme. Sibley, after the experience of the transcontinental telegraph, felt there was no reason the project could not be successfully completed within two years.[27] The Civil War prevented him from actively supporting the project until 1864.

In March 1864 Western Union took over the Russian-American line from Perry Collins in exchange for $100 000 and one-tenth of the stock issued for the undertaking. The Western Union directors meant to compete with the Atlantic cable, whose prospects they considered dim.[28] Publicly, Western Union denied any competition between the Russian-American line and the Atlantic cable. Privately, Hiram Sibley admitted to the competition. Meeting Cyrus Field in London, Sibley stated, "I would give $50 000 to *know* if you are ever going to succeed. I hope you will but I would like to be certain before we spend any more in Russia."

To finance the line, Sibley and his associates set up Western Union Extension Co. and issued 100 000 shares of stock at $100 per share; the stock was taken up immediately.[29]

At the same time a bill was passed by the Congress, with the support of Lincoln and Seward, providing assistance to the project. Western Union Extension Co. received rights of way, free timber and stone for telegraph stations, military protection of the line, assistance in surveying the route, and assistance in laying the cable. In exchange the government received priority rights for messages; rates were to be decided in concert with Russia and Britain.

The Extension company hired Col. Charles Bulkley to supervise the surveying and construction of the line. He sent out three exploring parties to survey the route: one to the Siberian section, one to Russian America, and the third to British Columbia. He planned to start building at several points

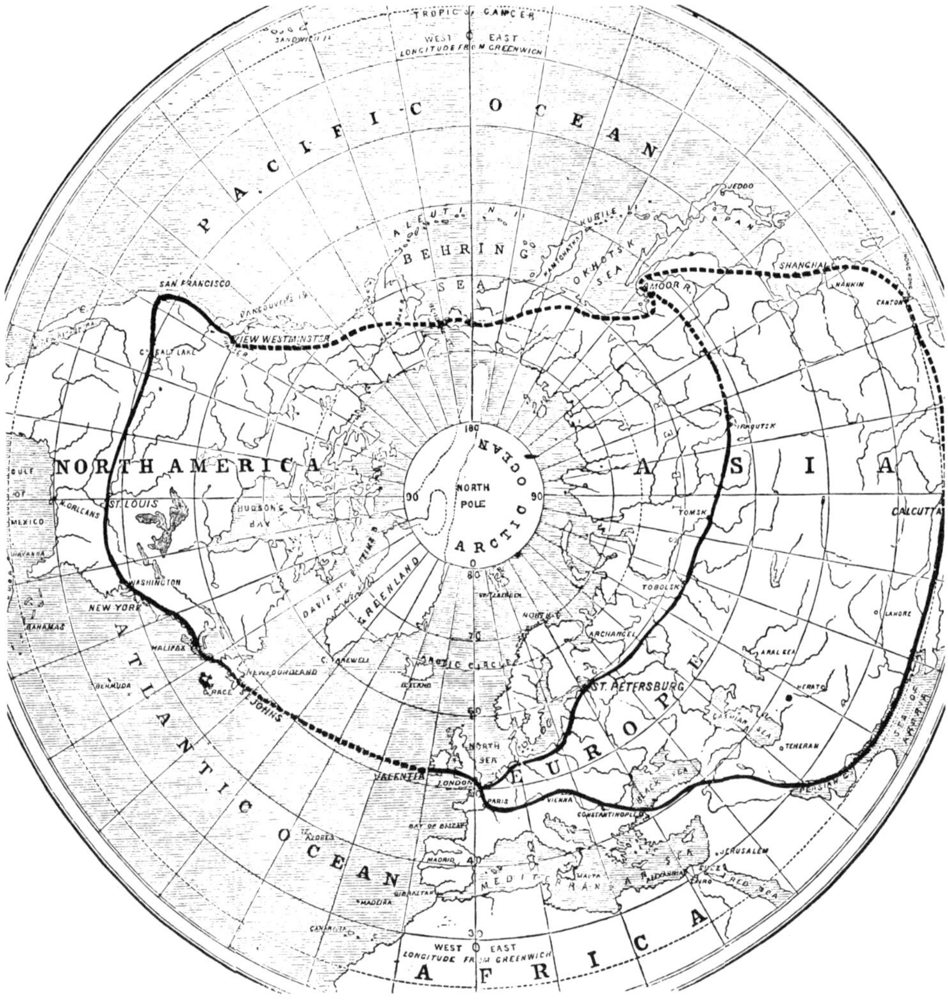

FIG. 2-4.—Linking the continents, 1865: Proposed routes of the Russian-American telegraph line and the Atlantic cable. (From Robert Thompson, *Wiring a Continent*, 1947.)

along the route, but construction on the line did not begin until 1866, and after the success of the Atlantic cable in August of that year the Russian-American project was gradually abandoned. Western Union Extension Co. continued to report progress on the line, while the Western Union leadership exchanged its extension stock for Western Union stock, leaving the small stockholders in both companies to absorb the loss.

When the Russian section heard in May 1867 of the success of the 1866 cable, they immediately ceased work on the line. Before leaving for America, they held a bazaar and sold what equipment they could. Before they heard of the cable's success they were confident that their project would defeat the Atlantic cable. One of the songs sung by the workers, to the tune of "When Johnny comes marching home," was:

In eighteen hundred and sixty eight
Hurrah! Hurrah!
In eighteen hundred and sixty eight
Hurrah! Hurrah!

In eighteen hundred and sixty eight,
The cable will be in a miserable state,
And we'll all feel gay
When they use it to fish for whales.

In eighteen hundred and sixty nine
Hurrah! Hurrah!
In eighteen hundred and sixty nine
Hurrah! Hurrah!

In eighteen hundred and sixty nine
We're going to finish this overland line;
And we'll all feel gay
When it brings us good news from home.[30]

Integration of an Industry

Field and the other promoters of the Atlantic cable also faced problems of competition within the submarine cable industry. When Field contracted for the Gulf of St. Lawrence cable in 1855, cable manufacturing was a relatively new business. The construction of cables was done by companies originally set up for other purposes: London Gutta Percha Co.—various gutta percha products; R S. Newall and Co.—wire-rope manufacturers for colliery; Kuper and Co.—wire-rope manufacturers for colliery; Siemens and Halske—design and manufacture of electrical instruments; and W T. Henley—manufactures of electrical instruments. The manufacturing process was fragmented; one firm made the insulated core, another the outer covering. Rapidly the manufacturer of the outer covering (therefore the finished product) became the party that laid the cable. This arrangement obviated the need for transferring responsibility for the electrical qualities of the cable to a third party.

The early days of the submarine-cable industry were highly uncertain. Cables cost a great deal and if they failed that investment was lost. There was no direct experience to build on, very little related experience from other industries, and a lack of systematic knowledge about cables. Under these circumstances the leading firms became those that were the most successful at improvising and insuring operational reliability. The earliest successful companies were those run by individuals of determined but venturesome character—men such as R. A. Newall and Richard Glass.

When Field and his associates presented the cable industry with an order for a 2500-mile cable in four months, it was obvious that no one firm could produce it. On the earlier Gulf of St. Lawrence cable, the contract had been let to Glass, Elliot and Co. Richard Glass, the managing director, had shown an active interest in the cable project since his initial introduction to Field by John Brett. In this case, the work was split—with London Gutta Percha Co. manufacturing the core and Glass, Elliot and R. A. Newall manufacturing the outer coverings—both for speed of manufacture and because of the threat of opposition from R. A. Newall if Glass, Elliot got the whole contract.[31]

The division of labor proved unsatisfactory; the two companies completed the work on time but had unfortunately applied the outer coverings with different lays—one right-handed, the other left-handed. After the cable was lost in 1857 Glass, Elliot was selected to manufacture the replacement. R. A. Newall never again participated in the Atlantic cable project.

A strong identification of Glass, Elliot with Atlantic Telegraph Co. developed. Glass, Elliot and London Gutta Percha and been active in the cable project from the beginning; Stratham of London Gutta Percha was an original director of Atlantic Telegraph Co., and Glass had been interested in the project since Brett first introduced him to Field. After the failure of the 1858 cable Glass, Elliot became increasingly important to the venture. It was their 1862 offer to construct and lay the cable as well as subscribe £25 000 that got the cable project moving again. For the first time a cable manufacturer had, as the Board of Trade Commission had recommended, a direct financial interest in the venture. The growing closeness of Glass, Elliot and Atlantic Telegraph Co. was legitimized in 1863 when the Atlantic's independent scientific consulting firm, set up pursuant to the Board of Trade Commission's recommendation in 1861, chose Glass, Elliot from eleven competitors to construct the new Atlantic cable.

When it became apparent that the demands of the consulting committee would require a significantly more expensive cable than had been anticipated, Glass, Elliot was forced to renegotiate the terms of its offer. During these renegotiations several members of Glass, Elliot; London Gutta Percha Co.; and Atlantic Telegraph Co., recognized the inability of any one or their companies to raise all the capital needed. They therefore decided to combine London Gutta Percha Co. with Glass, Elliot to form Telegraph Construction and Maintenance Co. (TC&M). With this merger the entire manufacture of a cable finally came within the province of one company. The originators of the merger believed such a consolidation would testify to the practicability of submarine telegraphy, a question still in doubt in the minds of many investors. Richard Glass was appointed managing director and John Pender, an original director of Atlantic Telegraph Co., its chairman. Soon after its establishment TC&M subscribed the necessary capital to lease the *Great Eastern* and to proceed with the construction of an Atlantic cable.

Despite the growing ties between Atlantic Telegraph Co. and the cable manufacturers, tensions arose between the two groups. After the cable broke in 1865 there was a dispute between the directors of TC&M and Atlantic Telegraph Co. over whether proper precautions had been taken.[32] The threat of a lawsuit was avoided when the Atlantic agreed to pay TC&M £500 000 (plus another £100 000 if successful) to build and lay a new cable and then to lift, splice, and lay the remainder of the 1865 cable.

The tension became more pronounced when Atlantic Telegraph Co. was prevented by the British government from raising their dividend to 12% in order to attract new investors.

Their only recourse was to set up a new company, the Anglo-American Telegraph Co., made up primarily of people from TC&M. TC&M led the subscriptions with £100 000. With this offer Brassey, Pender, Field, Gooch, Barclay, Elliot, Glass, Campbell, Bewley, and Smith needed to put up only £100 000 each, so that before subscriptions were opened to the public over £200 000 of the £600 000 had already been raised. Within fourteen days of the public announcement, the rest of the money was raised.

It was the nature of the agreement conditions that made people ready to invest in a project that had been struggling for twelve years. Anglo-American Telegraph Co., as agent of Atlantic Telegraph Co., would construct, lay, and operate the cables (the new 1866 and the repaired 1865 cables). In return they would receive £125 000 per annum from Atlantic Telegraph Co. revenues and £25 000 per annum from the New York, Newfoundland and London revenues. This £150 000 per annum equalled a 25% dividend on the company's £600 000.[33]

The movement toward consolidating all the aspects of cable financing, constructing, laying, and operating came to fruition with the success of the Atlantic cable, but first the growing tensions between the old Atlantic investors and the newer TC&M investors developed into open conflict. With the initial low volume of traffic across the cable, it became apparent that after the Anglo-American had taken its £25 000 there would be very little left for the Atlantic. Compounding the problem was the debt the latter had incurred in the erection of telegraph stations. The Atlantic was "at the mercy of its agents, without any efficient power of control."[34] Dissensions between the Atlantic and Anglo-American boards occurred more frequently.

Atlantic finally decided either to buy out or merge with Anglo-American. First they tried to buy it out, but with 25% revenue and £1 200 000 in mortgages, Anglo-American was unwilling to provide any feasible terms. Atlantic then sought an Act of Parliament authorizing it to raise more capital. TC&M and the Anglo-American (their boards highly interlocked) protested, but withdrew when clauses were inserted that required assent of three-fourths of the shareholders to raising more capital. When Atlantic seemed to have a possibility of raising the capital, a few of the Anglo-American directors began to purchase large blocks of Atlantic stock. By the time the necessary stockholders' meeting was held, Anglo-American had sufficient power to block raising of new capital.[35] In April 1868 Anglo-American took over Atlantic and control over the cables was consolidated.

In 1868 a French company, Société du Câble Trans-Atlantique, threatened the Anglo-American monopoly. The Société proposed to lay a cable 3300 miles from Brest in France, to St. Pierre, an island near Newfoundland, and then to Duxbury, Mass. Despite the fact that the cable was so long, the necessary capital ($6 000 000) was raised in eight days. Much of it came from Field's former supporters in England, especially TC&M investors.

The Anglo-American and the New York, Newfoundland and London Telegraph companies were opposed, but they did not control the landing rights in Massachusetts or St. Pierre. Peter Cooper got a bill introduced in Congress to extend the New York, Newfoundland and London's rights to

Massachusetts, but despite these protests, the project proceeded. TC&M manufactured the cable; the *Great Eastern* was hired to lay it. This development touched off bitter controversy within the management of the existing Atlantic cable companies.

But Field found a way to shortcircuit the emerging state of competition by bringing the French cable into the same revenue system as the Anglo-American cables. On 15 January 1870 an agreement was signed between the Société and Anglo-American making them partners in the control and management of the North Atlantic cable business.[36] Anglo-American received 63⅓% and the Société 36⅔% of the transatlantic revenues.

Thus, by reducing competition, maximum profit was assured for all concerned. The same pattern of consolidation was evolving in the domestic landline systems of the United States and Britain. In 1866 the Electric and the Magnetic were still the major companies in Britain; both had lines to Valentia that connected with the cable. Service was limited to the primarily urban areas of the country; large areas were without any telegraph at all. The total number of miles of wire had increased gradually until 1866, when the Post Office announced plans to buy up all existing telegraph companies.

Government regulation of the telegraph had been increasing in Britain during the 1860s. The Telegraph Act of 1863 was aimed at reducing the financial abuses and construction problems created by the private telegraph companies. By 1868 public dissatisfaction with the private companies had grown because of inequitable distribution of service, inaccuracies in transmission, frequent delays, and high costs. The press was particularly dissatisfied with the companies' handling of news. Following the Continental example, Britain in 1869 added the telegraph to postal service, insurance, and annuities as a state enterprise. The Telegraph Acts did not nationalize submarine cable companies; instead, cables were leased by the Post Office. Nationalization had little effect on the Atlantic cable, since by being connected to the Electric and the Magnetic, it already had access to all the major commercial centers.

In the United States, the trend toward consolidation in the domestic telegraph business had much more significance for the Atlantic cable. Consolidation meant the cable traffic would have much easier access to the inland cities controlled by Western Union and other competitors to the American.

With the outbreak of civil war in the United States in the spring of 1861, and the secession of the Southern states, the existing telegraph industry was thrown into a state of disarray. American Telegraph Co., whose lines ran from Maine to New Orleans, was hardest hit. Their southern lines were confiscated by the South and put under the leadership of Dr. William Morris, a former important stockholder in American Telegraph Co. It was not clear whether Dr. Morris was protecting the lines for the Confederacy or for the company. This disruption took on even more distressing proportions because of the burgeoning of its chief rival, Hiram Sibley's Western Union, most of whose lines ran east to west, north of the Confederacy. Moreover, the general manager of Western Union, Anson Stager, was named head of all military telegraphs.

The Civil War, in cutting the American's total mileage of lines in half, gave Western Union a chance to flourish. However, a new company, United States Telegraph Co., formed from companies opposed to Western Union's dominance, entered the industry. With peace in 1865, the three companies began battling for control of the domestic telegraph business. Peaceful conditions made it possible for the United States and the American to flourish.[37] American Telegraph Co.'s southern lines, albeit in disrepair, were returned to them; and increasing North-South business created a demand for more lines. The American also had a prominent, experienced management headed by Col. E. S. Sanford; Cyrus Field had rejoined the Board of Directors. The prospect of successful laying of an Atlantic cable and the acquisition of the Southwestern Telegraph (Kentucky, Tennessee, Mississippi, Louisiana, and Texas—4800 miles) put the American in an excellent position to contest Western Union's dominance.

Perhaps the most important element of the American's renewed strength was its low capitalization per mile of wire relative to Western Union's or the United States'. Figure 2–5 illustrates the

American's strong position at the close of 1865, when its capitalization was only $100 per mile compared to Western Union's $485 per mile.

Because of its high capitalization, Western Union was not doing well with the investing public. Hiram Sibley resigned; Jephtha Wade, his successor, moved to solidify its base. When the U.S. government turned over much of its war-inspired telegraph lines to the company, the directors felt their goal of gaining a monopoly on U.S. telegraphy was within reach.[38] The next step was its merger with the floundering United States Telegraph Co. on 1 March 1865.

By the beginning of 1866 the North American Telegraph Association was dominated by Western Union and the American. The American and the other remaining telegraph companies now had two choices: join Western Union or band together and form a competing national (north-south, east-west) network. "The advantages of consolidation were obvious—all would be benefitted by the added strength, security, efficiency, and economy which the establishment of a monopoly would give them."[39] On 12 June 1866 Western Union and American Telegraph Co. were consolidated. Thus when the Atlantic cable was laid in Ausut 1866 it connected with the largest private telegraph network in the world and the nation's first private monopoly.

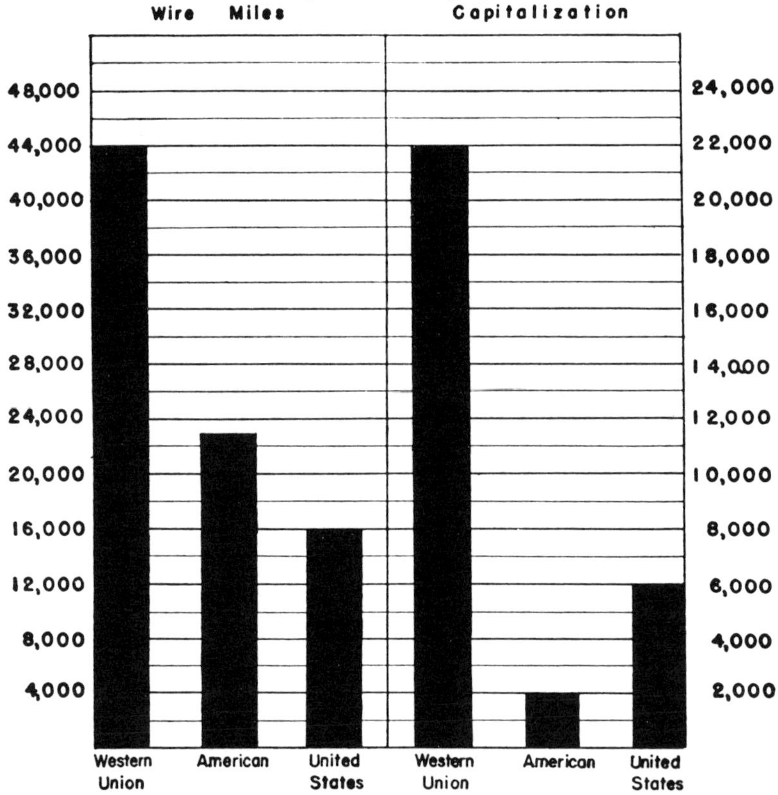

Fig. 2–5.—Wire miles of telegraph in operation and capitalization of big three, close of 1865. (From *Wiring a Continent*, 1947.)

Consequences of Integration

Field's and his associates' reactions to the competitive risks of their enterprise led to integration of a new communications industry. An indirect impact of the competition surrounding the cable was the American interest in Alaska. When Collins went before Congress with his plan for a line that could compete with the Atlantic cable, the possibility of opening Russian America to American influence was widely recognized. The Russian America Co.'s grant of privilege in the area was to expire in 1862. Surveying of the area in connection with establishing a telegraph line was considered beneficial to stimulating American commerce. According to Thompson, ". . . the enterprise . . . added greatly to the United States government's knowledge of Russian America, and was thus indirectly a factor in its later acquisition."[40] Russian America was purchased in October 1867 at the instigation of Secretary of State Seward, a strong supporter of the Atlantic and Bering Straits projects. In addition, the construction of a telegraph line in the Pacific Northwest and in British Columbia did a great deal to open up the area to colonization.[41]

The drive to reduce the competition also contributed to the creation of institutions that remained important to communications development. The special relationship between the New York, Newfoundland and London Telegraph Co. and NYAP established the Associated Press as a major newsgathering institution with exclusive access to foreign news. Although the NYAP dominance did not go unchallenged, its connection with the cable project stood it in good stead in those challenges.

The Atlantic cable project led directly to the vertical integration of the submarine cable industry. Financiers, core manufacturers, outer-cover manufacturers, and cable layers came together to form a major cable concern—the Telegraph Construction and Maintenance Co. The success of the 1866 Atlantic cable showed the value of combining the promotion, manufacture, and operation of a cable under the control of a single group.[42]

Finally, the Atlantic cable contributed to the legitimacy of horizontal integration in domestic and international telegraph business. The pooling arrangement with Société du Câble Trans-Atlantique was later copied by Pender's Eastern group and other cable companies throughout the world. In each case competitors found it more profitable to cooperate than to compete for a limited amount of business.

As to domestic telegraphy, the cable's impact on horizontal integration of the landline companies was more pronounced in America than in Britain where the drive to consolidate telegraph lines was not related to the cable. (Both of the major companies had access to cable traffic.) The demand for consolidation was to cut domestic costs, delays, and inaccuracies. The consolidation of American landlines under Western Union was done for much the same reasons—cutting costs, increasing efficiency, expanding business, and increasing profits. However, in the United States one company had exclusive connections with the Atlantic cable, and it was this circumstance that drove toward the final merger and monopoly.

A Natural Monopoly

The nature of telegraphic communication causes it to be an unstable competitive enterprise. Telegraphically encoded messages have only "symbolic" value in contrast to the "intrinsic" value of packages or goods. Because symbols can be encoded in a variety of formats and carried over virtually any transmission system, there is nearly complete substitutability among the services provided by competitive communications systems.

In addition, telegraphic communication benefits more from increasing market scale than does transportation, manufacturing, or utility services. For communications, the benefits of increasing size increase as rapidly as the increase in the number of *pairs* of potential customers to send and receive messaages—that is, as the square of the number of terminals in the total system. Therefore, combining two telegraph networks of equal size increase the potential customer pairs by a factor of four rather than by a factor of two.

By contrast, transportation costs increase with the distance over which goods or passengers are carried much more than do telegraphic communication costs. The costs of spanning distance tends to isolate transportation markets. Manufacturing benefits from economies of scale (in equipment purchases and marketing of distribution systems); but manufactured goods differ widely in quality, price, capabilities, and basic uses whereas communications is a generic service—in manufacturing terms, there is little product or service "differentiation" within the telegraphic industry.

A stereotyped product or service is provided by utilities, but these services differ from those of telegraphy because they do not depend on the *origin* of the product but only on the destination. It matters little to a customer whether electricity is supplied by a local generating plant or from a power pool drawing electricity from a distant hydroelectric plant. In communication the sender is as important as the receiver, so that increasing size of the system, by increasing the number of possible pairs, is doubly beneficial.

With this imperative for increasing size, and the substitutability of the service, competition becomes fierce and the market unstable; one competitor eventually comes to dominate the market unless there are extenuating circumstances. This is analogous to the Competitive Exclusion Principle in systems ecology (sometimes called Gauze's Principle): No two species that are ecologically identical can long coexist.[43]

Figure 2–6 is a graphical summary of the mergers, bankruptcies, withdrawals, and absorptions resulting from the entrepreneurial challenge surrounding the early years of telegraphy in Britain and America. The overall pattern is one of consolidation of a fragmented group of fledgling telegraph companies into a limited number of monopolistic blocks. Treaties, agreements, interlocking boards of directors, and government privileges were not sufficient to stablilize the competitive drives of the participants.

This pattern persisted throughout the telegraph industry. European nations all nationalized their telegraph systems. The British Post Office has a monopoly over all telecommunications in Britain to this day. John Pender, with his central control of the gutta-percha sources and vertically integrated manufacturing capability, came to dominate international submarine telegraph connections, as will be described in Chap. 5. The Atlantic cable ''Pool'' agreements were first challenged in 1884 but persisted until the cables were bought out by U.S. companies. Finally, Western Union was to become the first national monopoly in the U.S. Telecommunications in the U.S. continues to be a highly centralized industry to this day.

What prevented telegraphy from becoming an international monopoly? What eventually stabilizes competitive interactions? A full treatment of this question would lead too far afield from the purpose of this report, but one important finding warrants mention. The qualifications to the Competitive Exclusion Principle of systems ecology find direct analogs in the competitive interactions within the telegraph communications industry.

Table 2–6 lists eight conditions identified by systems ecologists as stabilizing influences on competitive interactions. Systems ecologists would use the term ''species'' or the more technical

TABLE 2–6.—Qualifications to the Competitive Exclusion Principle. (After E. C. Pielou, *Ecological Diversity*, 1975.)

Condition	Telegraph analog
(1) Conditions not shared with other competitors	• Landing rights, unique access to international connections, agreements with particular clients—governments, new services
(2) Community not at equilibrium	• Industry still growing so that demand exceeds supply
(3) Environment is changing with time	• New applications are arising or there is a need for diversity to protect against possible conditions (war, loss of a line)
(4) The competitors are still changing in time	• Technological innovation is still changing the nature of the competitors and their competition
(5) The competitors are not resource-limited	• The market for telegraph service is still expanding faster than competitors can crowd one another out
(6) The marketplace is variegated	• Different types of service require different types of organizations and technology—e.g., landlines versus submarine lines
(7) The marketplace is segmented by natural boundaries	• Geographical divisions form a base for each competitor's market—encroachment difficult
(8) Potential competitors cannot enter the others' marketplace	• Laws, regulations, international agreements preclude or forbid direct competition

"gamodeme" where we have used "competitor" and the terms "space, competitive environment, or the local ecology" where we have substituted the term "marketplace." For each of the eight qualifying conditions there are ample illustrations of analogous conditions within telegraphic competition. Notice that the dimension of time, implicit in qualifications (2), (3), and (4), applies to the formative years of the emerging telegraphic industry—growth, change, and technological innovation can hold off the competitive exclusion during the development phase of an industry's emergence, but not in the long term for the mature industry. Another illustration would be the concentration of automobile manufacturing.

C. The Political Challenge

In 1978 electronic communications with other nations appears indispensable to the conduct of public affairs in peace and in war. But from 1854 to 1866 neither the British nor the U.S. government took an active role in initiating or implementing the transatlantic cable project. It was conceived and carried through by private entrepreneurs and investors.

The entrepreneurs sought three kinds of assistance from the two governments: monopolistic landing rights, ships and soundings to assist in laying the cable, and contracts for a guaranteed minimum annual payment for services after the cable was in operation, at a rate that would provide a satisfactory return on investment. After several cable-laying failures, the entrepreneurs also tried unsuccessfully to get unconditional guarantees—that is, government subsidy for further attempts, regardless of the outcome of the venture.

Throughout this period, questions were raised in both Britain and America about securing the privacy of information going over the cable, and about the risk of false information being sent through the cable by accident or fraudulent design. Such questions from time to time led to suggestions of government monopoly, at least in Britain. But so strong and unquestioned was the tradition of limited government and laissez-faire that these proposals were never seriously or officially entertained. Instead, throughout this period in both countries, public-policy considerations focused on questions of aid or subsidy rather than government control.

In both countries there was a strong faith in science and technology as the mainsprings of progress and a rising standard of living; and in both countries, although one was the dominant nation of the world and the other a young and still developing agricultural nation, there was great emphasis on national pride and prestige coupled with strong admiration for the daring and aggressiveness of entrepreneurs—even those who were most ruthless and rapacious.

British Policy

The main issue was to reconcile Britain's commercial, colonial, and international communication needs with her commitment to free trade. How much government assistance would have to be given the private sector?

As we saw, the Newfoundland government had given Cyrus Field considerable aid, and Prince Edward Island, Nova Scotia, and the colonial government of Canada had also chipped in, all without objection from London. Field's request for soundings and for cable-laying ships was referred by the Foreign Secretary, Lord Clarendon, to the Treasury, whose financial secretary, James Wilson, came up with a proposal promising both, plus a schedule of subsidies and user fees that would go a long way toward providing financial assistance by the government without making it liable in case of failure—much like a mail contract. The proposal sailed through both Houses of Parliament with next to no objections. But after the 1857 and 1858 failure, Prime Minister Palmerston and Chancellor of the Exchequer Gladstone refused further government support, except for a raise of the initial annual subsidy, in exchange for which the promoters had to forego exclusive landing rights in Newfoundland. This subsidy was nevertheless sufficient to stimulate the flow of new private investment capital.

This policy must be regarded as judicious and successful: It encouraged the project by technical and financial aid but kept it in the private sector (which lacked neither will nor means to invest heavily in far-flung technological enterprises)—all without incurring liability and where possible creating

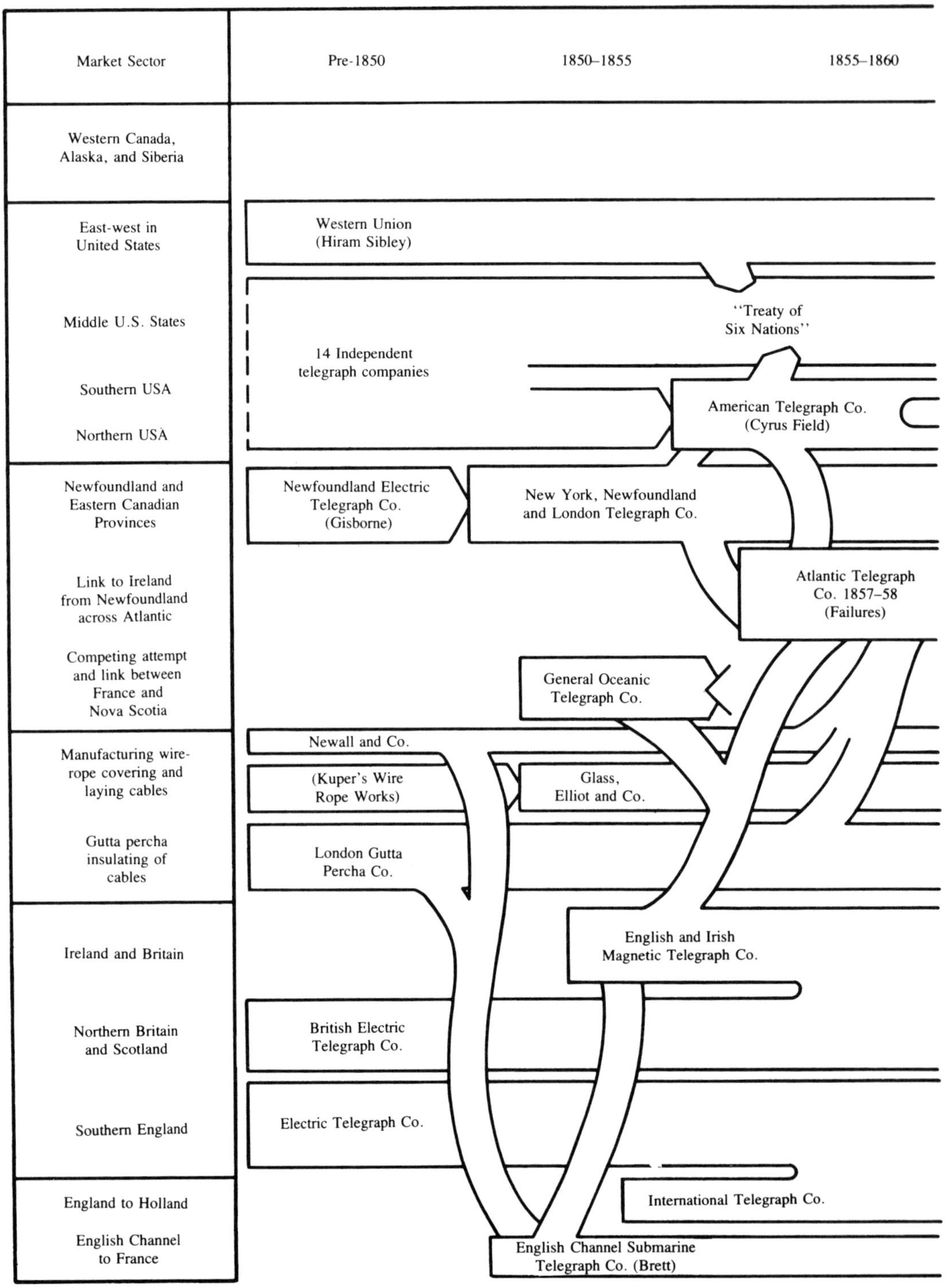

FIG. 2–6.—Telegraphy's progressive consolidation into monopolistic clusters.

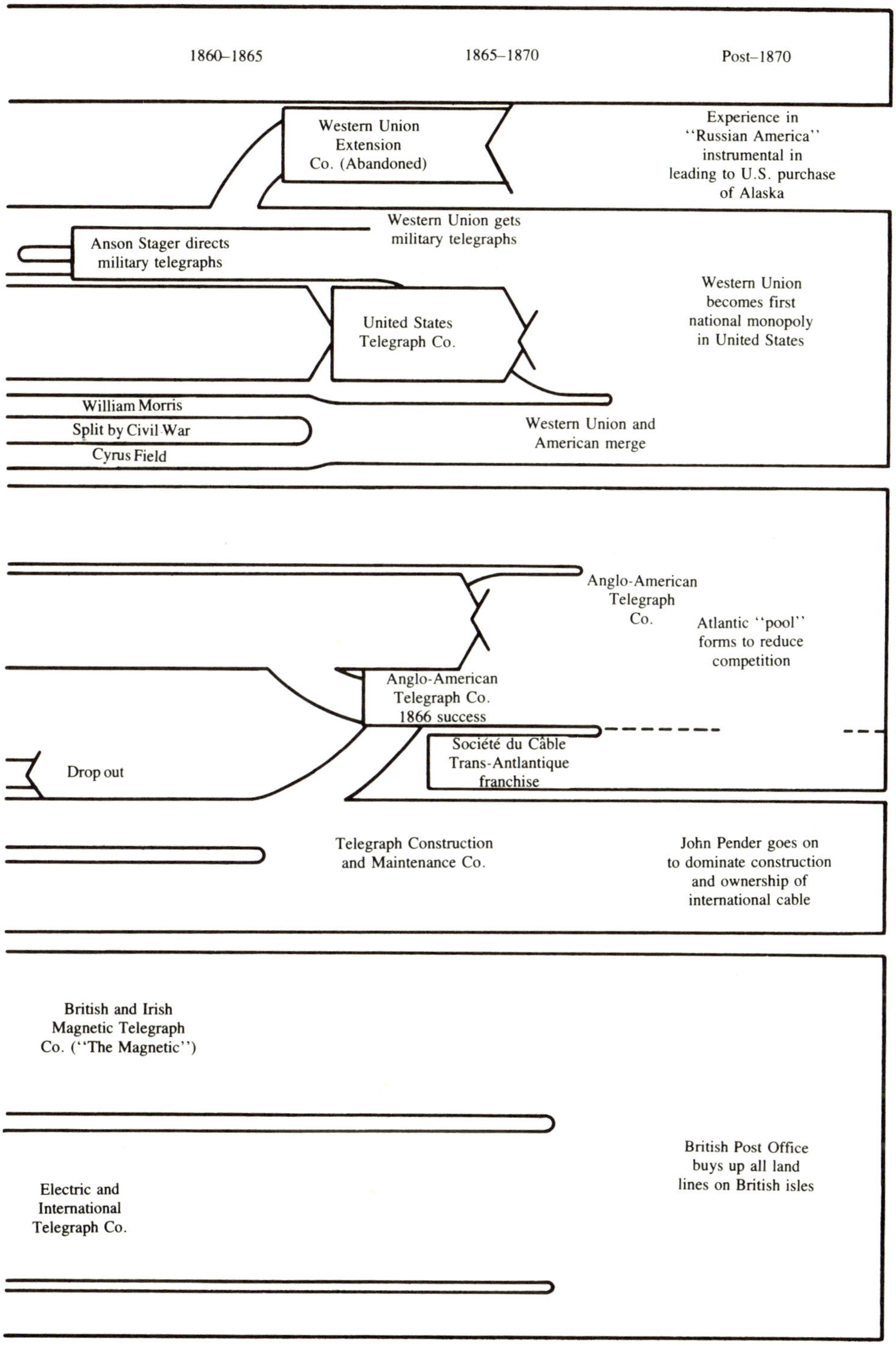
1860–1865
1865–1870
Post–1870
Western Union Extension Co. (Abandoned)
Experience in "Russian America" instrumental in leading to U.S. purchase of Alaska
Anson Stager directs military telegraphs
Western Union gets military telegraphs
United States Telegraph Co.
Western Union becomes first national monopoly in United States
William Morris
Split by Civil War
Cyrus Field
Western Union and American merge
Anglo-American Telegraph Co.
Atlantic "pool" forms to reduce competition
Anglo-American Telegraph Co. 1866 success
Drop out
Société du Câble Trans-Antlantique franchise
Telegraph Construction and Maintenance Co.
John Pender goes on to dominate construction and ownership of international cable
British and Irish Magnetic Telegraph Co. ("The Magnetic")
British Post Office buys up all land lines on British isles
Electric and International Telegraph Co.

Fig. 2–6 (Cont'd)

neither private nor public monopoly.

No government had been as yet seriously threatened with the use of information as a weapon by a foreign power. Other countries without Britain's advantage in trade, and therefore not convinced of the benefits of laissez-faire economic policies, used government monopoly of cables as an instrument of policy, an option not seriously considered by the British government during the period between 1854 and 1866.

American Policy

In America, Field had to contend with hostility and suspicion of what soon came to be perceived as an enterprise centered largely on Britain, the great political and commercial rival; and with the turmoil that preceded the Civil War. A Senate bill providing guarantees like those given by the British government squeaked through by one vote. Most Senators from Southern and Western States thought such aid would benefit the industrialized Northeast at their expense, by perpetuating monopolies and other regional inequities and increasing speculation "by which the industry of the laboring man who creates value," as Senator Toombs of Georgia put it, anticipating Marx, "is to be transferred to the pockets of those who merely deal in values."

The following were the dominant issues in the American debate:

Suspicion of the British, and especially of their policy in the internal American conflict	*vs*	The economic drive to supply British markets and to find other world markets;
and in the South: resentment of the increasing financial dominance of the Northeastern merchants who were seen as a cable's chief beneficiaries	*vs*	the fact that the cotton market was in England, and the demand for ready information which that implied.

As in Britain, the policy issues revolved around technical aid and a conditional guarantee rather than around either direct allocation of public resources or questions of control of communication channels. The latter questions were not raised in the public policy discussion.

One additional issue raised in the Senate was that of competing priorities—namely, establishment of better links with the western and northwestern portions of the American continent. However, since the bill called only for payment if the cable was successful and since it was generally assumed that the minimum payment guaranteed would not be excessive in terms of real usage, this was in a sense a fake issue.

In the House, where debate was superficial and brief, the bill was easily passed with a majority of 19 out of 234. In the Senate the bill was passed on Febraruary 25 with a vote of 20–19 and was then signed into law as the last official act of President Pierce. It empowered the Secretary of State under the direction of the President to contract with any competent persons or associations for aid in establishing and operating a telegraphic cable between Ireland and Newfoundland, and to pay for services an annual sum not to exceed $70 000 (£14 000) until the profits of such company reached the equivalent of 6% interest on investment, at which time the payment per year should not exceed $50 000 (£10 000). The contract could extend for 25 years and was conditional on a similar agreement by Britain (which Field already had), both countries to participate in tariff setting and to be on an equal basis with regard to priority of messages. The guaranteed minimum payment, it should be noted, was a maximum as well as a minimum.

Consequences of the Political Challenge

The commercial interests in America had carried the day, albeit by a narrow margin, and the U.S. had in effect signified its belief in a future as an industrial, commercial nation and an aspiring partner in a world economy. This result was probably more important as an indicator of the balance of power within the country and of long-range developments than it was as a determinant of the future of the cable project. There was very little gain in terms of American investment; the uncertainties and demands of civil war soon swept the nation, overshadowing any interest in laying the cable. But Field gained the symbolic participation of his own country in the project, and prevented the cable, once successful, from becoming a wholly British concern.

D. Lessons from the Challenges and Responses

There are many ways to view history, and many lessons are to be learned from it. These reflections on the process that eventually led to the successful laying of the 1866 transatlantic telegraph cables are guided by the desire to inform the process of technology assessment. Recapitulation of the lessons from the responses to the three types of challenge—technological, entrepreneurial, and political—are relevant to today's policy makers. Lessons are never learned once, for all time; each new situation poses nuances and peculiarities which require rethinking what was thought to have been understood.

Responses to technological challenges eventually led to a systematic examination of submarine telegraphy as a whole, combining the experience of several competing firms and individuals across proprietary boundaries and to systematic testing of alternatives before large investments were at stake. Throughout the industry of submarine telegraphy, there was a decided increase in the influence of engineers and scientists in what had been exclusively business decisions. Management and contracting procedures which reflected this shift allowed adequate project control and responsibility to fall to the most technically qualified. The ability to repair lost or malfunctioning cables radically reduced the risks of submarine telegraph construction and operation. Such innovations are often unsung among technical people but are crucial to investors and entrepreneurs.

Responses to the entrepreneurial challenge led to the consolidation of the telegraph industry under more centralized control, either public ownership or a private monopoly. Before telegraphy was thirty years old, this consolidation was complete, never to reverse. Only national jurisdictional boundaries stopped the consolidation process. Whether these boundaries will continue to be adequate as the nations of the world become evermore interdependent remains an open question. It is clear that the volatility of competition within the emerging telegraph industry was unprecedented in the experience of its entrepreneurial sponsors. The forces that drove telegraphy toward consolidation are present in today's telecommunications markets: the ease with which a single message can be encoded and transmitted over different systems, the overwhelming advantage of greater geographical scope, and the many dimensions of competition itself—for market areas, for investment capital, for connection privileges and landing rights, for government sanctions or subsidies, for interlocking agreements with users, and more. The conditions that led to momentary stability of this inexorable drive toward consolidation are directly analogous to conditions found in natural ecologies as exceptions to the Competitive Exclusion Principle. Systemic attributes of competition and requirements for stability appear to bridge disciplines.

Responses to the political challenge both reaffirmed established social values and revealed the balance of influence behind trends toward change or polarization in those values. Both British and American policy continued to reflect free-trade principles while encouraging the project with technical assistance. United States policy increasingly reflected the growing division between North and South; debate and vote on cable support was split along these political and ideological lines.

The urban, industrial, and commercial growth of the U.S. system of cities has been traced to the pattern of exchanges of information.[44] Urban centers attract innovations through their inventors who find the financial support and specialized skills necessary for development and implementation. In this way cities become the focus of the development process and are first to benefit from innovation. On the international scale, more highly developed and economically dominant nations may similarly attract and integrate communication patterns, skills, and financial resources. A transatlantic cable proved to be well beyond the risk capital resources of the then less developed United States. That British economic and technological leadership of the mid-19th centry came to dominate the emerging submarine telegraph industry is completely consistent with this generalization of the information-exchange theory of the development of systems of cities.

The technological lessons leading to the ultimate success of the 1866 transtlantic cables were paralleled by a social and institutional learning process as well. Simple free trade and laissez faire principles of entrepreneurial development possibly encouraged the creation of the telegraph industry. But these principles could neither maintain the stability of competitive interaction in the U.S. nor satisfy the public needs in Britain. Nationalization of the British telegraph system in 1869 and the creation of the transatlantic pool agreements soon modified British policy.

In this experience one sees an important distinction between scientific/technical knowledge and the refinement of the prescriptive norms of societal and institutional organization. The objects of interest in the physical sciences admit of experiments that do not depend on the historical time frame or experience of the examiner as long as the object of study can be isolated from exogenous influence and the observed result is clearly discernible within the precision of the instruments of investigation. Prescriptive norms embodied in legal precedent, accepted business practice, social/legislative/diplomatic protocol, and cultural traditions can be woven into complex deductive arguments as in the sciences. However, the underlying norms guiding action are constantly subject to revision and clearly depend on the historical context and the priorities of the participants in the decisionmaking process.

Since technological advance and societal organization appear to be based upon different types of learning processes and adaptive principles, it would be informative to examine the distinctions heretofore taken for granted among the technological challenge, the entrepreneurial challenge, and the political challenge. Table 2–7 compares these challenges according to their broad purposes, inherent limiting uncertainties, and general principles for approaching problems.The challenges described in this chapter suggest many illustrations, too numerous to recount here. Of course real decisions most often involve aspects of each type of challenge.

Cyrus Field's incessant drive to create a telegraphic link to Europe appeared essential to the success of the project. It is remarkable that an American was to be so prominent in what was destined to become predominantly a British enterprise. Unfortunately the demands of entrepreneurship—impatient investors, and contingent exclusive landing rights in British America—pressured Field to hurry the construction, laying, and operation of the 1858 cable, probably causing its demise. Similarly, the driving force of Field's approach offended Congressmen to the extent that he had to withdraw from active lobbying and allow his friends to garner support for the legislation to guarantee revenues for a successful cable. The personal qualities that nurtured success in one arena appeared to deny it in the others.

Entrepreneurs are always looking for the means of reducing risk in otherwise promising ventures. Telegraphic technology was used to obtain competitive leverage for the growing New York Associated Press through agreements with American Telegraph Co. Political channels were used by the promoters of the Red Sea cable to obtain British government guarantee underwriting the risks of construction; the public suffered the loss. Landing rights and other political privileges operate in a similar fashion. The proper formula for risk sharing among stakeholders has yet to be found. Certainly going to extremes as shown above is generally unnacceptable. Even today the negotiation of risk sharing is an often poorly articulated process in telecommunications policy.

By 1860 the distinction between technological problems and political ones surrounding submarine telegraphy seem to have been clear in the minds of the British decisionmakers. Postal contracts matched the legitimate powers of government to the requirements of guaranteeing returns to investors in a technology with obvious public benefits. Policymakers thus displayed a tendency which may be said to persist today: dealing with a new technology in terms of existing categories and precedents that have grown up around old technologies. The policy makers found it easier to treat new technology according to its superficial similarities with familiar technologies, leaving the real scientific and technological differences unconsidered. These procedures had the advantage of utilizing existing governmental coalitions and power relationships while leaving other problems to emerge as the unique character of the new technology inevitably burst through the limitations of old policies—for example, the demand for British nationalization of the telegraph, and the necessity of censorship in wartime (to be mentioned later in Chap. 3 on the social impacts). In an analogous situation today, geothermal energy is likely to be dealt with in terms of either water rights or mineral extraction rights in its first phases of development, instead of as a different, and unique, category of technology.

TABLE 2–7.—Distinguishing characteristics of different types of challenges.

Challenges	Broad Purposes	Inherent Uncertainties	General Principles
Technological Challenges	1. Satisfy a human desire with less human effort. 2. Extend man's control over nature and reduce man's dependency on nature.	1. Determining the limits on theoretical understanding of matter and energy. 2. Determining the properties of materials. 3. Anticipating further discovery and innovation.	1. Base designs on empirical understanding of physical relationships which are true for all time. 2. Divide complex projects into components and steps since these can be realistically understood separately.
Entrepreneurial Challenges	1. Create a productive exchange and make a profit. 2. Create an organization or gain recognition. 3. Establish a lasting enterprise —remain viable.	1. Assessing marketplace demand for goods or services. 2. Obtaining financial and other support in the face of risk. 3. The stability of relationships and agreements among market participants.	1. Be first to discover where gains are to be made. 2. Be constantly aware of the desires of each participant and be in a position to appeal to those motives. 3. Plan careful hedges against the risks in the venture.
Political Challenges	1. Increase national power and resources. 2. Establish, increase or maintain personal influence. 3. Gain recognition, acclaim, or satisfaction.	1. Shifting priorities according to unforeseen events and resource availability. 2. Changing the norms or values within the organization or society by which actions are legitimated. 3. The stability of the coalition which determines one's power base.	1. Perceive, express, and shape the values and norms which legitimate action. 2. Structure proposed actions to gain support of (or create) a dominant coalition of influence.

The distinctions between technical and essentially political criteria for decisions are never absolute or without ambiguity. The force of personalities, such as Whitehouse's application of very high voltages to the 1858 cable without approval of his colleagues, always enter into technical decisions, and diplomacy, nationalism, and threats of war always enter into the decisions to invest in technology; for example, the French cable of 1869.

Ultimately, retrospective technology assessment, the study of history, and related activities are the only means by which society can perceive the broad patterns of development and change brought about by the introduction of a new technology into a social, institutional, and international context. The perception of these patterns and the clarification of the distinctions among different types of challenges and appropriate responses is the only known systematic means of anticipating the future consequences of analogous technological advances.

Chapter 3

SUBMARINE TELEGRAPHY CHANGED THE WORLD

RAPID COMMUNICATION at any distance provides a way to manage affairs from a central point—it provides a control system. Information flowing from many directions gives powerful leverage to those at the center who coordinate, integrate, and hoard or disperse that information. Rapid communication also allows people in widely dispersed locations to influence or participate in decisionmaking by contributing ideas and information to, or pressing demands on, decisionmakers within the narrow time in which decisions must be made. Information widely and rapidly dispersed makes people sensitive and reactive to distant happenings that may influence their own lives which in turn affects the decisionmaking environment. Thus rapid communication both centralizes and opens up decisionmaking.

When for the first time in history people could exchange information almost instantly from continent to continent, they were given power over both time and distance. For now it was possible to watch events developing, to see trends building, to identify interactions formerly obscured by the lags in communication; and it was also possible to intervene in developing events. One could manage one's own affairs at a distance more effectively, and act in concert with others of like minds throughout the world.

But in the same way one is rendered vulnerable to the influence or control of others. Thus rapid communications can reduce uncertainties and create strong control systems; it can also create new vulnerabilities and new interdependencies.

When the societal environment changes, when perceptions of time and distance change, when the capabilitity to act and react changes, then social behavior and social institutions also change as they adapt and respond. But the way in which social institutions and behavior change, and the speed with which they change, are at best only partly predictable. We suppose that this depends on many variables, or many behavioral laws, of which we have little knowledge.

The Victorian observers of the new technology of submarine telegraphy were, as we shall see in the next chapter, able to anticipate with some acuity which of their institutions and behavior patterns would be affected by rapid communication, although they were less perceptive as to what those changes and adaptations would be, and about the resistance that would in many cases slow down the inevitable changes. It is we of the 20th century, fully creatures of our modern world and fully dependent on the converging technologies that produced it, who are surprised when we look back at how thoroughly electric communications changed society. This chapter briefly surveys the most significant impacts of submarine telegraphy on late 19th century society, beginning with those which were most direct and most quickly manifested and moving on toward those which were more subtle and complex, and perhaps in the long run most critical. An understanding of these sometimes contradictory effects of a major innovation may perhaps give us some useful understanding of how thoroughly now emerging technologies could change our own world.

A. Shipping

In the last quarter of the 19th century, the development of steamships revolutionized commerce. Steam engines permitted improved design of ships, more capacity, and versatility in the handling of specialized cargoes. With swift communication, the precise scheduling of service and multiport voyages became possible. Cable communication combined with swifter and larger steamships made it possible efficiently to coordinate and direct the purchase, transport, routing, insuring, receipt, and distribution of goods, and thus created the vast and complex network through which modern international commerce is carried on.[1]

The transition from sailing ships to steamships came during the two decades following the laying of the cable, especially in the British merchant fleet, which carried most of the world's commerce (Table 3–1).[2] Without submarine telegraphy, steamships would have been valuable for their speed, economy, dependability, and size, but commerce would not have changed so rapidly; the world would still have been made up of local markets kept separate by lack of prior information about supply and demand. On the other hand, cables probably could not have been laid without steamships, since sailing vessels lacked the controlled, steady pace and course essential to the task of paying out the cable. Thus the two technologies were closely complementary in bringing about changes in society.

Cables made it possible to know at once the movement of ships in and out of major ports. Lloyd's,[3] which had provided the latest shipping intelligence to traders since the 1740s, immediately made use of cables.

TABLE 3–1.—Ships registered in the United Kingdom, 1850–1910.

	Total Vessels		Steam Vessels	
Date	(Number)	(1000 net tons)	(Number)	(% of total net tonnage)
1850	29 984	3 565	1 187	4.7
1860	27 663	4 159	2 000	9.7
1870	26 367	5 691	3 178	19.6
1880	25 185	6 575	5 247	41.4
1890	21 591	7 979	7 410	63.2
1900	19 982	9 304	9 209	77.5
1910	21 090	11 556	12 000	90.4

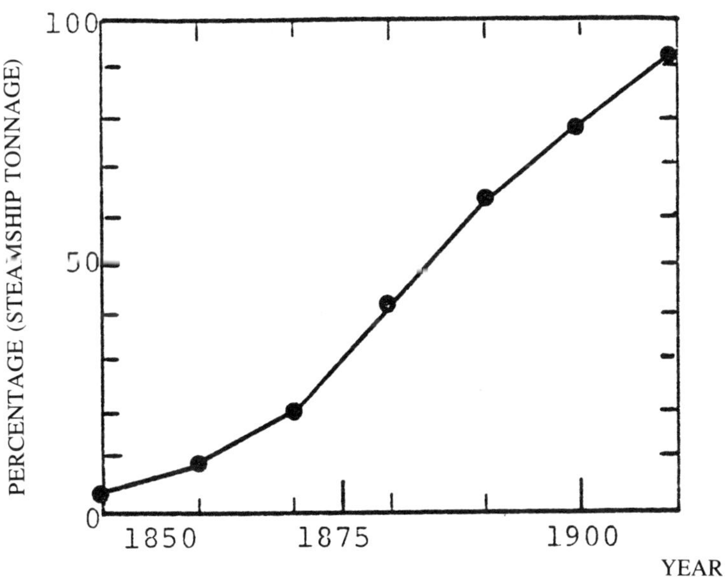

FIG. 3–1.—Portions of net tonnage as steamships.

Lloyd's had first used the telegraph in 1845 and first installed telegraph equipment in its offices in 1851. Two years later agents in major British ports were reporting ship movements three times daily, and Lloyd's agent in Austria was relaying reports from Trieste to an agent in Alexandria for forwarding to India and the Far East. By 1857 most Continental ports were reporting regularly, and when the Atlantic cable began operating in 1866 Lloyd's was one of the first users, sharing message costs with various marine insurance companies.[4]

By 1874 shipping intelligence accounted for the principal expenditure of the corporation (£2345 for intelligence by mail, £1230 for British telegrams, £5549 for foreign telegrams annually). In the 1870s 1200 Lloyd's agents in ports around the world generated a constant flow of information on ship movements, cargoes, wrecks, and casualties, covering 60 000 ships in North America, Africa, the Far East, and Australia, and transit points such as Suez and Gibraltar.[5]

Cables provided links only between ports, but in 1869 Lloyd's began building signal stations along the coasts of major shipping routes that could transmit messages by day or by night to passing ships by visual means. By 1884 there were 17 such stations along British coasts and six abroad, with the nerve center in London. Seven years later there were 40 stations in the United Kingdom and 118 overseas. (In 1898 Lloyd's began experimenting with Marconi's wireless equipment, and some years later radio was widely utilized by the signal stations.)[6]

The ability to communicate with a ship underway meant that the merchant was in constant contact with world market conditions. A steamer loaded with wheat in the United States might sail with its final destination undetermined; off Falmouth it could be signaled to deliver the cargo to Copenhagen. Grain ships from the U.S. Pacific Coast en route to Britain would be signalled from Cork in Ireland as to their final port, and ships destined for Europe from the East via the Suez Canal would receive instructions at Suez or off Falmouth. Ships en route from the South Atlantic would be signaled at Cape Verde or the Madeira Islands.[7] Besides receiving timely instructions ships sent word of their safety to shipowners and to relatives awaiting news of passengers. By 1884 it is estimated that over 90% of oceangoing ships communicated with signal stations before arriving at British ports.[8]

Before cable communications, shipowners relinquished control of their venture when the ship left port; the captain or supercargo was perforce responsible for the disposal of the cargo and for finding a return cargo and thus for the success or failure of the enterprise.[9] After cables were laid and signal stations established, shipowners had a source of information that put them firmly in control of their operations, directing the ship from port to port where market conditions were favorable and return cargoes waited.

In a typical voyage an owner might charter his ship to carry steel rails from Britain to Calcutta. While the ship was en route the owner, through his brokers, might arrange by cable for the ship to take on, in Calcutta, a load of gunnybags for Australia; there the captain would receive cabled instruction to proceed to New South Wales and pick up coal for delivery in South America. In South America the ship might remain in port for two weeks while the owner awaited a pending increase in freight rates, then take on grain for delivery somewhere in Europe. While replenishing its stores at the Cape Verde Islands en route for Europe, the ship could receive instruction to deliver the grain to Rotterdam, where a load of coal awaited delivery to Genoa, after which the ship might return to Britain. Thus financial risk was reduced and the efficiency of tramp operations greatly increased.

The founders of Royal Netherlands Steamship Co. (original prospectus issued in 1856) recognized that the new role of the steamship combined with telegraph connections between important ports and the resulting expansion of the credit system (to be discussed below) made "speed an all-important factor in international commerce."[10] The time was therefore ripe to establish new steamship lines. As Royal Netherlands Steamship Co. (and many other entrepreneurs) recognized, the shipping industry was in for a period of thorough and rapid change.

Ocean liners, unlike tramps, adhere to a regular schedule, which necessitates a sizable fleet and a large shore organization for coordination of cargoes and passengers for a fixed departure. Scheduled transatlantic sailing had existed for twenty years before steamships were scheduled on the Atlantic route in 1838, but liner operations expanded greatly after steam was introduced. The development of the compound engine permitted low-cost common carrier services, and efforts to improve overseas postal services by the British and American governments led to government support of scheduled shipping operations. Cables in turn stimulated the need for dependability and regularity in shipping goods to meet known market requirements.

Merchants could now anticipate and plan the movement of their goods with increased confidence, a particularly valuable development for the merchant or manufacturer dealing in small shipments and specialized nonbulk cargoes. By the turn of the century a case could be made that the great increase in the consumption of goods in less-developed areas of the world was partially due to the ability to provide a dependable and regular supply of goods more closely responding to the demand. For example, an enormous increase in consumption of western goods in India was attributed partially to the regularity and reduced costs of shipments made possible by 365 scheduled annual shipments between India and the United Kingdom.[11]

In the last quarter of the 19th century growing competition caused freight rates to fall. The large increase in capacity of the shipping fleet (Fig. 3–2) was a result largely of the transition to steam. By 1885 gross tonnage of steamships exceeded that of sailing vessels. The introduction of the compound engine, followed by the adoption of the triple-expansion steam engine, resulted in improved fuel economy and larger cargo capabilities. Incrased speed also made possible more trips per year. Between 1873 and 1897 voyage time from the United Kingdom to Australia was shortened by 13–16 days, to China by 11 days, and to India by 7 days.[12]

As we have seen, tramp steamers no longer had to hunt cargoes through the ports of the world. Instead, managers or owners operating through shipping brokers competed for cargoes over large areas. By the 1880s a manufacturer needing a vessel for his cargo might receive offers from several brokers. The result was an inflated view of the supply of ships as well as effectively increased capacity.

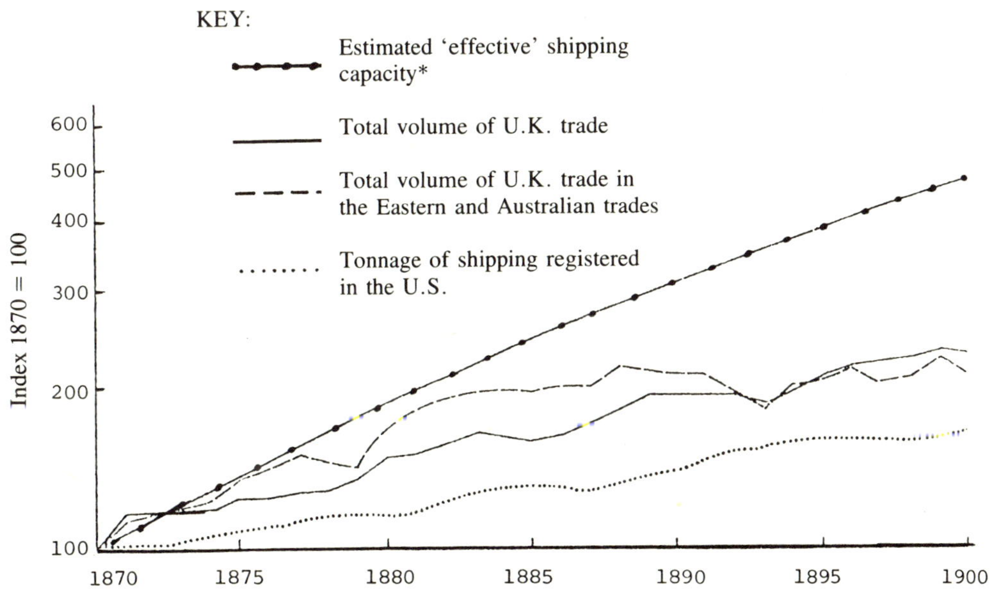

FIG. 3–2. Trends in the volume of trade, shipping tonnage and 'effective' shipping capacity in United Kingdom—Eastern and Australian trade, 1870–1900.

*Tonnage of shipping adjusted for changes in the average time taken for voyages in the Eastern and Australian trades. (Source: Deakin, *Shipping Conferences*, 20.)

As early as the 1870s, as competition drove freight rates down, shippers began organizing "conferences" on major routes as a way of setting rates and minimizing competition. A deferred rebate system insured that members of the "conference" would not undercut the rates that had been set. The shipping conferences or pools still exist in the 1970s and provide uniform rates in what is now a limited field of shipping competitors.

Thus cable communication by improving market information and the ability to transact business rapidly in response to changing demands, and steam navigation by permitting merchants to act on this information, converged to reduce uncertainty and risk. The establishment of shipping conferences permitted a standardization of rates that further reduced the risk of price fluctuation. Reliable shipping eliminated the need to stockpile goods and merchants operating on minimal capital developed an acute awareness of the reduced interest paid on their high-value goods when shipped on fast and reliable ships.

B. Trade and Commerce

The venture system of shipping, before the use of cable communications, was a form of speculation—the attempt to make a profit from fluctuations or changes in the value of a commodity, stock, or other form of property. The profit was viewed as just compensation for the risk involved in buying and selling goods in a condition of uncertainty as to their current or future value in the intended market.[13] Under the venture system the speculator dealt in real goods and speculation of this kind, being necessary to trade, was a respected occupation.

The risk was significantly reduced once a worldwide cable network was established. Speculation then became an important aspect of "futures" transactions in organized markets. But the speculator in the futures market deals in abstracts rather than real goods—in future demands—and the ownership and utilization of the commodity is of no real concern to him. The trader in futures was at first regarded as a nonproductive element in society, a parasite. In the 1870s, 1880s, and 1890s, Wall Street was condemned as "an evil in the land," the Chicago Board of Trade as a "den of speculators," and the produce exchanges as agents of misery and ruin.[14] After the turn of the century, however, a number of writers began to defend the modern exchange systems and the beneficial functions of speculation in organized markets,[15] as the restructuring of world markets (an important consequence of cable communications) came to be better understood.

Futures trading and the organization of markets. Trading in futures is the buying and selling of contracts to deliver or receive a commodity at some future time, with the time of delivery, quality, amount, and price of the commodity stipulated in the contract.[16] Futures transactions existed in transatlantic trade before cable communications, particularly in the cotton market. It came about when (with the introduction of mail steamers, telegraph facilities from Halifax to New York, and scheduled liner operations at limited points on the Atlantic coast) news began to cross the Atlantic faster than did shipped cotton.[17] This made it possible to buy and sell "arrival" or "transit" cotton.

A market for "arrival" cotton existed by mid-century in New York, where cotton could be purchased from samples for "arrival" in Liverpool and then shipped directly from southern ports to Liverpool. Futures trading took hold in the uncertain cotton market during the Civil War and immediately afterward when the demand for cotton far exceeded the supply, which made it difficult for manufacturers to purchase an adequate supply in the spot market (for immediate delivery) and desirable for them to purchase cotton for future delivery. Merchants were eager to sell at current high prices cotton that had not yet been shipped.[18]

With cable communications available after 1866 an extensive system of futures trading in cotton developed very quickly. Now speculation was also possible in a declining market, by selling short in anticipation of a decline in price.[19] Liverpool, the primary market for American cotton and the major world market for "spot" or "cash" cotton trading, saw a rapid decline in spot sales, which had peaked in 1861 at 1 329 100 bales. By 1873, seven years after the cable, spot sales dropped below 400 000 bales and by 1883 below 200 000 bales.[20] In 1871 spot sales in New York (Table 3–2) accounted for only 16.9% of the total cotton crop and this fraction continued to decline steadily to

TABLE 3-2.—U.S. cotton trading, 1869–1895 (in thousands of bales).

Year ending August 31	Total spot sales, New York[a]	Total futures sales, New York[b]	Total futures sales, New Orleans[c]	Total U.S. cotton production[d]
1869 (8 mo)	NA	102[e]	—	3011
1870	NA	592[e]	—	4352
1871	739	3 000	3	2974
1872	529	4 934	14	3933
1873	503	5 300	NA	4168
1874	461	6 188	NA	3836
1875	439	8 358	NA	4631
1876	373	7 234	NA	4474
1877	390	10 735	NA	4773
1878	328	12 973	NA	5074
1879	276	25 411	NA	5756
1880	320	34 007	5 207	6606
1881	315	29 801	12 608	5456
1882	365	33 077	15 706	6949
1883	241	26 544	No record	5713
1884	296	24 632	8 939	5682
1885	175	20 890	8 263	6576
1886	204	23 271	7 018	6505
1887	202	26 482	11 642	7047
1888	125	25 764	8 710	6938
1889	201	18 765	7 000	7473
1890	94	22 138	6 971	8653
1891	96	24 886	9 141	9035
1892	164	34 187	12 794	6700
1893	191	53 245	16 248	7493
1894	205	37 888	12 650	9901
1895	112	39 369	17 796	7162

Sources: (a) *Report of the Commissioner of Corporation on Cotton Exchanges* (Washington, D.C. 1909), pt. 1, 248–249; (b) *Annual Reports of the New York Cotton Exchange*, and the annual *Cotton Movements and Fluctuations* published by Latham, Alexander and Co.; (c) J. E. Boyle, *Cotton and the New Orleans Cotton Exchange*, 164; (d) U.S. Bureau of the Census, *Historical Statistics of the United States: Colonial Times to 1937* (Washington, D.C., 1960), 301–302; (e) G. W. Hoffman, *Future Trading*, 43, and Note (a) above, pt. 1, 55.

0.9% at the turn of the century—from 56 707 bales in 1871 to 16 446 in 1881, 1226 in 1891, and less than 1000 bales in the early 1900s.

Increasing futures transactions led to the organization of cotton markets in the United States between 1868 and 1872. The volume of cotton bought and sold in futures transactions considerably exceeded the actual supply of cotton.[21]

Futures transactions reduce risk in a fluctuating market. By 1868 Liverpool merchants realized that the cable permitted them to know cotton prices in Liverpool and the American South at the same time; if the price was favorable one could purchase cotton in the South and at the same time sell off part of the cotton at a a profit in Liverpool.[22] This scheme provided a "hedge" to the importer against a possible decline in price before he could sell the rest of the shipment in the Liverpool spot market. Thus hedging was first associated with speculative trade. But a grain dealer interested in the "trade profit" as opposed to the "speculative profit" could also use hedging to reduce his risk. If he customarily bought grain with 90% borrowed funds, he could now buy wheat at $1.10 per bushel and sell an equal amount at the same price in the futures market; if on delivery the merchant's wheat sold at a loss of 10¢ per bushel because of market fluctuation, this loss would be cancelled by the profit of 10¢ per bushel from his short-sale transaction on the futures market, so that he could sell his grain at the current market price without loss.[23] The flexibility provided by hedging acts to eliminate risk in handling a commodity between production and consumption. By the turn of the century hedging was generally a requirement for securing loans from British banks for purchase of cotton. Over 90% of U.S. cotton shipped to Liverpool was hedged, as well as 90% of the grain sold on major U.S. grain exchanges.

Since the cotton grown was bought and sold several times over in futures trading, a continuous market was provided for speculators, dealers, manufacturers, and farmers the year around.[24]Manufacturers could assure a future supply at a given price and therefore could accept contracts for future production of cloth. The farmer benefitted because otherwise the price of his commodity would have dropped at harvest time; buyers were now available who could store and sell it for future delivery.

Comparison of the period 1841–1860 with the period 1870–1921 indicates that the general effect of future transactions in the wheat market was a braking action on price fluctuation, which served to reduce the spread of high and low prices.[25] (However, futures trading depended on improvements in railroad and steamship services as well as on telegraph, cable, and later telephone communications.)

The increase in futures transactions led to organization and regulation of the cotton markets. In 1870, only four years after the transatlantic cable, the New York Cotton Exchange was formed, and the New Orleans Cotton Exchange the following year. During the 1870s cotton exchanges were organized in Mobile, Galveston, Savannah, Charleston, Memphis, and Houston; these cotton exchanges, unlike the first two, dealt only in spot trading.[26] Futures trading was primarily centered in New York and after 1880 developed in New Orleans.

These exchanges were part of a global system of markets dependent for its functions on landline and cable telegraphy. Table 3–2 represents an organized network of information collection and dissemination related to market variables, which quickly became essential to participants in the market. By the 1870s the New York Cotton Exchange was spending approximately one-third of its operating budget for telegraphic services.[27]

Developments in the grain market paralleled those in the cotton markets as U.S. exports grew and futures transactions increased. The Chicago Board of Trade, founded in 1848 as the major U.S. grain exchange, established regulations for futures sales by 1869. The Milwaukee Chamber of Commerce (established in 1858), the New York Produce Exchange (1861), and the Merchants' Exchange of St. Louis (1836) were all dealing in grain futures by the 1870s. Several new midwestern exchanges were established in the 1880s and western exchanges a decade later; all dealt in futures.[28]

The New York Coffee Exchange was established in 1882 and expanded as the Coffee and Sugar Exchange in 1914; futures transactions were important from the beginning. The New York Produce Exchange established a futures market in cottonseed oil in 1904. On the other hand the Liverpool and London cotton and grain markets did not have the intense trading in futures that was common in U.S. exchanges; the Bremen Cotton Exchange did not deal in futures, and the Le Havre Bourse developed only a limited volume.

Standardization and storing of commodities. Dealing in futures—without reference to a specific shipment of a specific crop—depends on the ability to standardize the commodity (that is, to assure delivery of an agreed-upon quality) and to store it safely. Many products of trade, such as lumber, cannot meet these criteria and highly organized speculative markets in these commodities did not develop.[29]

A standard system of grading cotton, grains, and other commodities (90% of which were purchased on credit) developed of necessity along with the futures market, as did accepted standard storage facilities and a system of warehouse receipts. An American Standard Classification for cotton was established in the 1870s, and by 1923 the U.S. Cotton Standards Act made all interstate and foreign commerce subject to a Federal grading system. This standard was accepted throughout the world, except in Liverpool, which had its own grading system.[30] Standard grades for grain, coffee, and other commodities also developed and permitted more accurate pricing and trading in world markets.

Relocation of markets. Between 1868 and 1870, as spot trading declined, cotton trade moved from Liverpool to New York and New Orleans, where futures trading was concentrated. Similar market relocations from centralized consumer areas to producer areas occurred after cables were laid, in commodities such as rubber, wheat, and wool. For example, the wool market shifted from London to Australia. For that to happen, an adequate auction system had to develop in Australia so that producers could sell locally, and traditional financial mechanisms had to be modified to permit brokers to operate in the colonial country rather than in London; rapid communication made both possible.

In the 1860s competition among wool manufacturers in Britain, and between British and foreign manufacturers, was intense and prices were depressed. It was important to purchage wool at the lowest price. The final cable link to Australia in 1872 (following the opening of the Suez Canal in 1869) meant that buyers could bypass consignment merchants and importers selling on the London market and buy directly in Australia, reducing middleman costs and shipping direct. An organized market came to maturity in Australia by the 1890s, with sales growing from 40–50 000 bales annually in the 1860s to 87 103 bales in 1872–73 and to 126 000 bales in 1874–75. By 1913, 70% of Australian wool was sold locally.

Reduction of middleman costs and merchants' inventories. By 1886 it was recognized that steam navigation and telegraphy had reduced middleman costs in most areas of overseas trade,[31] especially in the marketing of raw materials, where storage, credit, risk bearing, and sales were handled by middlemen. Between 1844 and the 1920s middleman costs in cotton trade between New Orleans and Liverpool dropped by 40%. (For example, ocean freight rates dropped from 1 to ⅓¢ per pound and marine insurance rates were cut to about half.) Reduced bagging, wharfage, and storage costs resulted from organization and relocation of markets, and reduced commissions resulted from increased competition between brokers and merchants. Previously, merchants imported cotton on consignment, receiving a 2.5% commission, and brokers handled the sales in Liverpool, receiving an 0.5% commission. After offers could be made and accepted by cable across the Atlantic, ending the consignment system, and after purchases were hedged in the futures market, allowing brokers with limited capital to compete with the well-financed merchant houses, the entire transaction carried a 1% commission. Middleman costs in marketing of Australian wool were 10.49% of gross price in 1866 but dropped to 8.23% by 1884.[32]

The need for merchants to hold large inventories disappeared with the expansion of steamships, the availability of rapid communications, and the development of futures transactions.[21] Small importers were able to enter the market, and lead time in the delivery of overseas raw materials to manufacturers was reduced, as imports were better timed to meet demands of manufacturers. Table 3–3 illustrates the 19th century trend in stocks of U.S. and East Indian cotton on hand in Britain—a reduction in supply on hand and a decreased yearly fluctuation within each decade. Consumption increased; that could have resulted in a shortage of storage facilities, but futures transactions provided an alternative. The effect on stocks of U.S. cotton is less dramatic because even before 1866 there was a relatively fast nine-day flow of information by scheduled shipping, landline telegraph for domestic dissemination of this information, and railroads to speed the flow of goods to ports. The impact of cable communications to India in 1865 was magnified four years later by the opening of the Suez Canal and subsequently by additional cable facilities, which resulted in a decrease from an average of 40.2 weeks' supply on hand in the 1850s to 24 weeks' supply in 1884. The lessened need to tie up capital in warehoused stock was obviously differentially important to small manufacturers and merchants.

Arbitrage. The mechanism for leveling prices between markets is arbitrage. Arbitragers buy in a lower, or cheaper, market and sell in a higher market, gambling against the tendency of market prices to level. In the simplest form, a pound sterling purchased in New York from London by cable might cost $1.80; if the rate in New York suddenly drops to $1.79 the arbitrager may be able to buy pounds

TABLE 3–3.—Stock in British ports and weekly British consumption of U.S. and East India cotton, 1811–1884, averaged over decades from annual data in Thomas Ellison, *The Cotton Trade of Great Britain* (London, 1886), Table 1. (Note low U.S. figures in Civil War decade 1861–70.)

	Number of weeks' consumption in stock on December 31				Average weekly consumption (thousands of bales)	
	U.S. cotton		East India cotton			
Years	Average for decade	Extremes during decade	Average for decade	Extremes during decade	U.S. cotton	East India cotton
1811–20	18.2	8–28	128.6	10–396	3.0	0.8
1821–30	23.9	13–38	128.5	47–197	7.8	0.8
1831–40	14.6	9–21	55.2	32–82	14.6	1.5
1841–50	21.4	15–33	64.6	29–131	23.2	2.7
1851–60	12.1	10–17	40.2	11–74	30.8	4.5
1861–70	11.2	6–32	25.4	11–59	15.1	15.8
1871–80	8.7	5–11	25.3	9–44	36.4	9.2
1881–84	11.8	N.A.	22.8	N.A.	47.5	7.0

in New York and sell in London before the market readjusts.[33] The demand in New York pushes that price up; the supply in London depresses the price there. Speculation of this sort is possible not only in foreign exchange but in such internationally traded commodities as raw materials, foodstuffs, bullion, and stocks and bonds.

A large volume of transactions in several commodities after the cable became operational made them important indicators of the marketplace. That was especially true of stocks and bonds; within a month of the laying of the cable a regular system developed for quoting London stock prices in New York, and arbitrage rapidly leveled the European and North American markets. In the 1870s arbitrage transactions accounted for 87% of the speculation on the London Stock Exchange and arbitrage transactions between London and New York totalled over 100 000 shares on some days.[34] A typical transaction would have an operator transmitting a message to the Atlantic Coast, a second transmission across the ocean, a third from the Irish coast to the London Stock Exchange, and a confirmation through the same three steps in reverse; the entire operation took only 3–4 minutes. This service obviously greatly reduced the risk to the speculator. Arbitrage transactions between the London Stock Exchange and the Paris Bourse were less efficient and even in the 1900s sometimes took 20 minutes; sometimes London messages were sent to Paris through New York to save time.

Four years after the cable was laid New York brokers were spending $1 million annually for communication with London. High cable rates (at first $100 for 20 words, reduced to $5 by 1901) led to the development of codes to make a few words do the work of many. By 1902 95% of all messages were in code; by then Wall Street generated 1000 cable messages per day and 30% of all cable traffic originated in New York.[35]

Although New York had become, along with London, a focus of world finance, the introduction of the "ticker" in the New York Stock Exchange a year after the cable and in the London Stock Exchange in the 1870s allowed investors all over the two countries to participate in the market.[36] There was an enormous increase in the number of transactions,[37] most of which were arbitrage transactions.

Foreign exchange and credit. Commercial credit, essential to foreign trade, was provided in the 19th century by banks and merchant houses by means of bills of exchange. They accepted trade bills or drafts against consignments, allowing 2–6 months for payment, and also issued bills under similar conditions to cover credit for one seeking to purchase commodities. Thus at least two sets of bills were required for a transaction, one allowing the purchaser to buy and the other allowiing the seller relatively fast use of his newly acquired assets. When wool was shipped from Australia to Britain, for example, the bill was sent from the seller (after the grace period of 2–6 months) to the London bank that had extended credit and arrived about 100 days after its issue, sufficient time for the wool to be transported and sold on the London market and the money to cover the draft to be deposited in the seller's London bank. After 1872 cable transfers were used that required immediate payment. Banks and merchants had to increase their cash reserves.

Cable transfers were increasingly popular even though they reduced float[38] because it was not necessary to take discounts and costs such as stamp duties into account. U.S. bankers, in purchasing bills in Britain, had run some risks, since they did not know what discount rates were being used. The cable transfer was a kind of international check, and by being faster reduced the risk of bad or unhonored credit, which had caused merchants and banks to allow credit only to long-established and highly reputable firms. Especially in the Far East, because of the long times involved, only major merchant houses had been able to absorb the risks involved in trade. The development of international banking by means of cable transfers gave access to foreign markets for smaller merchants and manufacturers.

In Anglo-American trade a few firms had dominated the foreign-exchange market. A typical merchant banking house, such as Brown Brothers, restricted its operations to the purchase and sale of sterling, advances against cotton consigned to Liverpool for sale, credits to merchants importing British goods to the U.S., and letters of credit for travelers. When granting credit to merchants, Brown Brothers controlled the goods until delivery and then provided the merchant with credit for their purchase, thus shortening the period of the debt; they insisted on providing all the financing for any merchant with whom they dealt so that they could monitor his financial status and protect their investment. With cable communications and the increase in trade, the volume handled by a merchant often was too large to be safely handled by one banking house; in any case, Brown Brothers and other merchant banking houses met increased competition from other kinds of banks.[39] In 1870 fifty-six private houses and four banks participated significantly in the New York foreign exchange market; the number of banks increased steadily until after World War I they dominated the market.[40]

Arthur H. Cole has studied the sterling exchange rates during the 19th century and found that they fluctuated erratically between 1825 and 1860, which he attributed to inadequate development of the New York money market and "the difficulties of communication."[41] Between 1865 and 1913 seasonal variations of exchange rates, and differences between highs and lows of the monthly averages, narrowed significantly. He attributed that to the closer relationship of the New York and London money markets as a result largely of cable communications, along with the use of cable transfers and the increasing use of an international loan fund.[42] Cable communications and steam shipping also contributed to stabilization of the exchange rate by decreasing the cost of shipping gold; they allowed quicker response to the fluctuations in exchange rates which dictated the shipment of bullion in and out of the country (until 1913), and as a result the "gold points" or prices at which gold was imported or exported converged steadily.

Increasing stability of the foreign-exchange market in turn reduced the profit margin exacted by foreign-exchange firms. Before the Civil War, for example, Brown Brothers had to demand a margin of up to 1% on daily exchange operations to assure a net profit of 0.25% after losses resulting from slow payment and uncertain discount rates. By 1900 the firm was content to charge 0.25% as these uncertainties were reduced by rapid communications.

Competition. The major British commercial organizations and merchant houses had championed and financed cable communications because they had vast overseas interests and a strong need for a rapid and reliable flow of market information. Ironically, 1872—when basic cable routes between Europe and the Far East were completed—also proved to be the peak of British domination of foreign trade. Newly emerging industrial powers—especially Germany and the U.S.—penetrated British overseas markets and smaller British merchants made inroads into what had been the private preserve of the major merchant houses. The merchant houses had dominated foreign trade through the consignment system because only the large houses could finance all phases of marketing, accept the financial risks and the long credit periods, and invest in large accumulations of stocks. With cable communications, the consignment system disappeared—abruptly in the U.S. cotton trade and more gradually in other areas, such as the Australian wool market.

Trade with the Far East had always been difficult because of the 2–3 months needed for communication between Liverpool and Shanghai. The opening of the Suez Canal, improved steamships, and cables were expected to bring great benefits to the merchants. But the natural monopolies of the large houses eroded as local, foreign, and smaller British merchants went into the markets. For a time British exports to China even declined as a result of competition, from almost £9 million in 1872 to between £7 and £8 million in the late 1880s. Treaty ports were flooded with goods while the interior remained relatively closed as a market.[43, 44] A Royal Commission on Depression of Trade and Industry investigated the state of British trade with the Far East in 1885–86 and placed most of the blame on foreign competition resulting from the Suez Canal, but several observers, such as Francis Strange of the Peking legation, stressed the role of "regular steam and telegraphic communications with Europe," which had altered the role of the middleman and ended the era of large profits by a few firms having a virtual monopoly.[45]

In Latin America the British mercantile houses had many of the same problems but a substantial number of them hung on by bringing together a large variety of goods at favorable prices. Some manufacturers retained their traditional tie with the wholesale merchant rather than entering into direct trade with retailers,[46] but on the whole cable communications had wrought dramatic changes in the structure of Victorian trade and commerce. D.C.M. Platt points out that the "intense specialization" of Victorian commerce when manufacturers produced, and left the shipping, marketing, and credit functions to others, gave way during the late 19th century to direct selling—ignoring the skills of the traditional "merchant."[47] Rapid communication made this streamlined procedure possible.

C. Weather Forecasting

Mankind has always wanted to know what the weather is going to be like tomorrow or next week, especially for planting and for shipping. Foklore is full of techniques for predicting weather, from bodily clues (such as rheumatism) to observation of the behavior of animals, magical spells, and divinations. By the 19th century attempts at forecasting the weather were much more "scientific" and systematic. Systems for collecting information about weather trends had developed, and landline telegraphy was immediately utilized for this purpose. But most nations lacked sufficient land area to make national weather forecasting effective: international cooperation was necessary to collect and integrate signals of changing weather from beyond national boundaries.

This factor was especially important for Britain, a small country with a large shipping fleet very vulnerable to the vagaries of weather. By 1863 Britain had a system of telegraphic weather reporting utilizing cables from Ireland, the Isle of Man, Jersey, and Heligoland, and from France, Portugal, Denmark, and Norway. But there were many problems, including the high tariff rates for cables, the need for international cooperation, and the uncertain and uneven development of the state of the art of meteorology.

Britain particularly needed warning of storms in the Atlantic. Ships arriving in Boston and New York after crossing the Atlantic would by 1885 submit weather observations that were consolidated and cabled back to Paris and London, but the high cost of the cables and the uncertain results led to early abandonment of this attempt. In the 1890s Britain got daily consolidated weather reports from various Continental locations and from the Azores. But cable stations were not necessarily available at the critical location for weather observation, especially on the open seas. Time and again international meteorological conferences discussed the better use of cables for weather observation. Cables linked to ships stationed off the Atlantic Coast were suggested. The Prince of Monaco recommended a system of weather observatories at the Cape Verde Islands, the Azores, and the Bermudas. Another proposal was a special cable system to serve observatories in Iceland and the Faroe Islands at a cost of £20 000 annually, to be divided among ten nations in proportion to the relative importance of the service to them. (Britain's share was to be £9855.) But such special systems were never built and weather services continued to be dependent on existing cable routes, which were not planned to meet the needs of weather forecasters.

In some areas, however, cables had a major impact on weather forecasting capability. The first U.S. cable reports on hurricane formation came from Cuba in 1873 and accurate tracing of West Indian hurricanes started in 1878. The need for facilities on foreign soil was a major problem. A well-organized hurricane warning system was not fully developed until 1901, when the U.S. Weather Bureau was receiving cabled reports from Havana during the hurricane season which consolidated information cabled from stations throughout Cuba and in Jamaica, Santo Domingo, Puerto Rico, St. Kitts, Trinidad, Barbados, Curaçao, and Dominica, supplemented by observations from British stations at Turks Island, Nassau, and Bermuda. Even when this system was in the early stages of development it was cost-effective. After 1890 no hurricane reached a U.S. seaport without a warning having been issued well in advance. This service was reported to have greatly reduced damages to property; in one case it was estimated that $38 million in property was removed or secured without damage as a result of the warning.

Forecasting of weather forming over the North Atlantic and understanding of its effect on weather conditions in Britain and on the Continent were slow to develop. Real improvement had to wait for marine radio, which permitted rapid reporting from many areas of the ocean simultaneously.

But the use of cable communications for weather reporting had greatly stimulated the development and standardization of equipment and techniques for observing and measuring meteorological change. It had also stimulated international cooperation in weather observation and forecasting, beginning with the first International Meteorological Conference in 1872. A more subtle consequence was the growing realization that though weather could not be controlled, it could be reliably anticipated and its impacts moderated—another step toward planning and scheduling and another aspect of the importance of rapid communications in reducing uncertainty.

D. The News

As soon as the first landline telegraph was built, newsmen seized on this marvelous way of being first with the news. New organizations sprang up for collecting the news and getting it to both newspapers and commercial interests. Among the first were Correspondance Garnier (founded in 1832) and Havas (1835) on the Continent; by the 1840s Havas operated a regular service between London, Brussels, and Paris[48] that used pigeons to bridge the gaps between telegraph connections. Bernhard Wolff established a news agency in Berlin in 1848 to capitalize on the new Prussian telegraph line from Berlin to Aachen, and Julius Reuter began to operate a news service on the other end of the line. As landlines between Berlin and Paris expanded, Reuter switched from pigeons to horse relays. By 1850 Berlin and Paris were directly connected by telegraph. Out of business, Reuter turned to supplying news between Britain and the Continent; in 1851 the Dover-Calais cable was laid and Reuter built offices at Calais and Ostend to provide cable news services. At first all these organizations dealt primarily in commercial information but also carried official government news, local newspaper items, and advertising, providing the necessary translation services.

The high costs of cable transmission and the competition that developed meant that profits were low. In 1856 Reuters in Britain, Wolff in Prussia, and Havas in France agreed to share their commercial information business, and three years later their political news services. In 1869—six months before the Franco-Prussian War—the three agencies divided up the news world. Wolff was to serve Germany, Austria-Hungary, Russia, and Scandinavia. Havas was to control the French Empire, Italy, Spain, Portugal, and South America. The rest was left to Reuters. But the latter's portion included Britain, and British-financed and British-laid cables soon linked the major commercial and political centers of the world with Europe. Reuters' offices at the ends of these cable lines seemed to reflect the dominant British position throughout the 19th century world. A Reuters branch was established in Bombay in 1866, soon after the opening of the Indo-European landline and cable, and by the end of that year there were offices in Calcutta, Madras, and Rangoon. Karachi, at the terminus of the Persian Gulf cable, became the distribution center for news from the East. A Reuters office opened in New York in 1867, and Reuters followed the cables to Singapore, Java, and Shanghai in 1871, to Australia in 1872, and to Yokohama and Nagasaki in 1873.

The public press had expanded in both the U.S. and Britain as technical advances were made in paper and printing, literacy increased, urbanization grew, and new advertising techniques were developed. By mid-19th century there were many daily newspapers, but international news was featured in only a few. London newspapers, primarily *The Times*, could be distributed throughout Britain on the same day that the foreign news was received; provincial papers (none of which was a daily) therefore did not carry foreign news.[49] On 11 December 1849 the London *Times* had astonished speculators at the Paris Bourse by utilizing railway and steamship to deliver the London paper, with its commercial news, to them at 1:30 P.M. on the day it was printed.[50] But landline telegraph lines meant that for the first time provincial newspapers in Britain could compete with the London *Times*, and the public press soon took full advantage of it.

In the United States, because of the larger size and more uneven development of the country, the New York papers had not dominated the public press as did the London papers in Britain. The early and rapid development of landlines immediately stimulated a cooperative newsgathering organization, the Associated Press (AP). Organized by major New York newspapers in 1848, it expanded to provide news—especially foreign news—to a large number of subscribing newspapers throughout the country. Competitors challenged the AP's control of foreign news sources, patents, and telegraph facilities,[51] but by the later 1860s AP and Western Union Telegraph Co. had a firm and mutually beneficial alliance through which they dominated domestic wire news services.[52]

In Britain the first domestic news services were provided by the telegraph companies themselves. The *Times* management soon realized that there was no practical way to prevent telegraph companies from copying *Times* news reports and selling the information to the provincial papers; in 1855 they made the best of this situation and reached an understanding with Electric Telegraph Co. by which the latter received *Times* dispatches prior to publication and in return provided the newspaper with full telegraph service in the United Kingdom and over the Channel cable.[53]

The provincial papers could not do without the wire services sold by the telegraph companies, but they suffered from poor service.[54] Dissatisfaction by the newspapers with the inept management of the telegraph companies contributed strongly to the movement to nationalize the British telegraph, and in 1868–70 the assets of the telegraph companies were purchased by the Government and a state monopoly was established. A Press Association owned and operated by the provincial newspapers was established to collect and distribute news, relying on Reuters for foreign news. This close relationship led to control of Reuters by the Press Association after World War I.

By contrast to landline telegraphy and domestic news services, ownership of international news services and cables remained separate. In 1858, when the first Atlantic cable appeared to be successful, Edward Taylor of *The Manchester Guardian* had sought the assurance of Atlantic Telegraph Co. that it would stay out of the news business.[55] At this time Reuters was just moving to establish an internatonal news service, and had the first cable continued to operate, Atlantic Telegraph Co. might well have decided to compete with Reuters. But eight years later, when the second cable finally succeeded, Reuters, Havas, and Wolff were well established and the opportunity had passed.

Before electric communications news was a valuable and salable commodity but only in a limited market. Once printed, news is no longer exclusive; control over it is lost. With the telegraph and cables, news became a perishable commodity that could be distributed and sold simultaneously over a large area to many customers. Hence a high investment in efficient newsgathering could be maintained.

The increase in volume, regularity, and speed of news transmission over a part of the route increased the importance of speed in handling the news throughout its transmission. Reuters, Havas, and other news organizations used pigeons and horse relays to expedite the flow of information at the end of the early telegraph lines. Troops of uniformed messenger boys were organized at telegraph terminals to provide the final ink.

Before the development of telegraphy and scheduled steamer service, news had filtered into North America from Europe through a number of eastern seaboard ports; inputs were random, unpredictable, frequently duplicated but often inconsistent, and not selected or edited specifically for American customers. As the channel of news flow became narrower, with the cable, efforts to retrieve the news became more systematic as well as more intense.

Between 1851 and 1858 AP had already managed to develop an effective monopoly on foreign news. After the failure of the 1858 cable AP leased Cyrus Field's idle Newfoundland telegraph system for five years or until a successful cable was laid and intercepted ships off Cape May to retrieve news: 13 ships in 1859, 31 in 1860, 34 in 1861, and 46 in 1862.[56] On the other side of the Atlantic, Reuters and *The Times* competed for the latest news from North America, first by using fast boats to intercept ships off Southampton, and when a cable was laid to Ireland in 1853 by intercepting ships off the south coast of Ireland. Still later Reuters built a landline to the extreme southwest tip of Ireland to get a jump on their competitor.[57]

Mowbray Morris, editor of the London *Times*, had called the Atlantic cable a "great bore," saying that the public would expect it to be utilized for the latest news no matter how trivial, and the newspapers would be obliged to shoulder the expense.[58] But Morris as well as Reuters and AP utilized the short-lived 1858 cable even at the high cost of $10 per word.[59] The same rate applied in 1866, and all news agencies realized that they had no choice but to use the cable, yet such high costs could make competition among agencies mutually disastrous. A news "war" broke out in 1866–67 between Western Union (which dominated the landline system serving the cable) and AP, whose "insensitive" treatment of western papers had led to a revolt against New York domination of AP. Western Union challenged AP's monopoly. Before the brief "war" ended both sides were near bankruptcy. New York papers, which had been spending about $800 a week each for their share of AP's cable tolls, had to contribute $2100 without the western subscribers. The settlement decreased the western papers' share from 33% to 22% and AP agreed to use only Western Union facilities where they were available in return for an unchallenged monopoly on supplying foreign news.

During the news war the western papers had purchased Reuters dispatches; AP had continued to rely on its London agent's transmitting items after they were published in the London papers.[60] The Reuters news was thus fresher but the AP representative was able to select items of most interest in America. After the press war, however, AP continued the use of Reuters dispatches.

The AP "monopoly" did not mean that subscribers depended entirely on the news service; the New York *Tribune*, for example, supplemented its subscription to AP, which cost about $14 000 per year, with a permanent staff in Europe.[61] But AP maintained its London office around the clock, processing 35 000 words a day by cable by the late 1860s. AP's cable costs were sometimes $2000 a day and never less than $300 even after cable rates began to decline.[62]

Extensive coverage of European events as they happened rapidly became the standard. In 1866, before the cable was working, Henry Villard of the New York *Tribune* missed the events of the Austro-Prussian war because the "Prussians moved faster than the correspondence," since there was a ten-day delay for transatlantic news. In the Franco-Prussian war four years later George W. Smalley led a team of *Tribune* reporters reporting continuously by telegraphy, a technique developed during the American Civil War.[63] During the Spanish-American war AP had a fleet of boats operating along with the U.S. naval forces off Cuba and rushing news dispatches to the nearest cable ending, spending $300 000 on this operation; and during the Santiago campaign 2000–6000 words were filed daily by newsmen.[64]

The reporting of the Russo-Japanese war cost AP more than $300 000. During World War I the association spent an average of half a million dollars annually for coverage, with an expenditure of $801 195 in 1918 alone.[65]

The foreign news being reported in ever-increasing detail through wire services to local newspapers had important repercussions. Dorald L. Saw, in a study of "bias" in the reporting of presidential campaings in Wisconsin daily papers from 1852 to 1916, concluded that increased use of telegraphic news during this period was inversely proportional to the number of "biased" articles printed, and that this relationship was most evident after 1880 with the increasing dependence of local newspapers on wire service news (Fig. 3–3 and Table 3–4). Shaw defined as biased those articles which "contained value statements in such a way that the over all impression created upon today's reader was a positive or negative feeling toward the referent."[67]

The systematic and often cooperative use of the telegraph for reporting and distributing news made it available to many more local papers, but the increasingly centralized function of news processing made it more vulnerable to control. The increasing size of the potential audience meant that there was a strong incentive for governments to guard, limit, or distort the flow of information to fit national objectives—in other words, to censor the news or to use propaganda to influence public opinion.

During the Crimean War in 1853–56, the British government first showed serious concern over the potential of the enemy's benefiting from what had previously been harmless public information. The enemy (Russia) was in telegraphic communication with London and Paris as well as with the Crimean battlefields. Britain and France established communications with the front lines through a

FIG. 3–3.—Percentage of election-year stories judged biased (solid line) and percentage of stories sent by wire (dashed line), 1852–1916.[66]

TABLE 3–4.—Percentage of election-year stories judged biased, wire and nonwire, summarized for 1852–1880 and 1884–1916.[66]

Election years	Per cent judged biased of: Wire news stories %	Wire news stories *n*	Nonwire news stories %	Nonwire news stories *n*	Total number of stories
1852–1880	23.2	56	53.0	264	320
1884–1916	1.4	653	34.7	124	777
Entire period	3.1	709	47.2	388	1097

series of landlines across Europe and a Black Sea cable. Reports of British correspondents published by the London press were available in St. Petersburg on the same day they were published in London. Lord Raglan in the Crimea and the Duke of Newcastle at the War Office were soon protesting the publication of reports on the movement of artillery, munitions, and troops; such information was even found in private letters from soldiers to their families, which were published in newspapers. But not all government officials immediately recognized the implications; Lord Hardinge, the Master-General of Ordinance, was regularly supplying *The Times* with detailed information about troops being sent to the Crimea and the War Office was pressuring the newspaper to exert self-censorship.[68] In the face of growing public criticism about the conduct of the war, the government did not dare attempt to impose official censorship.

Earlier, during the American Civil War, newspapers had published reports helpful to the enemy, and during the Franco-Prussian war Prussian army observers had monitored French news cabled to London and then cabled this information directly to Northern Germany.[69] When the U.S. Navy, in the Spanish-American war, desperately wanted information about the Spanish fleet in the Caribbean, the AP reporter aboard the flagship cabled AP correspondents in major cities around the world. From Madrid he received information, published in local newspapers, that the Spanish fleet had entered Santiago harbor the day before. American naval forces immediately deployed to Santiago, setting the stage for the crucial battle.[70]

By World War I governments had fully assimilated these lessons. Slogans such as "Loose Lips Sink Ships" passed the lessons on to the public. Information, however, can be not only denied public dissemination by self-imposed or government-imposed censorship; it can also be used for persuasion. The telegraph network coupled with organized news services provided a channel that could be utilized and was too important to be ignored.

International news services, like the cables themselves, enjoyed something of an "international" status although each was privately owned and yet closely linked to a European power (Reuters to Britain, Havas to France, Wolff to the rising Prussian State). Relations between news agencies sometimes reflected international political alliances; in the 1880s, Stefani in Italy, Wolff in Germany, and the Korrespondenz-Bureau in Austria-Hungary (that is, the Triple Alliance powers) sought to reduce the territory of Havas. Reuters resisted pressure to join this alliance and by 1890 all were again on friendly terms.[71]

In 1893 AP sought and gained exclusive rights to the Reuters news service in the Western Hemisphere, thus becoming a member of the powerful Reuters, Havas, and Wolff news combine. AP served only North Ameria, and later Central America. This was a major triumph for AP over its only important domestic rival, United Press (UP). It meant that all AP-gathered domestic news transmitted for foreign consumption went through Reuters, and all foreign news destined for AP distribution went through Reuters also. Combined with the British-dominated cable arteries of the world, this made London the undisputed center of world newsgathering at the turn of the century. Within a few days after the outbreak of World War I Britain cut the German-owned Emden-to-New York cables and throughout the remainder of the war all news to the Western Hemisphere from Europe went through London and Paris.

London and Paris censored all traffic on the transatlantic cables during the war; the Central Powers could get only a trickle of news uncensored by the Allied Powers into the U.S. through the wireless.[72] During this period messages direct from Germany accounted for 6% of the war news printed on the front and editorial pages of the New York *Times*. News from Berlin via London accounted for 3.5 to 8%, from neutral nations 6% or less, and London-censored news from 57% (in the first year of the war) to 46%. Before the German cables were seized or cut in 1914, 80% of the news printed in the U.S. about Germany and Austria-Hungary originated in those countries, but in the first month after the cables were cut only six of the 96 New York *Times* items dealing with Germany originated there and 73 were written in London. By contrast at least 70% of the war news referring to France and Britain originated in those countries. Thus the Allied powers were able to dominate the perception of the World War by the U.S. public.

When the French refused to allow German war reports to go over the cables to South America in 1914, AP considered moving into this Havas-controlled territory. But the only direct cable line between the U.S. and South America followed the west coast and could not compete with the direct London-to-Buenos Aires cable; and a break in the alliance between AP and Havas and Reuters would have prejudiced collection of news in Europe and increased the cost of collecting foreign news. When the U.S. entered the war in 1917 there was government pressure on AP to expand to South America, and State Department influence reduced rates on the U.S. cables below the 25¢ per word rate Havas was paying on the transatlantic cable. If the AP had accepted this opportunity the French government was prepared to support Havas by underwriting all cable charges to South America. But AP refused to be a vehicle for U.S. news as Reuters and Havas were for their governments.[73] Eventually AP did expand without government aid to become an international news service, and in 1934 the tie with Reuters was severed and the independence of the American news service was assured.

By this time the economic limitations, so important to the cooperative efforts of the early international news agencies to carve up the world into spheres of influence, no longer existed. Cable tolls had decreased to the point that competition could be sustainined. Also, the objective of foreign news services had expanded from simply acquiring foreign news to dissemination of U.S. news abroad. The U.S. had become sensitive to the fact that U.S. news reaching South America through Europe tended to emphasize lynchings, murders, and embezzlements.[74] To an emerging international power such as the U.S. such a world image was unsatisfactory.

The first effort to provide the world with the American image of itself began with the U.S. entry into World War I, when the Committee of Public Information was established to provide a channel for the dissemination of the American position to the world. Since Britain and the Continental powers controlled the cables, wireless proved to be the answer for an American voice to the world.

The growing importance of propaganda was a direct result of the growth of news-conscious publics and the increased participation of the general public in the political process in modern industrial countries. By World War II all countries recognized the importance of the news media in achieving domestic solidarity and national objectives, and in influencing world opinion over the heads of national governments. Cable communications, fifty years before wireless capability, stimulated and fed public demand for news of world events as they happened, and public consciousness of having an immediate and direct stake in those events.

E. Diplomacy

Diplomacy is "the conduct of relations between governments by officially accredited representatives" or foreign service officers.[75] With each breakthrough in communications capability come predictions that the foreign service will be rendered obsolete. For example: "[When] the most distant post can be reached immediately by cable or radio . . . [the diplomatic representative] will be merely an automaton carrying out the will of the foreign office."[76] Or: "When the techniques of long distance telephone (and television) communications have been developed a little further [the foreign officer] . . . will fade, like the Cheshire Cat, into a new social grin: for foreign ministers will then be able to conduct their business directly with one another as they sit at their office desks."[77]

Such predictions were common in the early years of the cable; one example will suffice. Samuel S. Cox, a U.S.Representative from New York, observed in 1874:

> Telegraph, steam, with their prompt communications; newspaper enterprise ever in advance of diplomatic dispatch; these and other elements of progress, have rendered ministers abroad trifling, expensive, and useless for every purpose of national comity, interest, or glory.[78]

There were scores of such predictions, and they had a ring of truth. Telecommunications facilities provide physical capabilities which if accompanied by mutual international or bilateral agreement as to their utilization could conceivably support diplomatic intercourse without a system of overseas representation. This situation never developed. The cable instead was used to serve the existing system of diplomatic reporting and exchange. Telegraphy was a far more rapid means of communications than was required for most diplomatic uses until nation states acquired capabilities in other areas which rivalled electronic communication in speed. The U.S.-USSR "hot line" is an indication that the traditional mode of diplomatic communications through representatives will not suffice in the modern era. But in the 19th century there were relatively few situations in which such swift reactions were neccessary or desirable. Communications of a personal nature between individuals and transactions between buyer and seller, or between commodity markets, are based on underlying common relationships in which intermediaries may impede or detract from the efficiency of the process. But in diplomacy intermediaries serve to protect the head of government against embarrassment, too hasty judgment, and the necessity for consistency—they provide a valuable buffer. Transactions in diplomacy tend to be defensive and conserving in nature, a searching out of mutual advantage or vulnerability; they depend heavily on precedent, experience, understanding, and flexibility.

The predictions of obsolescence of traditional foreign-service functions were accurate in reflecting the capabilities of the technology but not in assessing the ability, willingness, or desirability of the institution to utilize the technology to its full potential. The predictions assumed a swift molding of institutions to fit the technical innovation; but in practice its impacts were indirect and reached major importance only after additional innovations in institutional structure and behavior developed to fit the new technical capability.

In three other areas the consequences of rapid communication for diplomatic practices, and for international relations, were much more significant: the shift to a continual-response mode, the opening up of new areas of cooperation and new interdependencies, and the involvement of the general public in foreign affairs. The development of a "continuous response mode," in which governments can (and therefore must) react to events as they are happening, has two consequences which are at best in precarious balance: it tends to exaggerate urgency and magnify crises, but it also requires a continual give-and-take between governments which probably acts to modify policies and to keep them flexible, tentative, and pragmatic.

Rapid communications also made it possible for nations to cooperate in some areas where quickly shared information was essential; for example, in apprehending fugitives, in controlling epidemics, and later in coping with natural and manmade disasters. There instances were not so dramatic as the interdependencies created in commerce, currency, and warfare, but they gave rise later to international organizations and mechanisms that no doubt have a stabilizing effect in international relations.

On the other hand, the involvement of the general public in military and foreign affairs decisionmaking, which is to be discussed more fully in another section, creates new constraints and pressures on diplomacy that may be either stabilizing or destabilizing, and may most often push a government into extreme and aggressive positions.

A detailed study of communications between the U.S. Department of State and its mission in London for the first forty years of cable telegraphy, and a sampling of the communications since then, provides insight into this evolution.

Pre-World War I use of the cable. In 1866 five cable messages were exchanged between the Department of State and London. For the next 43 years, the volume of correspondence between the two posts rarely reached 150 messages a year (Fig. 3–4). Between 1909 and the outbreak of World War I (1914) the volume increased somewhat, with a peak of 552 messages in 1910. Comparison of the cable traffic with the volume of letter instructions and shipped dispatches shows that the diplomats depended for the huge bulk of communications on letters, the traditional mode. The 1000 words transmitted by cable in 1880 or the 9000 transmitted in 1890 hardly had a major impact on diplomacy.

Cablegrams were used to complement letters—to convey or to request a specific bit of information—which was developed in detail by letter. For example, on 27 September 1878 the Department of State cabled to London that instructions regarding a dispute over fishing in North American waters had been sent by pouch the same day, and that a copy was to be delivered to Lord Salisbury immediately on its arrival. The U.S. minister was to cable the results of the discussions to

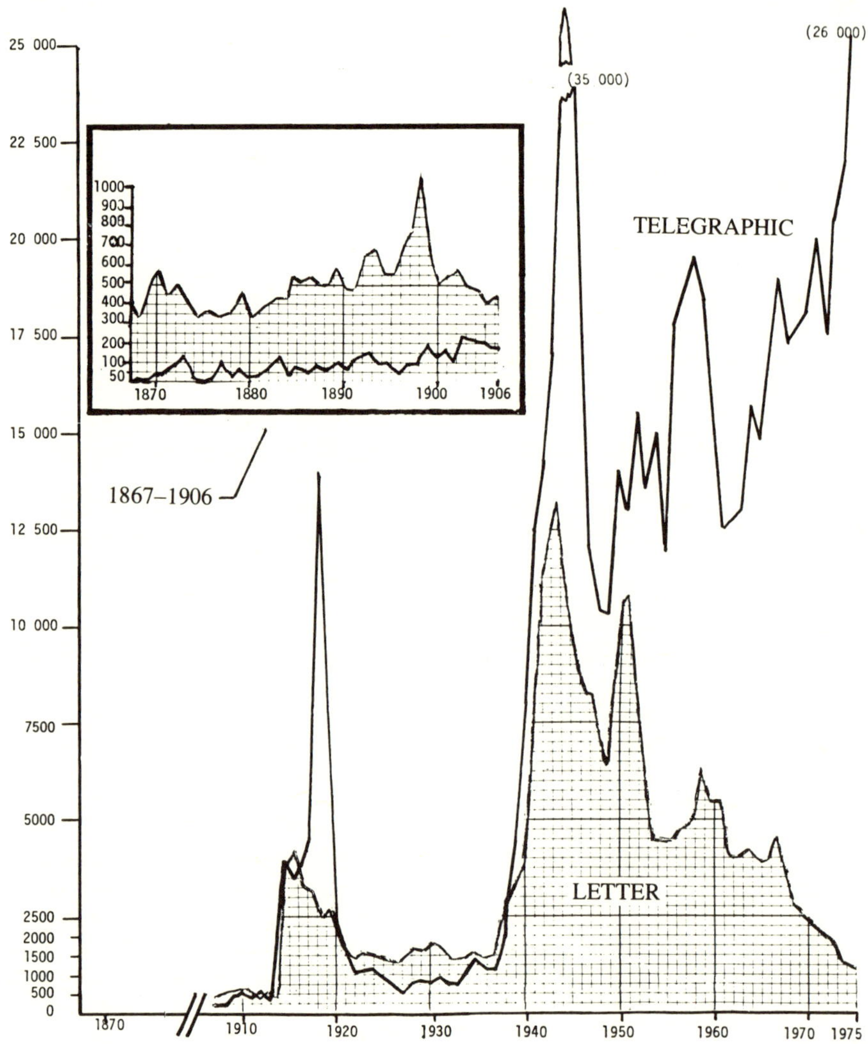

FIG. 3–4.—Total pieces of letter and telegraphic correspondence between the U.S. Department of State and its mission in London, by year, for 1867 through 1975.

Washington immediately by cipher. This was a typical example; the cable was used merely to expedite business that could have been conducted entirely by cable.

In a crisis a series of short telegrams in both directions would be generated, but generally they were used to request or send information about the immediate status of a politically sensitive, short-lived event (perhaps an American thrown into jail in Ireland); continuity and broad policy instructions were transmitted via mail pouch.

The *Alabama* Claims negotiations of 1872 were the first diplomatic ''event'' in which there was a surge of cablegrams between the U.S. and British governmentt. (The U.S. government claimed damages from the British for allowing *Alabama* and other Confederate raiders to operate against Union shipping from British ports during the Civil War.) Transmission errors were few, but there was a problem in keeping track of the rapid exchange of cablegrams; they sometimes arrived out of sequence and there was confusion over what information had been received at each end when responses were sent. The British Foreign Secretary, Granville, expressed great frustration over using the telegraph in negotiations,[79] and Robert C. Schenck, U.S. minister to London, was relieved that the negotiations were carried on in the U.S. and there was no need ''to correspond through [the] difficult and unsatisfactory medium of the telegraph.''[80] To reduce the confusion, a system was adopted of listing each cablegram in the text of letter instructions, but once the negotiations were completed this practice ceased (although it reappeared on later occasions). Telegraphic messages were not serialized by the State Department until 1911, although they were usually duplicated in letter correspondence, which was serialized.

Another problem that plagued the negotiators was security; Secretary of State Fish warned the U.S. minister in London to use ciphers because cables between them were being published in U.S. journals and in the London *Daily News*; the leak was thought to be in the telegraph office.

After the *Alabama* Claims were settled cable traffic dwindled. It surged again in 1874 (an abnormal volume of extradition proceedings), in 1882 (U.S. citizens in Irish jails), in 1892 (a fishing boundaries dispute, and efforts to promote a monetary conference), and in 1898 (the Spanish-American war and entanglement of U.S. shipping in the Boer War). But in no year prior to World War I did cablegrams exceed letters, and in no one event or series of negotiations was the bulk of information carried by cable.

In general, cable traffic between the State Department and the London mission grew along with the American presence overseas; London was the center of world shipping, commerce, and finance. But use fluctuated: London received 51 cabled instructions in 1872, four in 1874, 29 in 1881, 53 the next year, and only 16 the following year. Figures 3–5 and 3–6 categorize the various uses of cables. Once urgency had been established by use of the cable, the reply generally also went by cable, often with an interim cable stating that the response would arrive in a day or two. Some messages were long and detailed (in the early years they often used the flowery and circuitous phrasing of diplomatic letters), but as a rule those pertaining to negotiations were brief; background material was provided in the continuous dialog of the pouches. Although the high cost of cable service pushed toward brevity, no uniform system of abbreviations developed and many excess words were still employed, reflecting a fear that extreme economy would sacrifice clarity.

At the turn of the century, standing instructions to U.S. diplomats continued to warn that use of cables was ''not permitted in the ordinary business of the mission or in communications with the Department of State except when justified by important and urgent cases.''[81] But from the first the cable was not in fact reserved for such urgent matters. Cables were used for appointment of negotiators or representatives to conferences; assignment of military attachés and observers; procuring copies of agreements, maps, or other bits of information; and permission to issue passports. These functions accelerated the tempo of diplomatic exchange and many of these matters were not found in pre-cable diplomatic letters. The existence of instantaneous transtlantic communication was already generating new requirements for information on which to base actions. For example, in control of epidemics international cooperation had not been practical until there was swift communication; now cables were used to coordinate efforts to stop the spread of hoof-and-mouth disease, a

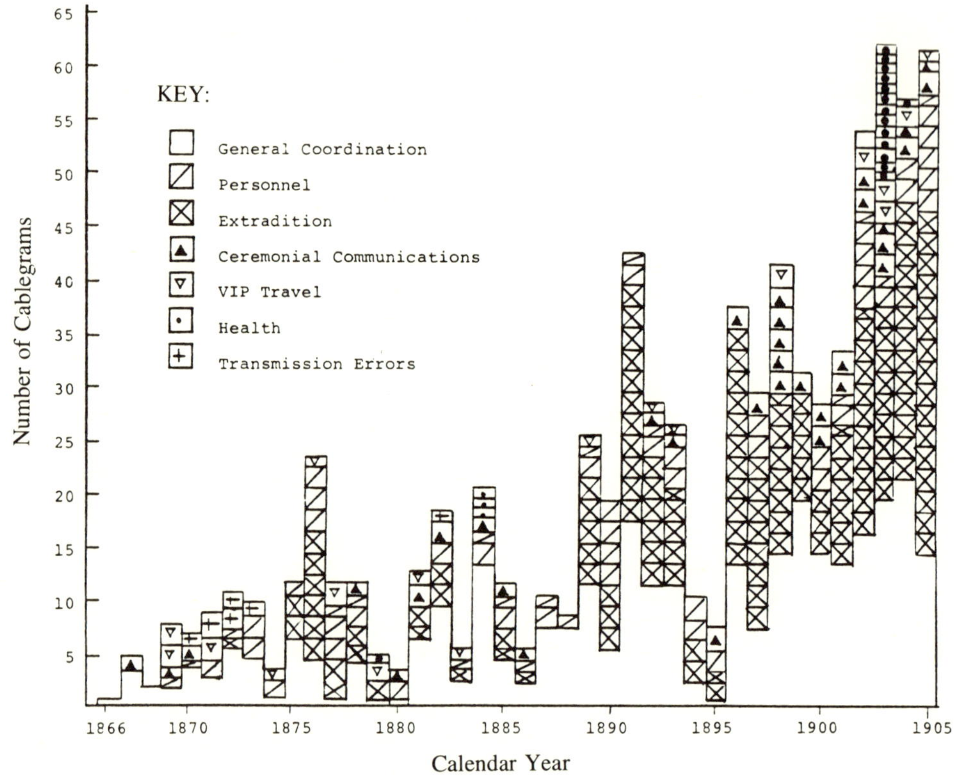

FIG. 3–5.—Composition of nondiplomatic telegraphic instructions received in London from the U.S. Department of State.

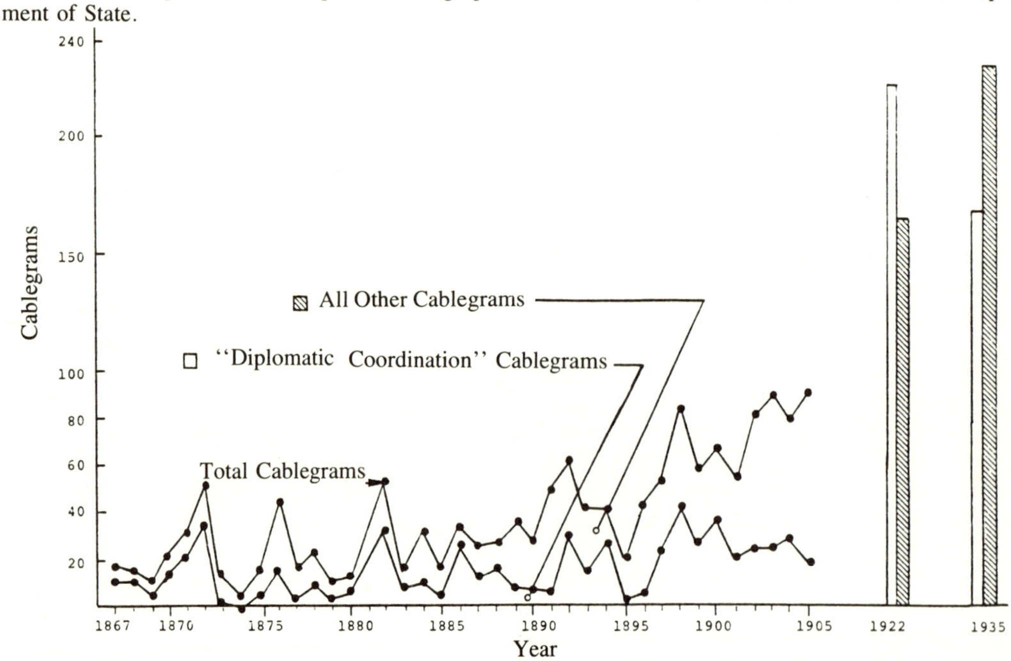

FIG. 3–6.—"Diplomatic coordination" and all other cablegrams received in London from the U.S. Department of State, 1867–1905, 1922, and 1935.

matter of considerable economic importance. A second example that developed rapidly was the use of cable communications to request arrest of fugitives arriving by ship, and their extradition. Thus the cables made possible several new forms of practical international cooperation, which later gave rise to new international organizations to carry out these functions.

Considerable cable traffic also developed for ceremonial and "personal comfort" purposes—congratulations and condolences, arranging greetings for important persons on arrival in the country, and announcing births, deaths, marriages, or appointments to high office.

Use of the cable during and after World War I. For the first half of 1914 cable traffic between the U.S. State Department and London was like that of the previous years, about 150 messages each way. With the outbreak of the war in Europe it increased about tenfold and continued high for the three years of neutrality. When the U.S. entered the war cable traffic increased still further, and continued high during the peace treaty negotiations.

Around 1922 cable traffic between the State Department and London stabilized at 500–900 messages annually during the 1920s and 700–1300 during the early 1930s. U.S. involvement in Europe never returned to the prewar level, nor did cable traffic. Letter volume had approximately equaled cable traffic during the three years of U.S. neutrality and was far eclipsed by cables after the U.S. was involved in the war and in peace negotiations; however, by 1922, the number of letters and dispatches was again much higher than the number of cables. Yet by the end of the World War I the cable was for the first time a fully integrated, "normal" mode of diplomatic exchange.

In 1938 traffic volume between London and Washington again began to increase rapidly—jumping by 50.5% each year until the peek of 35 000 cablegrams in 1945 at the height of the U.S. participation in World War II. (The message traffic build-up into World War I averaged a 52% increase per year over the interval 1913 to 1918.) Manual coding and decoding in the peak years of World War II were often as much as two days behind; at times it took five to seven days to process a routine message in-house (an interval roughly equal to an Atlantic crossing by ship). Congestion threatened to defeat the assumed advantage of electrical communication—speed. A decline began immediately after the war, but the lowest annual volume in postwar years (10 000 messages in 1949) was four times higher than the highest annual volume before the war (2298 messages in 1938). After 1938 the volume of diplomatic correspondence by letter never again exceeded cables in number, and letter correspondence has declined steadily since 1951. The increase in volume of cable communication throughout this period is greater than counting the number of messages would indicate. The information transmitted in each message was much larger (Fig. 3–7).

Since World War II, there has been considerable fluctuation in traffic volume; the changing status of Great Britain as a world power meant that the London post was not directly involved in some events important to the U.S. in recent decades (e.g., the Korean War), but participated in others, such as disputes about the Suez Canal and the Middle East.

Cable costs and traffic volumes. The five cablegrams between Washington and London in 1866 cost roughly $2400; at $10 a word, cablegrams were not routine events. By 1872 rates had dropped to $5 per word, soon after to $2.50. By 1884 the cost was 50¢ per word, by 1885 40¢, by 1888 25¢, and in 1904, 20¢ per word.[84] Rates had dropped steadily at an average rate of 15% per year for 20 years (Fig. 3–8).

In that same time frame, State Department expenditures for telegraphy climbed rapidly, at 9% per year (Table 3–5). In comparison, total operating expenditures, excluding salaries, increased at only 2.5% per year. Since the actual message volume between London and Washington increased at less than 10% per year during this period, much of the overseas telegraphic budget must have gone into landline connection with London and other U.S. foreign missions.

However, the large declines in cost during the first 20 years of transatlantic telegraphy were not accompanied by comparable increases in message traffic. A change in the way telegraphy was utilized within the State Department was less a function of economics than of organization and habit. At the opposite extreme, the gravity and increased pace of world events, and the growing involvement of the U.S. on the international scene, especially during the World Wars, forced the integration

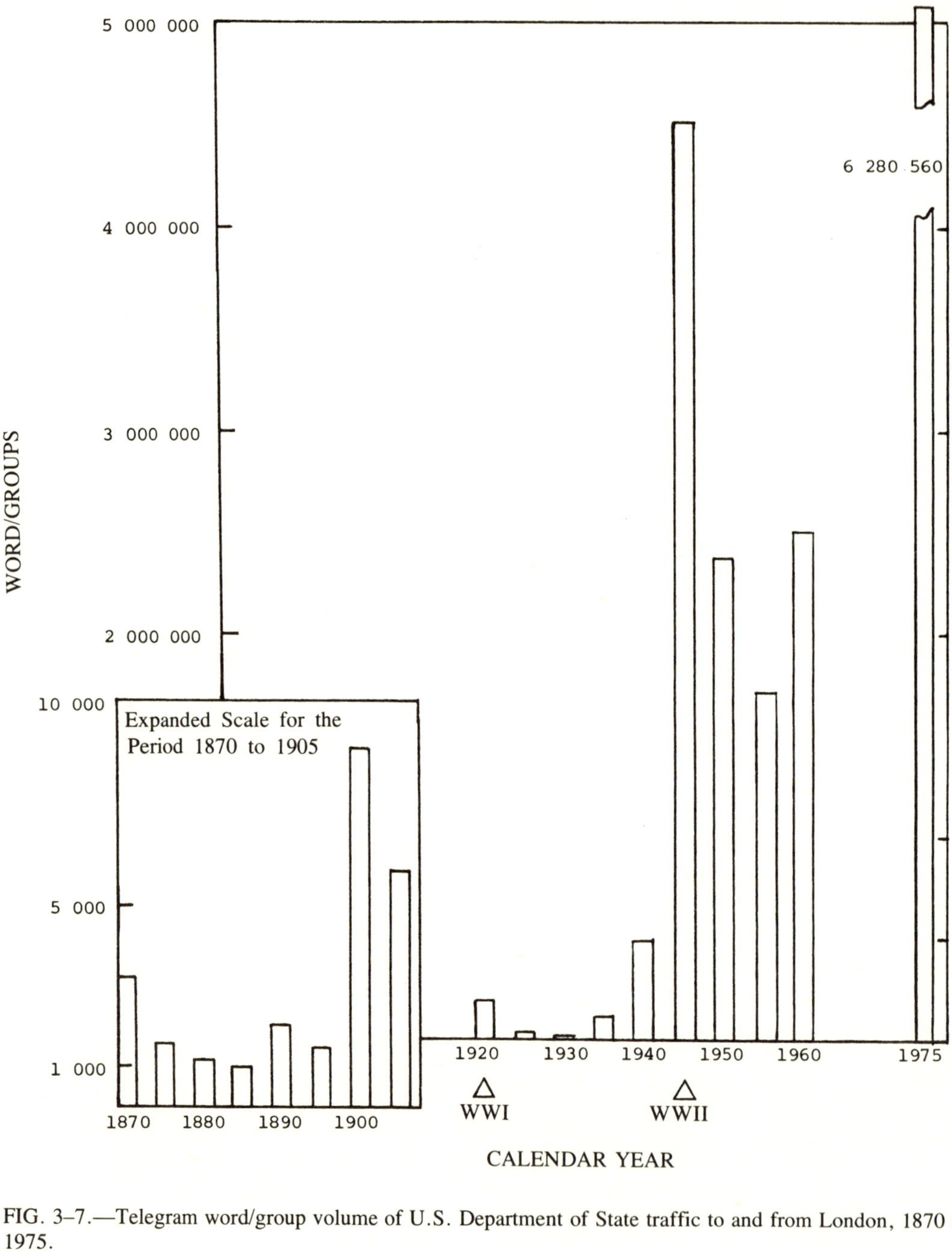

FIG. 3–7.—Telegram word/group volume of U.S. Department of State traffic to and from London, 1870 to 1975.

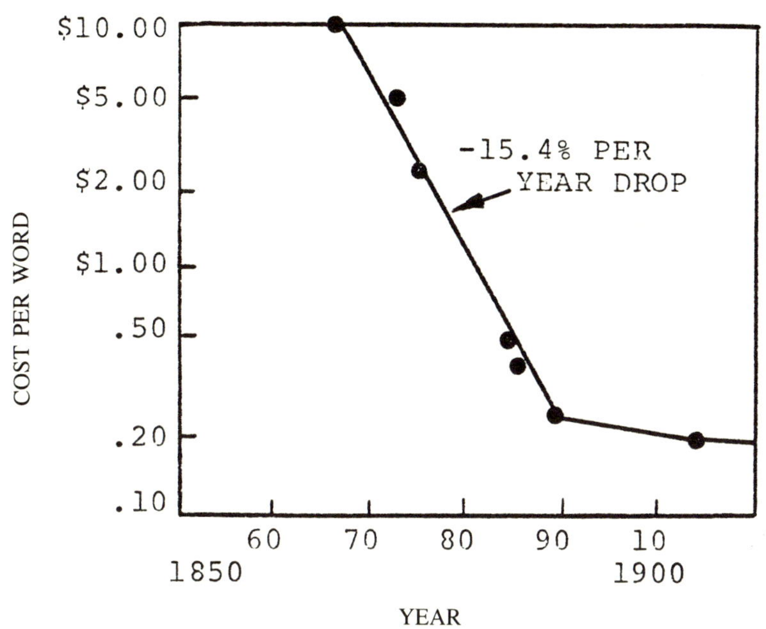

FIG. 3–8.—Decrease in cost of transatlantic telegrams, 1865–1905.

TABLE 3–5.—Increase in U.S. Department of State expenditures for telegraphic communication with its foreign missions, compared with increase in total operating costs excluding salaries, 1866–1917.

FY	Total telegraphic expenditures	Total contingent expenditures
1866 (pre-cable)	$ 1,397.83	$305,828.02
1870	4,354.06	389,450.58
1875	2,670.86	422,378.26
1885	11,227.57	440,246.66
1907	75,597.01	555,620.11
1910	170,472.30	793,888.07
1916	147,365.53	825,777.54
1917	179,240.72	957,291.56

of telegraphy into diplomatic practices in brief periods of intense adaption and stress.

The Department of State first leased wires on the transatlantic route in 1939, with two telegraph printing machines operating over a direct circuit to London. From then on, the greater the utilization of the system, the more economically efficient its operation. It was during the following period that letter volume significantly declined relative to cable traffic.

But well before this time procedures within the institution of diplomacy had begun to change. The Department of State in 1868 hired a temporary clerk whose duties were to send and receive cable messages for the Secretary. Messages were transmitted via a private telegraph line between the Department and the Western Union office 13 city blocks away to be processed. Forty-two years later (1910) the Department's telegraph room had five clerks and handled about 1000 messages per month.[83] By the 1930s 60-word-per-minute teletype machines were in operation, along with the traditional telegraph key; nine operators were on duty in a round-the-clock operations, averaging 200 messages a day. During the 1940s and 1950s the Department further developed its system; telex was introduced and by 1961 a "crisis response center" was set up to take care of more critical events. These developments reflect the change which was occurring, within the Department and within the practice of diplomacy, from leisurely and reflective development of response to international affairs to a continuous mode of response to events throughout the world.

Codes. Codes were developed to meet the needs both for economy and for privacy and security. They began in the commercial sector. British businessmen in the Far East trade were particularly sensitive to having trade information intercepted as it flowed across continental telegraph lines to and from the East. Price quotations were especially valuable; in the 1870s there was a Parliamentary investigation of Indian telegraph operators falsifying opium price quotations. Diplomats had similar problems, beginning with the "news leaks" during the *Alabama* Claims negotiations. On the other hand, Reuters News Service developed effective codes less for security reasons than to win a competitive cost advantage by reducing the size of cable messages.

In 1867 the Department of State first provided its missions with a specific cipher for transmitting sensitive information over the Atlantic cable. The Department's first code book (the "Red Code") was used from 1876 to 1918. As the volume of traffic increased, the manual system of encoding and decoding became intolerably slow and cumbersome compared to the speed of the communication itself. During World War II the Department's clerks were often two days behind, and at times to took 5–7 days to get a routine message coded. In 1945 the Department began using electro-mechanical crypto-mechanical equipment which was about ten times faster than the manual method.[84] By the early 1960s full automation permitted all traffic to be securely transmitted without delay for encoding.

As the technologically advanced nations upgraded their secure communications systems, they also increased their capability to intercept the messages of the less-developed nations that lacked such advanced systems. This created a disparity, or magnified an old disparity, in the conduct of international affairs.

There was considerable disparity in the generations of technical equipment available in the Department of State at any time. After 1920 wireless communications were adopted as the primary mode of communications to some areas where cable service was marginal or not available, and also provided an alternate route for all communications. Cables continued to be the primary communication mode with Europe and with certain other areas (such as Ethiopia). Upgrading of telegraph equipment was not uniform; teletype machines operated side by side with manual Morse equipment until about 1954, when the manual circuit to Mexico City was finally abandoned. Older equipment was often retained after more advanced equipment was introduced, as backup or in some cases as a less expensive mode. Satellite communications have added a new dimension of versatility.

The slow acceptance of the cables was later paralleled by resistance to the telephone as a diplomatic tool. Secretary of State Stimson in the 1930s frequently used radiotelephone for discussions with the London or Paris embassy, with advisors using headphones ready to answer questions when needed. But that proved to be an exception. The radiotelephone was not widely used for diplomatic purposes, except as the cable had been used in the early years, either in second-by-second crisis situations or to collect isolated bits of information—for two reasons: the difficulty of secure voice communications and the absence of a record of what had been said.

The slow change in diplomatic practices. Lord Lyons, British minister to the U.S. during the Civil War, remarked in later years that had there been an Atlantic cable at the time of the *Trent* Affair (when the captain of a Union warship intercepted the British steamer *Trent* and seized some Confederate officials from it), the U.S. and Britain would certainly have gone to war; as it was, he and his counterpart in London, Charles Francis Adams, maintained silence for several weeks until the public uproar in both countries had died down and tempers had cooled enough for normal relations to be resumed.[85]

Today it is startling to think of a several weeks' delay in responding to a critical diplomatic issue or perceived threat to national security. The capacity for rapid communications seems to necessitate rapid response—in hours or minutes, not weeks—whether or not a wise policy can be formulated within that time. The ability, and the perceived necessity, to act upon events as they occur changed the tempo of world events and therefore of diplomatic practices. But this change was surprisingly gradual, compared to changes in commerce and trade practices.

The British Foreign Office had telegraphic access to the Continent by 1851. In 1859 resident clerks were first hired to handle telegrams after normal business hours,[86] and a number of high government officials had telegraph facilities installed in or near their country homes. The U.S. Department of State employed its first telegrapher in 1867 and linked the Department, the White House, and the homes of the Secretary of State and Chief Clerk by telegraph. The British Colonial Office employed resident clerks to handle telegrams after working hours in 1874.[87] After 1871 the British Ambassador to St. Petersburg had two members of his staff living in his residence because "telegrams came in the night and it is difficult to find a secretary unless he is living in the house."[88]

From this ability to transmit and receive information without regard for the local rhythm of life new organizations developed; for example, the "watch officer," or today's 24-hour Operations Center in the Department of State. Earlier around-the-clock readiness, such as ship's watches and police operations, were meant to guard against physically present threats; the new institutions were designed to act on information about far-distant events.

Through the mid-19th century the British Colonial Office had functioned smoothly through procedures geared to the "monthly mails."[89] A message from India took twelve weeks to arrive; from Australia, nineteen weeks. A response could be prepared as "an abstraction from the reality of time and place, and by the time the governor received it, it was perhaps irrelevant as his own message had been at the time of its consideration in London."[90] Now there was "a revolutionary alteration of the base of communications itself,"[91] and during the 1870s the smoothly functioning, carefully staffed Colonial Office practices were under severe strain because of increased overseas activity. The workload, measured by pieces of correspondence (letters, telegrams, cables), increased from 25 000 in 1870 to 46 000 in 1890, while the staff increased only from 62 to 65. Cables were routed directly to the Colonial Secretary and created a sense that crises were occurring on a continuous basis. In the past all decisions were made by top officials; now "the clerks . . . were put to work."[92] Telegraphy both increased centralization, from the colonies to headquarters, and diffused responsibility within the Colonial Office staff.

Yet the impact of cables was gradual and slow; in fact, high costs and unreliable facilities prevented greater utilization of telegraphy.[93] In 1870–71 only £800 (about $4000) was spent for Colonial Office telegrams. In ten years, in spite of reductions in rates and the use of codes, the amount had risen to £2800.[94] But most of the correspondence was still in the form of dispatches. In 1896 the Colonial Office handled 56 000 pieces of correspondence and 3.6% were telegrams; in 1899 the count was up to 75 000 (6.5% telegrams), and in 1902 at the height of the Boer War, 116 000, of which 8% were telegrams. Through these times of the vigorous direction of Joseph Chamberlain the Colonial Office, like the U. S. Department of State, thus relied primarily on written messages.

Nevertheless, the capability to intervene in distant events immediately, and therefore the necessity to do so, had been created, which implied an eventual change in the relationship between the man at the center and the "man on the spot," the distant representative. The man at the center had access to information from all over the globe and hence a larger number of variables to consider. In theory, governments always exercise control over foreign affairs; with the new ability to receive information and transmit instructions implied a necessity for doing so; that is, a pressure toward centralization of decisionmaking.

However, as long as the distant diplomat or administrator had first access to, and control over, the information transmitted he continued to play a major role in making and implementing policy. Moreover, headquarters was subjected to a deluge of information with which the existing system could not cope efficiently. The traditional mode of operating was to place trust and confidence in the man on the spot, who had been trained and selected to act independently and responsible. British Colonial administrators therefore continued to act with considerable independence, and the Colonial Office was given a long respite in which to develop means of handling the new information capability. For example, during riots in Barbados in 1876, Governor Pope-Hennessy refused to initiate telegraphic communications with the Colonial Office until he had firm control of the local situation.[95] At other times colonial administrators deliberately used the cables to force the Colonial Office toward actions they wished to promote. Robert V. Kubicek, describing Joseph Chamberlain's administration of the Colonial Office in the 1890s, says:

> If anything, the telegraph gave local administrators the opportunity to create a sense of the immediacy of a problem or crisis with an intensity that dispatches, which took weeks or months to reach London, could not. The telegraph . . . made the Office more acutely aware of the need to support rather than curb or direct the actions of the man on the spot.[96]

Kubicek thus concludes that the telegraph "intensified involvement but denied [Chamberlain] its corollary, control."

In short, the British Colonial Office at the end of the 19th century was an institution under siege from an expanding workload, an exploding information load, changing institutional functions, and a fixed staff. Thus the centralization of decisionmaking that had been widely predicted was extremely slow to develop.

The same difficulties can be traced in the response of the British Foreign Office and the U.S. Department of State. The increased volume of information complicated rather than improved channels of communication from headquarters because of a lack of qualified staff to process the information for the decisionmaker. After World War I diplomats sometimes resorted to use of the cable not because their message was urgent but because there was no other way to get it to the attention of the minister or the Secretary of State, and some diplomats report that out of frustration they would give information to news correspondents hoping that the responsible official at home would read it in his daily newspaper.[97]

Yet eventually the centralization of authority did occur, in part because the constant flow of information from all over the globe had to be integrated and cross-checked at a central location and therefore decisions and policies must also be evaluated there; in part because, as will be discussed below, the public also had immediate information about developing events and constantly demanded a response from the government. Once that response is publicly given it is immediately available to other governments as well, without the mediation of an official representative.

The adaptation of diplomatic institutions can be illustrated by a look at the pattern of growth in the U.S. Department of State (Table 3–6). In 1870 Secretary Hamilton Fish carried through a major reorganization of the Department. Twenty-five administrators with eight specialists and a small staff

TABLE 3–6.—Increase in U.S. Department of State personnel, functions, and organizational units. (Personnel figures from O. L. Nelson, *Report on the Organization of the Department of State*, 1946; separate-unit figures from J. L. McCamy, *The Administration of American Foreign Affairs*, 1950. "Separate units" are defined as any distinct offices, bureaus, or divisions, as indicated on the organization chart.)

Personnel	1870	1909	1922	1938	1943	1946
Supervisory staff	25	71	64	21	211	491
Geographic	7	19	59	112	274	582
Economic	1	16	28	127	419	880
Information	—	21	30	98	249	1628
Intelligence	—	4	5	—	18	815
Miscellaneous	3	5	160	162	521	704
Administration	16	73	285	443	1063	2523
Total	52	209	631	963	2755	7623
Separate Units	16	23 (1913)	30	36 (1931)	62	113 (1948)

carried out all functions. By 1922 the growing number of specialists in economics, information, and intelligence indicates the response to the flow of information reaching the Department, and the growth and diversity of specialized functions within the Department continue to grow throughout the period, paralleling the increase in message volume (Table 3–7).

Telegraphic messages were only one factor in the increasing volume of information. Table 3–7 compares the Department's correspondence in 1910 with a sampling in 1955, with particular attention to telegrams. The number of copies of telegrams circulated within the Department in 1955 and 1960 (Tables 3–7 and 3–8) gives some hint of the increased necessity for coordination—that is,

TABLE 3–7.—Comparative estimates for 1910 and 1955 of U.S. Department of State correspondence.*

	FY 1910	FY 1955
Volume		
Total number of incoming and outgoing correspondence-including telegrams	121 000	949 907
Total number of incoming and outgoing telegrams	12 000	169 794
Workload		
Number of incoming and outgoing correspondence/number personnel in Department	578	258
Number of incoming and outgoing telegrams/number of personnel in Department	57	46
Internal Complexity		
Average number of copies made of each telegram	20	75
Number of copies of telegrams in circulation	240 000	14 000 000
Number of copies of telegrams/number of personnel in Department	1148	3800

*The 1955 figures contain all correspondence including publications and unrecorded mail. The 1910 figures are approximations listed in Department of State, *Outline of the Organization and Work of the Department of State*, 79–81. A detailed listing of what is included in the 1910 correspondence figures is not available.

TABLE 3–8.—Sampling of number of copies of correspondence at U.S. Department of State in 1960.

Type of correspondence	Total number	Average number of copies per item	Number of copies
Instructions (out)	30 640	91	2 484 538
Dispatches (in)	103 897	70	7 237 755
Airgrams (in and out)	39 558	94	3 718 452
Telegrams (in and out)	432 997	75	32 474 775
Total	607 092	—	45 915 520

the growing complexity if the decisionmaking process. If this ratio can be taken as an indicator of workload, it must mean that the average decisionmaker in the Department is receiving many more inputs of information in later years.

Information is also presented to the decisionmaker at a different pace or rhythm; before electronic communications, information arrived at longer intervals but each message covered a number of topics, with considerable background and context provided in the same message. As the interval between messages shortened, the message became narrower in scope, both as to topics covered and the amont of contextual material included. The recipient of the information may have before him a complete file of messages on a question of interest comprising a continuous dialog of question, answer, and response over time; the integration of this information, whether performed by one person or a group of specialists, becomes a complex activity that is performed at the center of control rather than in the field. The institutional response to this situation for one 25-year period is indicated in Table 3–9: in 1930, of the total personnel at headquarters and in the field (excluding international activities), 15% were in Washington; by 1955 the figure was 40%.

TABLE 3–9.—Personnel distribution, 1930–1955.*

	1930	1940	1950	1955
Department of State	703	981	4 930	3 686
Foreign Service activities	3 852	4 129	9 410	5 935
International activities	171	334	12 109	2 971
Total	4 726	5 444	26 449	12 592

*U.S. Department of State, *The Department of State 1930–1955: Expanding Functions and Responsibilities*, Department of State Publication 5852, Washington, D.C., 1955, 60.

At the same time, U.S. operational activities overseas have expanded far beyond the diplomatic function; that is, more and more government agencies have functions and operations outside the U.S. This trend began in 1930 and expanded rapidly after World War II. It further adds to the complexity of conducting foreign affairs. Cable communications cannot be considered a direct cause, yet the ability of multiple agencies to function overseas at a variety of operational levels and to be coordinated at the center of government is closely linked to the capability to communicate rapidly and cheaply on a worldwide basis. This development illustrates the way in which an innovation in technology is first used to carry out existing activities more efficiently but later permits new activities and functions as an institution adjusts to new capabilities.

As already pointed out, cable communications allowed nations to cooperate effectively for the first time in some operational activities such as apprehending fugitives. A secondary consequence was the growth of international organizations. Table 3–4 above shows the increased involvement of the U.S. in international activities in terms of manpower. In the 1930s the U.S. was already participating annually in about 75 international conferences; this number rose to about 300 in the 1950s, and by 1955 about 30% of the State Department's budget was contributed to international organizations.[98] Before cables, international or multinational cooperation activity tended to focus on periodic conferences; the management of continuous efforts was hindered by cumbersome communication.[99] International organizations, made practical by efficient communications, are one aspect of the growing interdependence of nations made possible by cables. It also contributes to the need for centralized national decisionmaking in foreign affairs. Lord Strang, in *The Foreign Office* (1874), comments on the relationship between "the trend toward universalization in the international field" and "the trend toward centralization in the conduct of diplomatic relations by individual national states."[100]

Electronic communications also make it possible for geographically remote nations to participate more fully in such international activities; Burke once observed that nature prevented colonial representation in Parliament.[101]

The continuous exchange between nations and centralized decisionmaking tends to produce policies that are pragmatic, realistic, and flexible; but it also means that local problems and crises must be dealt with in a context of world affairs. G. N. Clark observed that "the greater the agreement and unity of purpose at [the] centre, the less would be the sympathy and harmony between the centre and the outlying territories. . . ."[102] Solutions to local problems can no longer be developed in isolation, which again adds to the complexity of managing foreign affairs.

Thus, in creating new interdependencies among nations, and in centralizing decisionmaking within nations, cable communications created new vulnerabilities that may or may not exceed the capabilities of new mechanisms for dealing with them, which cable communications also created.

F. International Law

No matter who has laid or financed a cable, only the cable end which is on a nation's shore is under the real control of that nation. The other cable end, on foreign shores, cannot be controlled by the "home" government except through the fiction of extraterritoriality, and to such a claim no strong government will consent.[103] The status of transnational cables is therefore peculiarly a matter of reciprocity, and therefore international law or agreement. One of the most important principles of international law is the "freedom of the high seas," and from this recognized right to traverse international waters was derived the concept that States are entitled to lay submarine cables from their own shores to those of other States which agree to the connection. Until 1884 the status of cables was uncertain under international law. The Paris Convention of 1884 established this and other general principles that were applicable in time of peace; after 1890 there were continuing efforts to establish principles that would also be respected in time of war.[104]

In 1871 the International Telegraph Union met in Rome to adopt uniform operating procedures for European land and cable telegraph systems. Cyrus Field attended the meeting, but the United States was not a member of the Union and did not sign the resulting agreements because U.S. telegraph lines were owned not by the government but by private companies.

In 1875 the International Telegraph Union met again in St. Petersburg and twenty countries, most of them European, signed further agreements, including a clause recognizing the absolute right of a government to censor or withhold messages. The U.S. again did not participate. Subsequently there was an attempt to embody the substance of the international agreements in a treaty to be approved by the U.S. Congress, but this attempt was opposed by the U.S. telegraph companies because of the censorship clause. (Only two cables serving the U.S. ended on U.S. soil.)

The International Telegraph Union (ITU) met once more in Paris in 1884. This time the United States, Russia, Germany, and Austria-Hungary (the last two having abstained from the 1875 meeting because of the weak state of their national telegraph unions)) were among the 26 convention signatories. The treaty embodying this agreement was approved by Congress in 1888 despite the censorship clause.

The Paris Convention of 1884 provides that the right to lay and maintain submarine cables is placed under the collective guarantee of the signatory states. All nations have absolute jurisdiction over waters within their three-mile limit; they may grant or refuse to grant licenses for cable landing to another nation or to a foreign subject under any conditions which they choose. U.S. licenses are granted by action of the President subject to further control by Congress.[105] In 1879 President Grant, for example, refused to allow a French cable to be landed at Cape Cod because France would not give American companies permission to land a cable in France. During time of war licensing one of the belligerents to land a cable on the shore of a neutral country would jeopardize that nation's neutral status.

The right to take necessary actions to maintain and repair cables is generally recognized. The International Law Commission and the Conference on Law of the Sea in 1956 and 1958 merely updated the provisions of the Paris Convention of 1884:

> *Subject to the right to take reasonable measures for the exploitation of its continental shelf and its natural resources*, the coastal State may not impede the laying or maintenance of such cables or pipelines, the State . . .

shall pay due regard to cables or pipelines already in position on the sea-bed. In particular, possibilities of repairing existing cables and pipelines shall not be prejudiced. [Emphasis added.]

The United States suggested in 1869, and the Institute of International Law considered in 1879, a proposal that wilful damage of cables on the high seas be considered an act of piracy. The Institute rejected the proposal, following a trend away from classification of such acts as piracy and toward substitututing legal suits for damages to compensate for economic loss. The International Convention for the Protection of Submarine Cables was signed in 1884 by 26 nations; other nations subsequently adhered, and in 1919 the Treaty of Versailles made this Convention applicable to the Allied and Associated Powers and to Germany. The Convention provides that "breakage or damage of a submarine cable wilfully or by culpable negligence is punishable without prejudice to a civil action for damages except in the case of damages caused in the effort for self-preservation." Cable ships are also protected during the laying or repairing of cables, and other ships must keep at least a mile away from such ships. Culpable negligence is defined as the absence of elementary precautions and of a fault of ordinary seamanship. Suits for damages are to be tried in the national courts of the infringing vessel. This convention was regarded as a significant advancement in international law.

The St. Petersburg Conference of 1875 established the absolute right of all nations to correspond by submarine telegraph, in secret codes if they so chose, but only if a State through which the messages passed did not feel the messages threatened its national interests. A nation could suspend any or all international telegraph service or stop any message passing through its territory "subject to notification . . . to the other contracting governments." While the telegraph was being widely heralded as a harbinger of peace and brotherhood, rather than guaranteeing the absolute right to communicate in time of peace, nations thus decided to retain the sovereign right of censorship they had always exercised over the mails. The U.S. accepted this position in ratifying the international treaty of 1888, over the objection of the American private telegraph companies. (Most foreign governments by this time owned their national telegraph systems.)

In time of war belligerent nations formerly sent their dispatches by ships, which were liable to visit, search, and capture. (Even the mail of neutral nations is subject to capture if carried by the ship of a belligerent.) Cables cannot be visited and searched, and therefore nations have been reluctant to renounce the "right" to cut cables on the high seas in time of war.[106] The status of cables during wartime has thus not been agreed upon by international convention.[107] The only general principle is that interruption should be as little as possible, consistent with the necessities of war, particularly with reference to states not engaged in the hostilities.[108]

The rules of land warfare established at the Fourth Hague Convention in 1907 provide that when a belligerent occupies enemy territory "submarine cables connecting occupied territory with a neutral territory shall not be seized or destroyed except in cases of absolute necessity; they must also be restored and the indemnities for them regulated at the peace" (Article 54).

In trying to work out the status of submarine cables in international law in time of war, several theories were advanced:

- Cables are like mail ships—it is not lawful to sink a mail steamer because circumstances prevent verification of her contents; therefore it is not lawful to cut a cable although it cannot be ascertained what messages are being sent. (Cyrus Field so argued at international conferences, and this was the position of the Buchanan Administration.)
- Cables are like bridges connecting two territories and liable to damage and destruction under the same circumstances as other property in time of war.

The prevailing view merely recognized that (a) neutrals have a prima facie right of access to belligerents, but (b) the object of naval warfare is to stop the circulation of the enemy's commerce and to cut his means of communication; if "necessity dictates" cables will be cut. Neutrals are therefore not entitled to an indemnity for damage suffered through lawful acts of war by property which they have connected to territory under military blockade or occupation.[109] This principle is held to apply even when a cable owned by a neutral country is cut on the high seas; there seems to be no other situations where neutral property may be destroyed in neutral territory without violation of international principles or agreements.

In the case of a cable between two neutral territories, by which there is "knowing carriage of or repetition of a message of a belligerent"—that is, performance of an "unneutral service"—the other belligerent is very likely to cut, or attempt to cut, such cables. If the neutral owners of the cable can later show that they were not in fact performing "unneutral services" they are in theory entitled to compensation.[110]

The policy of the United States is that in time of war cables between enemy territories, between two points of one enemy's territory, or between occupied territory and neutral territory, can be cut; cables between two neutral territories "should be held inviolable and free from interference."[111]

During the Spanish-American War the U.S. cut the cables connecting Cuba, Manila, and Puerto Rico with the outside, although they were neutral (British) property. The Manila-Hong Kong cable was of particular value to British commercial interests. The U.S. refused to make any compensation to the British cable companies, and the British-American Claims Arbitration Tribunal in 1923 disallowed British claims on the grounds that the action was "implicitly justified by that right of legitimate defence . . . of any belligerent nation," and that "the degree of Spanish control . . . was such that [the Manila-Hong Kong cables] were invested with the character of works of public utility."

Britain cut all three German cables in World War I. The Treaty of Versailles provided that Germany renounced title to the cables in favor of the Allies, and their value was credited to Germany in the reparations account. (This credit was not compensation for the cutting but for the future ownership.)[112] Britain also cut enemy cables in World War II, and Italy cut British cables in the Sicilian Channel in 1940.

Thus general principles of international law were incrementally adopted and expanded to cover the new technology of communications in time of peace, but in time of war this technology like others is subject to the laws of necessity—no rules are effective or enforceable. The opportunity to create by agreement new rights that had not applied to ships or mail was not seized.

G. War and National Security

In the first quarter of the 19th century, the Napoleonic wars had radically transformed warfare. Wars had been only one part of a complicated bargaining process between countries, usually between sovereigns, in which commercial relations, credit, exchange of territory, dynastic marriages, and often whim played a part.[113] War was not an interruption of but an intermittent part of the process of international relations, and it was seldom a decisive factor. The armies were

> . . . so restricted by the difficulty of communication and supply, by the limited capacity of weapons, by the appalling incidence of sickness, and above all by the exigencies of public finance and administration, that warfare although almost continuous . . . was seldom decisive.[114]

With the revolutionary, and later nationalistic, wars that began with Napoleon this situation changed radically. Napoleon demonstrated that an unprecedented proportion of the male population could be called up, controlled, and maneuvered by one leader—who was sometimes the political as well as the military leader. War became an interruption of, and a substitute for, political intercourse. This led to the doctrine of the "decisive battle," i.e., that the enemy could be, and must be, overwhelmed and subdued. After such a battle the issues were settled and political intercourse began on a new footing. The doctrine dominated military thinking for 100 years, until the first World War.[115]

Moreover, 19th-century technology had also changed the military situation. Railroads provided a conduit for constant replenishment of resources:

> So long as railways kept the armies supplied with the blood and treasure of the nations the armies could not be defeated until the nations themselves were exhausted and begging for peace.[116]

Railroads "naturally tended to make strategy run on strait and straightforward lines."[117] The use of railroads to move supplies led directly to an enormous expansion in the size of armies, but tied them tightly to the rail lines. The railroad actually tended to reduce mobility, as became apparent in the

American Civil War; the South exploited this weakness of the Northern armies until Sherman daringly cut loose from his supply lines and began his march to the sea. As the railroad led to larger, less mobile armies, and made warfare a long-drawn-out process of exhausting the enemy's supplies of men and materials, the telegraph "linked commanders in the field to centres of political and military control where a different perspective obtained over what was going on at the battlefield."[118]

It was the landline telegraph that made it possible to maneuver very large armies, which had become very dependent on the rails for supplies:

> The other new invention . . . which simply revolutionized the conduct of military operations was . . . that of the electric telegraph. Without it the coordination of the rapid movement of increasingly massive armies would not have been possible.[119]

Although the importance of the telegraph and the impact it had on military strategy and tactics are generally recognized by military historians, there has been relatively little detailed analysis—and next to none of submarine telegraphy—for at least two reasons. Telegraphy's impacts are part of the larger picture of change resulting from several converging trends in mid-19th century: professionalization of army control and command; emergence of the railroad as the logistical system for modern armies; development of massive armies, beginning with the revolutionary and nationalistic wars (especially the Napoleonic wars); and the impacts of steam, both for vessels and for manufacturing. A second factor is, according to military historian McElwel, that "in this case (as contrasted with the railroads) . . . development was far more evenly distributed, and no particular general staff was able to snatch any important advantage over its rivals."[120]

The telegraph, both landlines and submarine, had another major impact as well, which became apparent even in the Crimean War, 1854–56:

> And the Press began to exert an influence which made statesmen and generals tremble. It was the Press in England which clamored for the siege of Sebastopol, thus stirring the imagination of the people and in turn influencing the government. For the first time in the history of warfare newspaper correspondents accompanied an army to the front, and as witnesses of the appalling condition of the hospitals and the insufficiency of the supply services there can be no doubt they were instrumental in forcing a change.[121]

By World War I (as we have shown), this lesson had been learned:

> If the centre of enemy power lies not in his armed forces but in his civilian population, then that population must be attacked directly. It must be softened and subverted by propaganda. It must be starved and enfeebled by blockade. It must be remorselessly bombed by air. Its morale must be undermined to a point where its capacity for armed resistance is fatally weakened.[122]

The railroad and the telegraph thus supported the movement toward "total warfare," which began with the revolutionary and nationalistic wars of the early 19th century and was strengthened by the harnessing of steam, development of modern industrial processes, and the development of moden weapons. In 1897 a Polish banker, I. S. Bloch, published a three-volume treatise on *The War of the Future in Its Technical, Economic and Political Relationships,* in which he stressed the interdependence stemming from the inventions of steam engines, the railroad, and the telegraph, and predicted that ultimately wars would end only with "national bankruptcy."[123]

Until the 17th century territorial expansion primarily meant land warfare on the European continent. In the next two centuries, as nations increasingly came to depend on raw materials and foodstuffs from abroad, Britain developed an overseas empire. Emerging from the Napoleonic wars dominant on the seas and in control of world markets, Britain also eliminated European rivals as serious contenders for world power.[124]

Until World War II, with its emphasis on air power, control of the seas was essential to world power. By then, "The aeroplane was rapidly restoring the world of Ptolemy, a small and shrunken world, not of oceans containing huge island continents, but of great stretches of land divided by gigantic lakes."[125]

Therefore, to tease out the consequences of submarine telegraphy for military security and the conduct of war, we begin with its implications for naval warfare. Steam-powered navies and telegraph cables extended to global conflict the capabilities which railroads and landline telegraph conferred on continental warfare. We therefore look first at the effects of submarine telegraphy on mobility and centralized strategic control in naval warfare.

The development of total warfare, we have argued, was both driven and facilitated by the development of rapid communications. The ability to respond instantly to threat created the necessity to do so. The major nations of the world quickly adopted a state of readiness and constant preparation for defense—a state that itself posed a constant threat to the security of their rivals.

The value of information and the speed of its transmission grew with the interest in constant readiness, and military intelligence became as vital and as competitive an element of national defense as armaments.

Submarine telegraphy played a major role in the growth of the importance of constant intelligence to national defense. Created in war, as in peace, new control systems and new vulnerabilities made the world both more interdependent and more dangerous for every nation (Fig. 3–9).

The submarine telegraph was conceived and applied in an era in which "command of the seas" was the critical element in national defense for Britain. Great Britain's dominance in the cable industry reflected the country's maritime, technical, and industrial base. The cables were a way of maintaining British control of the seas for both commercial and military purposes. Did submarine

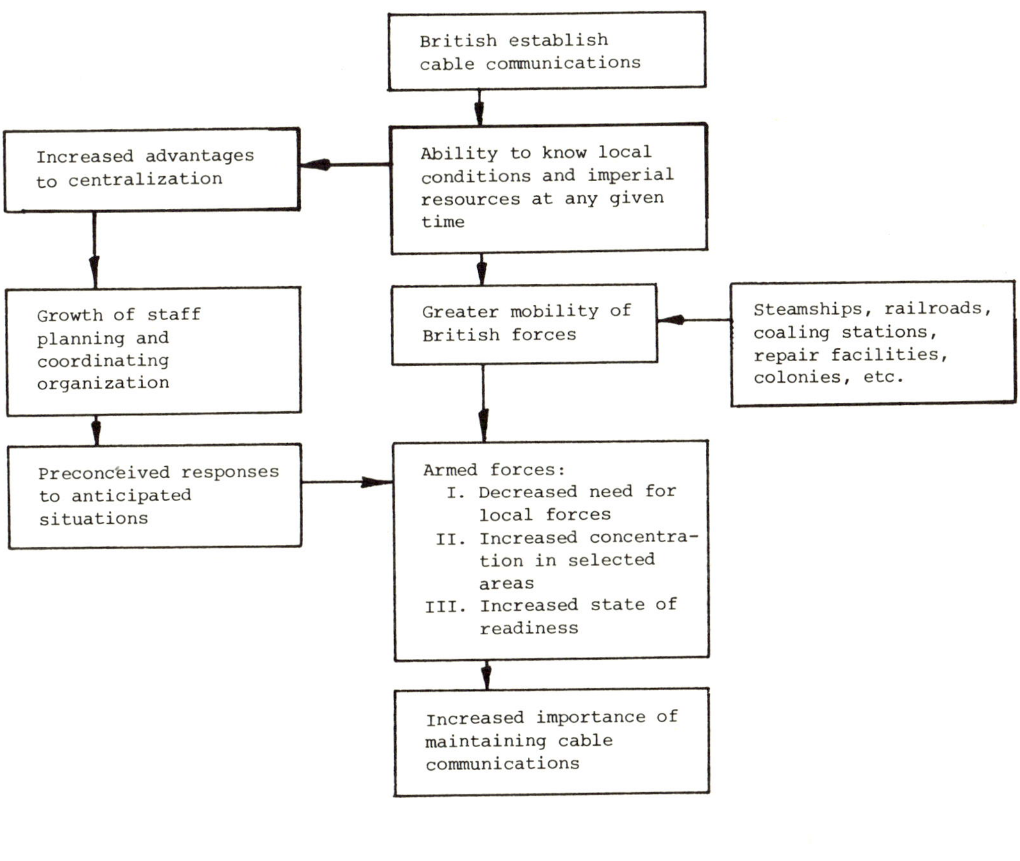

FIG. 3–9.—Potential impact of cables on Imperial defense in 19th century.

telegraphy serve to strengthen British hegemony? Or did this technology actually undermine both the British commercial monopoly overseas and its military superiority?

In identifying the defensive posture of the British Empire as it related to submarine telegraphy, three distinct periods stand out: (1) the pre-1870 era of immature cable technology, during which it had a negative effect on the defense of the Empire; (2) the 1870–1910 period, during which submarine telegraphy was the only means of worldwide communications and remained primarily under British control; and (3) the period after 1910, in which wireless offered alternative means of worldwide strategic communications free from the necessity of having to protect the route of the information.

In the 1850s and 1860s, until long-distance, deep-water cables could be developed, telegraphy was tied to the land masses of the world and cooperation among several countries was needed to make services possible. From the first permanent Channel cable in 1851 until the laying of the Atlantic cable in 1866 and the England-Gibraltar-Malta cable in 1870, the United Kingdom had to depend on European cooperation for all external telegraphic communications. Britain was at the end of the European telegraphic network. No European state depended on a strategic telegraph route through the British Isles. During this period of British dependence on Europe, the situation never became intolerable because of the relatively tranquil state of the overseas empire and the preoccupation of the European powers with Continental events. Nevertheless, Britain actively promoted the development of as many alternate telegraph routes across Europe as possible. Security was in multiple routes.

A pillar of British political and military posture was the geographical independence of the United Kingdom from the European continent. The British line of defense and communications depended on the unrestricted flow of goods, passengers, and information on the seas. Submarine cables laid along the sea routes could only strengthen British dominance on the seas. The Admiralty was the one government department that had supported the Brett brothers' first proposals for a Channel cable in the 1840s: ". . . their Lordships cannot but applaud the active and enterprising ingenuity which has advanced the project to its present state."[126]

In 1870 an all-cable system from Britain to India (via Gibraltar, Malta, Suez, and Aden) freed the Empire from the Continental telegraph network. Within ten years the major worldwide cable routes were established and by 1902 only a few minor British possessions were not telegraphically connected with Great Britain.[127] By then nearly 170 000 nautical miles of cable were in operation, 90% established by private enterprise and about 10% under the administration of governments. British capital financed roughly 75% of the private ventures and about 120 000 miles of the cable in existence was manufactured by one British firm—the Telegraph Construction and Maintenance Co.[128] In short, Britain dominated the cable routes of the world and London was the center of this network.

Although most cables were privately owned, in the event of war, rebellion, or other emergency, the British government had the power to take possession of and work the lines on its own account for as long as necessary. Under normal operating conditions government messages received priority handling and often a reduced tariff rate.

State financing of this world cable system would have been prohibitively costly; it was a burden that Germany, France, and other states undertook on a less grand scale for cables they deemed strategically essential. It was the British ability to blend strategic and commercial requirements in cable communications that permitted their domination of international cable routes. Somerset R. French, the Postmaster General of Cape Colony, pointed out: "A war may possibly only take place once in twenty-five years, whereas we have the necessity for good cable communication in connection with the commercial development of the Empire always with us."[129] Only Britain had a large enough overseas empire to support such a system. Commercial and strategic requirements did not always coincide, but in these instances, cable routes were established through state subsidies or, on occasion, by state ownership.

Emerging powers such as Germany and the United States had to depend on British-controlled cable routes for most overseas communications. For example, during the Spanish-American War, a considerable amount of intelligence and naval coordination on both sides went via the London cables. At the turn of the century neither the United States nor the Germany navy—nor any navy for that matter—could operate effectively without the cooperation of British-controlled cable facilities.

Mobility and Flexibility

It was the India Mutiny of 1857–58 that convinced Britain of the need for a rapid communications capability overseas. On 9 April 1857 the first telegram regarding the disaffection of Sepoy troops near Calcutta reached London. On May 2 the Secretary of State for War arranged for four regiments to leave the United Kingdom for India "without delay"—but the suggested departure date was to be "no later than June 20." Events in India were developing rapidly while London remained relatively unconcerned with what appeared a minor incident.

But the events of May 10 and 11, which included the seizure of Delhi, exposed the seriousness of the situation.[130] On May 16 Sir Henry Lawrence at Lucknow telegraphed the Governor General in Calcutta, "All is quiet here but affairs are critical; get every European you can from China, Ceylon, and elsewhere; also all the Goorkas from the hills; time is everything."

Time was everything, and a message was dispatched from Calcutta to London on May 18 describing the critical state of affairs. On May 27 the message reached Bombay and from there a packet steamer carried it to Suez. The Consulate at Alexandria placed the message on a steamer for Trieste on June 21. At noon on June 26 the telegram was sent from Trieste and after approximately twenty hours in transit via the European landlines and the Channel cables it arrived in London on the night of the 26th or the morning of the 27th of June.[131] Time was everything—and the time from Calcutta to London was forty days.

The reaction to the Mutiny in London was to reinforce the India garrisons; the choices were:[132]

From England to	Steamers	Sailing ships	Average number of days from England
Calcutta	9		82
		22	116
Madras	2		90
		2	131
Bombay	5		76
		9	118
Karachi	3		91
		10	128
India via Suez			33–40

The mere effort to coordinate activities in India from London was cumbersome and slow. The correspondence between the Secretary of State for the Colonies in London and the Governor of Mauritius, who was supporting British efforts in India, took 109 days' round trip—a delay that kep Britons in a state of tension for the remainder of 1857 and well into 1858, unable to appraise the true situation in India until long after the events occurred.

Had this crisis occurred in the last quarter of the century when a cable network existed, the events of 10 and 11 May 1857 would have been evaluated in London on the same day and local garrisons from Mauritius, Capetown, Australia, Ceylon, and elsewhere would have been dispatched immediately. Most would have arrived in India the same month. Troops from Malta and those en route to China would also have arrived within a matter of weeks. The garrisons abroad weakened by such troop movements would be in continuous contact with London and India in case problems developed locally. Moreover, the troops dispatched from England via the Cape by steamer and rail would still have been available for recall if a conflict developed in Europe. At the time, the strength of army units remaining in the United Kingdom was dangerously low in face of potential French hostilities, and the more than 31 000 troops sent to India via the Cape were beyond reach or recall for 3–4 months.

The first recorded contribution of cable communication to increasing the flexibility of military forces was the use of the short-lived 1858 Atlantic cable as the India Mutiny ended. The order to recall two regiments from Canada for service related to the Mutiny was countermanded from London by cable; the estimated £50 000 savings was well publicized by the cable promoters.

We can only estimate the indirect consequences of increased mobility for military expenditures. In 1870, if Gibraltar were invaded by an enemy, it took ten days for London to get news of it. Assuming a fleet was ready to sail, another month would pass before it arrived at Gibraltar, which had therefore to be capable of "sustaining siege and investment for at least forty days."[133]

Once steamships and telegraph became available, word of the siege would reach London in minutes and the relieving fleet could be at Gibraltar in four days. If the cable to Gibraltar were cut the delay time would double to approximately eight days. Consequently, a safe estimate of the resources to expend in the defense of Gibraltar would be based on the premise that relief could be provided in eight days.[134]

From about 1870 to World War I the concepts of imperial defense were the subject of extensive debate, much of it centered on the relative roles of the army and navy and the question of fortifications versus mobile naval forces. The navy won the debate by the 1890s, at least in terms of fiscal expenditures, but the increased emphasis on shipbuilding was in large part the result of growing competition from foreign navies; submarine telegraphy complemented the traditional role of the Royal Navy in defending the Empire.

Communications capability, beginning in the late 1860s and maturing by 1880, may be related to changing troop distributions and perhaps even to troop strength (Table 3–10). Had submarine telegraphy been the only factor operating, its impact would be measurable, but there were other

TABLE 3–10.—Strength and distribution of British Army, 1851–1913.

Location	1851	1861	1876–77	1880–81	1890–91	1900	1907	1913[c]
A. Home	24 175[a]	97 959[b]	96 275	101 541	115 554	131 802	120 806	125 209
B. East Indies	27 144[a]	92 838[b]	63 105	62 588	72 429	73 157	75 658	76 528
C. Colonies (plus Egypt)	42 498	44 481	23 373	27 519	35 069	51 204	51 249	44 125
North America	8 230	4 268	1 806	1 843	1 493	1 791	—	—
Australia and New Zealand	3 413	6 263	—	—	—	—	—	—
South Africa	5 050	4 788	2 501	5 386	3 324	8 827	17 533	6 826
West Africa	700	980	645	625	953	1 849	2 689 (2 438)	2 514 (2 111)
West Indies	6 019	4 140	2 216	2 361	3 011	3 278	1 344(946)	1 029(621)
Ceylon	3 553	2 226	1 251	1 242	1 416	1 776	1 046(65) [765]	1 290(842)
Mauritius	1 941	1 990	425	458	881	3 838	1 244(194)	1 369(135)
Bermuda	1 335	1 071	1 981	2 158	1 391	2 055	1 312	1 307
China	—	—	837	1 210	2 989	4 737	4 091(457) [1630]	9 502(420) [4582]
Straits Settlements	—	—	1 209	1 024	1 509	1 669	2 251(184) [733]	2 649(101) [843]
Cyprus	—	—	—	492	553	135	1 161[1503]	118
Malta	3 331	7 116	5 292	5 483	8 796	10 663	6 881(471)	6 934(437)
Egypt	—	—	—	—	3 389	4 363	5 692	6 115[200]
Ionian Islands	3 491	4 233	—	—	—	—	—	—
Gibraltar	3 618	6 001	4 949	5 009	5 196	5 428	4 327	3 877

a. 1850 figures
b. 1858 figures
c. ( ) indicates Colonial troops; [ ] indicates Indian troops.

trends: expansion of imperial territorial interests in China, the Straits Settlements of Malaya, Cyprus, Egypt, etc.; withdrawal of army units because of local stability or colonial self-government; and the changing composition of troops (Indian private armies, the East India Company army, colonial armies, hired irregulars, etc.).

On a local level the redistribution of troops resulting from the availability of rapid communications can be more readily identified. One example was the West Indies, where as early as 1861 consideration was given to the impact of availability of cable facilities on local troop distribution. The Duke of Newcastle, Secretary of State for the Colonies, observed that "If our West Indian possessions could be united by an inter-insular telegraph, the concentration of troops would be desirable."[135] Most imperial troops in the West Indies were ultimately concentrated in Jamaica.

The number of colonial and Indian troops deployed outside their home areas is indicated in Table 3–10 for the years 1907 and 1913. The fact that this information was recorded implies a growing awareness of the contribution of other than Regular British Army units to imperial defense. It also hints of an increased ability to direct and utilize military resources of varied origin under the control of self-governing colonies. Such coordination of Indian, colonial, and British forces suggests dependence on an improved means of communication not available in the first half of the 19th century.

An example of such coordination was the landing of major contingents of Indian troops at Malta during the 1878 crisis with Russia. Prime Minister Disraeli was thus able to dramatize his position to Russia in a manner not possible without rapid telegraphic communications. The British public viewed the new mobility of troops with delight and humor, as in this parody of G. W. Hunt's famous music-hall song of the day:

> We don't want to fight,
> But by jingo if we do,
> We'll stay at home and sing our songs
> And send the mild Hindoo.[136]

The new mobility of troops was also applied to British army units. Before the expansion of cable communications, troops in transit had to be discounted as effective forces because once committed to relocation they were in effect beyond convenient communications; the duration of this suspended state depended on the speed and reliability of transportation. Steam and telegraphy minimized the inactive status of troops in transit. Table 3–11 attempts to identify the changing trend in the movement of troops. These data on the number of British soldiers embarking and disembaring in British ports are available only for the years 1871 to 1887. The total strength of the British Army during this period remained relatively constant, as did the ratio of troops deployed overseas to the total army strength. In 1871, 20.6% of the total British Army force overseas departed from or arrived in Britain. In 1875 troops in transit during the year represented 24% of the total overseas; in 1880, 38%; and in 1885, 52%. Only the last year, 1887, saw a noticeable reduction to 27%. Although the sample is too short to be conclusive, it is highly probable that the transportation of troops increased considerably with the introduction of steam, cable, and the Suez Canal.

Naval units are by definition mobile, but steam introduced a new need for fuel and increasing sophisticated repair facilities; thus rapid communications were important in expanding the range of naval vessels.

The distribution of British naval vessels throughout the world during the 19th century (Table 3–12) lacks detail for the last 25 years of the century, but little change occurred during this period despite major changes in the design of naval vessels and armament, including the transition to steam, iron hulls, and screw propulsion. Cables did not alter the worldwide distribution or concentration of the Royal Navy noticeably between 1870 and 1900 because there was no serious naval threat to the Empire. At the turn of the century all changed.

In 1904 the first major revision in the distribution of the fleet was announced. The Board of the Admiralty identified certain new conditions that necessitated a review and readjustment of ships and stations:

TABLE 3–11.—Number of British soldiers embarking and disembarking in United Kingdom (service in colonies, Egypt, and India), 1871–1887.

Year	Total Embark*	Total Disembark*	Total Troops Overseas	Total Strength of Army
1871	9 998	8 067	86 908	192 665
1872	12 825	8 823	87 853	192 007
1873	15 235	9 280	87 796	188 379
1874	13 407	11 792	88 654	186 389
1875	10 398	10 671	87 476	184 669
1876	14 426	10 462	87 287	184 433
1877	15 386	12 738	88 598	190 502
1878	17 913	11 611	91 557	200 758
1879	26 310	17 159	103 075	191 290
1880	18 323	18 097	97 099	188 986
1881	17 686	16 543	96 852	188 798
1882	36 066	30 636	94 725	189 229
1883	19 808	23 644	91 890	181 971
1884	24 182	13 117	93 000	198 064
1885	31 889	23 518	106 485	203 805
1886	23 783	20 683	105 748	209 574
1887	13 231	14 098	102 807	211 105

*Figures from *Annual Returns of the British Army (Parl. Papers*, 1871–1888).

TABLE 3–12.—British Navy on-station strength (Ships/Men), 1847–1904. (From *Return Showing Number of Ships on Different Stations, Parl. Papers*, 1867–68, 45: 167, and 1876, 45.)

	1847	1857	1867	1874	1904
China/East Indies	28/5536	37/6845	38/4447	20/1852	
East Indies (separate for 1865)	—	—	7/1275	9/1128	
Australia (included in East Indies prior to 1857)	—	2/230	6/1367	9/704	Far East Fleet (Singapore)
Pacific	16/3864	12/2794	14/3321	9/117	
Southeastern coast of America	11/1670	8/1353	9/115	5/313	Cape Fleet (Capetown)
Cape of Good Hope	11/2098	9/1214	(3/446) ( )	11/1199	Atlantic Fleet (Gibraltar)
West Coast of Africa	32/3523	19/1904	(19/1894)		
North America/West Indies	14/2508	27/6241	29/5488	16/1696	
Mediterranean	18/2865	45/8909	19/5207	16/2035	Mediterranean Fleet (Malta)
British Isles	51/6232	42/6236	53/5608	59/6391	Channel Fleet (Dover)

The principles on which the present peace distribution of His Majesty's ships and the arrangement of their stations are based, date from a period when the electric telegraph did not exist and when wind was the motive power, and it is a wonderful testimony to the strategic and political soundness of those principles that they have stood the test of time and met all the needs of the service up to the present moment.[137]

The redistribution of the fleet reflected the Japanese Alliance of 1902 and the French Entente of 1904, which permitted a reduction in British fleet strength in the Far East and the Mediterranean, and a concentration in home waters to counter the growing German fleet. Five large fleets were organized to control the key points of Singapore, the Cape, Alexandria, Gibraltar, and Dover, which permitted rapid concentration. The Pacific and South Atlantic squadrons were eliminated and the China,

Australia, and East Indies squadrons were condensed into an Eastern Fleet at Singapore. The Cape squadron was responsible for the West African Coast and the East Coasts of North and South America. A Channel Fleet guarded the home waters and could be reinforced from the Atlantic Fleet stationed at Gibraltar, which could also support the Mediterranean Fleet centered at Malta.[138] The new distribution recognized that Britain could not police the entire world and had to concentrate on a growing threat in Europe. This was the first explicit incorporation of a rapid communications capability into a total strategic concept of imperial defence.

Admiral Sir John Fisher noted:

[T]elegraphy has been enormously developed, hence transmission of orders, and mutual conference of thought enormously bettered. Instead of a number of isolated squadrons acting under different heads, all actuated from different views of the same state of conditions, each independent of the strategy pursued by the other, we will have a coordinated whole. Unity of purpose is strength, ever so much more so in unity of strategy.[139]

In 1897 the Colonial Secretary, Joseph Chamberlain, supported the concept of employing the ships of the Australian squadron at considerable distances from Australia:

No organized expedition could be sent either from Japan, or from the United States, or from France without the full knowledge of the Admiralty. . . . If it were organized, our whole strength would be directed to defeating such a movement.[140]

Both Chamberlain and Newcastle suggested similar strategic concepts, that a rapid worldwide communications capability was an important element in the strategic posture of the Empire once non-European powers developed a potential naval threat to the Empire. Before cables, emphasis would have been placed on the independent defense of isolated areas with increased fortifications and less dependence on concentrated fleets to check potential enemy fleet movements; cable communications permitted increased security for the Empire by the availability of information.

On the other side of the coin, Newcastle in 1861 warned that the concentration of naval forces was essential because of the telegraph. He noted that with the electric telegraph "it would be an inevitable consequence that an enemy's fleet would take care to avoid those colonies with sufficient forces and attack those with few or none."[141]

Sir Henry Holland, Colonial Secretary in 1877, had a similar view:

. . . steam power affects the question of Imperial defense . . . By its means, rapid and certain naval combinations can be made, so that the liability of the outlying ports of the Empire to sudden attack, no longer impeded by wind and weather, is vastly increased. Similarly, the development of the telegraph cable systems of the world has further facilitated rapid and unexpected measures of aggression.[142]

The move to concentrate forces was closely associated with the effort to reduce the British military commitment overseas for fiscal reasons. Cables, the concentration of forces, and reduced military expenditures complemented each other. Chamberlain promoted a similar concept in the 1890s in response to the problems of limited imperial resources and the sensitivity of self-governing colonies—particularly Australia—to the depletion of local fleet units.

A new and definite stage has been reached in that evolution of the modern steam navy which has been going on for the last thirty years, and that stage is marked not only by changes in the materiel of the British Navy itself, but also by changes in the strategical position all over the world arising out of the development of foreign navies. In the Western Hemisphere the United States are forming a navy the power and size of which will be limited only by the amount of money which the American people choose to spend on it. In the Eastern Hemisphere the smaller but modern navy of Japan has been put to the test of war and has not been found wanting. The Russian navy has been greatly increased . . . The French navy stands, as always, in the fore-front. The new German navy has come into existence, it is a navy of the most efficient type and is so fortunately circumstanced that it is able to concentrate almost the whole of its fleet at its home ports. *Distribution and Mobilization of the Fleet* (Cd. 2335; Parl. Papers, 1905, 28, 1).

Military Readiness and Central Control

The flexibility and speed in collecting and utilizing military forces demanded improved control from the center. There was no conscious, explicit, or consistent effort to exercise greater control from afar, but the adjustment to changing conditions is clear in retrospect.

For example, during the Mahdi uprising in the Sudan in 1884 the War Office cabled the General Officer Commanding in Egypt exact instructions on how to relieve the garrison at Tokar, including the names of the officers to command the various contingents of the relief force, the selection of the units in Egypt to make up this force, how they were to proceed, weapons to employ, equipment to take, camels and camel-drivers needed—including guidance on whether to purchase or rent the camels and how to go about it. Also, each soldier was to be issued 250 rounds of ammunition in addition to the 70 rounds in his pouch. In conclusion the War Office stated, "All confidence felt in your judgment and experience to settle necessary details."[143] Twenty years earlier, this message would have been a slap in the face to a commanding general in the field. By 1884 such guidance appeared appropriate, although more routine situations may well have called for far less detailed instructions from the War Office.

The coordinating effort from London took into consideration the military situation beyond Egypt and the Sudan, and tapped a considerable source of knowledge beyond that immediately available to the local commander. The eventual result of this ease of communication between the field and the center is well illustrated in the correspondence of the American Expeditionary Force (AEF) during World War I.

Major-General John J. Pershing and his staff arrived in France on 13 June 1917. Within the staff structure were two divisions—the Cable Division and the Code Division—responsible for the handling of records, correspondence, orders, and communications of the General Headquarters. By 1918 they represented the efforts of approximately 70 clerks and officers primarily concerned with operating a system of communications with Washington.[144]

A series of cablegrams between Chief of Staff and AEF began in the form of weekly summaries that discussed "military operations on all fronts, including reports on political, social, and economic conditions in the belligerent countries." Eventually the summaries were made on a semiweekly basis and, beginning in March 1918, daily.

The detail of military communications noted in Egypt in 1884 had become institutionalized by the time of World War I, with a definite tendency toward central control of military activity, derived in part from 19th century Prussian military experience.

The successful campaigns against Denmark and Austria in the 1860s and the major victory over France in the Franco-Prussian War of 1870–71 were largely attributed to the efficiency of the Prussian General Staff under the elder Helmuth von Moltke. The function and importance of military staffs were increasing. General Paul Bronsart von Schellendorff of the Prussian General Staff pointed out that when armies were small and movements and formations were fairly standardized and rigid, the commander usually provided the details of execution in his plan; but by 1875, centralization was a must:

> The enormous numerical strength of modern Armies, and the way they must be organized to meet the constantly changing requirements of war, render an immense amount of detail necessary in carrying out military operations even under apparently similar circumstances of time and place.[145]

By midcentury "command and staff functioning had developed to the point where the commander was seldom a participant in battle." Moltke demonstrated central command in making his operational decisions in a command post 200 miles from the front lines when defeating the Austrians at Königgrätz in 1866.[146] Sixty years earlier, Napoleon had directed his wars from the battlefield. With rapid communications military endeavors could be orchestrated at any distance from the actual event.

The success of the Prussian General Staff was due less to brilliant battlefield tactics than to the central control of a large complex military machine with involved logistical systems. The entire complex effort of an industrial state at war, from supply movements at the railhead to sorties against enemy positions, was coordinated my means of the telegraph. Peacetime endeavors of the General Staff were equally important to wartime success. The peacetime duties were to prepare for war, i.e. to draw up and prepare

. . . schemes for strategical concentration of the Army in certain particular directions by road and rail, with collecting and estimating the strength, etc., of the various European Armies, with the study of theatres of war, and with the preparation of military maps.[147]

In short, the General Staff organization in peacetime was responsible for a degree of readiness enhanced by working out "all matters of mobilization, movements, quarters, manoeuvres, railways and telegraphs."[148] By the time of World War I,

Mobilization was a recognized act and condition; it meant not merely, as the word suggests, making the armies mobile, but also calling up reserves, canceling leaves, transferring troops to battle stations, imposing martial law in frontier areas, manning fortresses, taking over rail transport, and a thousand other measures that transformed the standing army into a much larger national force ready for action. It involved very careful and detailed plans in advance which, like all military plans, were practically impossible to alter, and difficult to arrest, once they began to be carried out.[149]

This emphasis on "readiness" was important in the immediate causes of World War I. Once a potential enemy mobilized, others also had to mobilize for self-defense. Russia mobilized its forces on 24 and 25 July 1914 as a deterrent against possible Austrian aggression in the Balkans. Germany, exposed to this Russian readiness, issued mobilization orders, which in turn necessitated French mobilization. Tension mounted, war broke out, and Germany executed the Schlieffen Plan, a prepared strategy devised to wage war on two fronts, against Russia and France. Belgium was a casualty of the plan, which led to a British declaration of war.

Had telegraphy not existed, preparatory planning of mobilization still would have developed, given improvements in transportation, nationalism, and advances in instruments of war. Telegraphy perhaps only made a difference of degree—although it is difficult to visulalize a major mobilization without telegraphy for the coordination and timing of every phase of the operation.

Improvements in transportation permitted the concentration and supply of large armies. Telegraphy contributed both to the efficient operation of transportation systems and to the coordination of the entire effort. Development of the general staff was important. Byproducts of such strides toward central control included the increased complexity of the national defense effort (also a byproduct of industrial society as a whole and of massive armies) and an emphasis on timing and speed.

Speed in diplomatic exchanges—a result of telegraphy—was an important ingredient in the escalation of events following the murder of the heir to the Austro-Hungarian throne at Sarajevo on 28 June 1914. Europe was engulfed in war within little more than a month; a rapid diplomatic exchange served to escalate rather than defuse the events of Sarajevo. "Mail pouch" diplomacy might have lessened the prospects of war. Yet we cannot know the number of conflicts *averted* by the timely use of the telegraph.

In British Empire defense, central control was slow to develop because of the conflicting interests of self-governing colonies, the lack of agreement on a concept of defense that satisfied both the War Office and the Admiralty, and the physical impossibility of organizing a farflung, diverse empire. Nevertheless, the need for a collective approach was appreciated especially beginning in the 1880s, when the vulnerability of the Empire to attack became a serious issue.

As late as 1909 General W. G. Nicholson, Chief of the General Staff, complained to colonial representatives that there were 1 200 000 officers and men to call in the event of war,

but there is a need to produce an army which will be effective, capable of promptly bringing its full power to bear at the decisive point, capable of its parts being able to employ in any theatre of operation where their services are required. Only a fraction of Empire forces can do this today.[150]

Part of the British Army's problem was the firmly entrenched views of the "Blue water" school which entrusted the defense of the Empire to naval superiority and a big-gun navy capable of dealing with potential foreign naval threats to the vulnerable shores of the overseas Empire. The fear of enemy cruisers creating havoc in the shipping lanes was discounted.

The shift from sail to steam had in one sense greatly reduced the mobility of modern warships. Ships needed coal or, later, oil; and Britain controlled the bulk of the strategic overseas coaling stations. Sophisticated repair facilities were also needed to maintain the increasingly complex vessels of war. Only Britain had a network of overseas shipyard facilities—and submarine cable facilities.

Admiral Sir John Fisher's redistribution of the fleet in 1904 was a major expression of the "Blue water" strategy. Moreover, he emphasized a new demand for instant fleet readiness for war. The number of warships in drydock at any time was carefully controlled, as was the condition of the reserve fleet.

Readiness and centralization contributed to military efficiency, but also to discontent and uncertainty in colonial areas such as Australia, where local defense considerations were now secondary to a total concept of imperial defense orchestrated from far-away London.

> . . . To destroy (an army's) lines of intercommunication—by which orders and reports pass—is to paralyze its sensory organization, the essential connection between brain and body. (B. H. Liddell Hart, *Strategy*, 1954, p. 199.)

Military Intelligence

It is ironic that the strategic potential of submarine cable was not fully realized until the age of radio. The great advantage of the cable over its successor was relative security, at least when combined with cryptography. Governments to this day generally prefer using cable to radio for transmission of sensitive information. According to David Kahn, "The telegraph created modern cryptography; the radio, modern cryptanalysis." Before 1914 no technology was available that could tap a cable once under water. Listening devices had to be installed at landing points or at places where submarine cables crossed land. There was one exception: the British grappled for and picked up the cable off Egypt in their naval operations off Alexandria in 1882, and thus were able to use it for direct communications with London to report the progress of their activities and to receive guidance.[151]

That no nation, particularly Imperial Germany, otherwise so ingenious in electrical technology, developed a submarine cable-intercept technology, was one of the mysteries of the First World War. It seems chiefly explained by the German reliance on radio and an unjustified faith in the impenetrability of their codes.

Before 1914 the fact that being the center of the world cable net gave the British an advantage in global warfare was just coming to be appreciated by the military planners, policymakers, and some segments of the informed public of Britain, the U.S., and Imperial Germany. In the Franco-Prussian War of 1870–71, which was studied until 1914 as the only modern war between two major European powers, an incident occurred that must have left an imprint on the military imagination. The German armies cut off Paris from the outside world by 20 September 1870. Before the blockade was complete, however, French authorities had with great secrecy laid a cable along the bed of the Seine outward to territory still controlled by French armies, its existence known only to a handful of officials of the Postes et Télégraphes and a few members of the Cabinet. It began transmitting on the 23rd, but the next day it was badly damaged, and on the 27th the Prussians succeeded in dredging it up. Unable to decode its messages, they destroyed it,[152] but it was a portent for the future.

A brief look at the use of cabled secret intelligence by the U.S. Navy illustrates some of the factors that slowed the more general impact of submarine telegraphy on military intelligence, as well as factors that pushed its development. Before 1900 it was common practice for the U.S. to allow foreign attachés to visit shipyards, and friendly nations such as the U.S. and Japan routinely exchanged information and even blueprints of their latest battleships. Only in the most advanced technological developments were U.S. naval attachés directed to obtain specific information. Such instructions would generally be sent by letter and the intelligence reports returned via diplomatic pouch. As late as 1904 the American naval attaché observing the Russian operations in Manchuria sent his reports by ordinary mail, through the German postal service operating in Chinese ports.[153] In the U.S., of course, although military intelligence was gathered, foreign wars were not regarded as imminent. But in other countries, too, the use of cables to carry secret information was slow to develop. One reason was that much military intelligence was traditionally in the form of maps, illustrations, and prints, which were not easily transmitted by cable.

Nevertheless, the U.S. Office of Naval Intelligence (ONI), established in 1882, used some cables from the beginning. By the 1890s the Navy encoded many of its transmissions to naval attachés in London, Paris, Berlin, Tokyo, and St. Petersburg, and by 1898 the U.S. Navy employed "the most secure and advanced code system of the day."[154]

Cabled secret intelligence came into real strategic importance with the full development of contingency planning.

At the outbreak of the Russo-Japanese war, President Theodore Roosevelt sent a coded dispatch to the American naval attaché in Tokyo directing him to cable semiweekly reports of Japanese preparations for war against Russia. The Japanese had begun the war on 8 February 1904, by torpedoing the Russian Pacific squadron at anchor, and immediately imposed strict censorship on cable traffic from Japan. The letterbook of the Office of Naval Intelligence for 1904–1905 shows the difficulties and pitfalls in this pioneer effort at conducting an intelligence operation by cable. Without any precedents to follow, things were always going wrong. Attachés forgot to register confidential cable addresses as they moved from place to place, cables went astray, and in one instance the attaché observing Russian efforts against the Japanese in Manchuria had the code, but not its key, which providentially happened to be in the hands of an American consul in China.[155] Since the Russians felt the U.S. favored the Japanese they limited the movements of the U.S. attaché in Manchuria and he was virtually unreachable via the Embassy in St. Petersburg by either cable or letter, though his dispatches reached Washington.

Approximately 30 cables were sent out by ONI from March 1904 to July 1905, perhaps a third of them in code. The messages in clear are characteristically short, occasionally cryptic one-liners, making assignments; and only occasionally was information asked for by cable. Messages sent in code were generally longer and more substantial, such as reporting the orders from the Russians that foreign naval attachés depart the Pacific area following the destruction of Rodzhensky's fleet.[156]

Washington was completely surprised by the annihilation of the second Russian fleet by the Japanese on May 27–28. On April 22, Schroeder, the chief intelligence officer, wrote McCully in Manchuria that there really did not seem to be any reason for him to stay: "I have some confidence for Rodzhensky and his fleet." Three days after the battle (May 27–28) Schroeder wrote Tokyo "We [in Washington] are still completely at sea as to the possible cause of that tremendous destruction of Russian vessels."

That does not mean that ONI's intelligence-gathering apparatus was a failure, but only that the Japanese were successful in masking their strategy and superior condition of their fleet from the American naval attaché in Tokyo. This superiority may have surprised the Russians and even the Japanese as it did the U.S. The Navy Department's last communication to its Russian attaché contained a classic understatement: "As the Russian Pacific fleet has been annihilated there does not appear to be any scope for further effort on your part."[157]

Submarine cables came to ONI's notice as objects of strategic interest in their own right. The chief intelligence officer wrote the naval attaché in Vienna in March 1904, enclosing a sheet of information about cables serving the Ottoman Empire:

> This represents exactly the kind of information, among others, that this office is desirous of acquiring, and it is thought probably that exhibits similar to this may be existent for countries other than Turkey . . . obtain such information as you can about all countries.[158]

The naval attaché at Paris noted that a submarine cable was shortly to be opened between Brest and Dakar for French Imperial communications. The Chief Intelligence Officer directed him "to obtain all information possible regarding that cable, especially in regard to its route at sea, on which side of various islands it is laid, etc., and also its landing place." Similar information was requested for the cables Réunion-Madagascar, Réunion-Mauritius, and Saigon-Pulo Condor and Pontianak.[159]

These messages reveal ONI's interest in the strategic importance of submarine cables and the Department's lack of knowledge about some of them.

World War I

In the early morning hours of 4 August 1914 the British cable ship *Telconia* carefully cut, one by one, the five German cables that ran down the English Channel to Vigo, Brest, Tenerife in Africa, and to New York. This was the first offensive action of the British Empire in the First World War.

One cable still connected Imperial Germany with the western hemisphere, a strand that ran between West Africa and Brazil and was largely American owned. Germany could radio to West Africa and have the message transmitted to the new world by cable. The British were loath to risk the displeasure of the Americans, but eventually persuaded Eastern Telegraph, a British-owned company, to wind up 30 miles of the cable it owned into the tanks of a cable ship. The cable blockade of the Central Powers was now complete. Austria-Hungary and Germany had no cable over which they could send and receive information in security from overseas. They were forced to resort to various expedients, which proved ultimately fateful to their cause and changed the course of history.

The British officer in charge of cable censorship, Col. George Cockerill, visited W. H. Page, U.S. ambassador to Britain, early in the war and found that "upon this very question of cable censorship, he had shown himself to be eminently reasonable." It is not likely that President Wilson, who championed the principles of open seas and open diplomacy, shared this attitude. J. W. Gerard, appointed U.S. ambassador to Berlin in 1913, attempted to obtain permission from the German authorities to land a Western Union cable that would run from the United States to Berlin. Permission was refused, Gerard believed, owing to the opposition of the German overseas cable companies, which were to be put out of business a year later in the opening hours of the war by the British. Had the Americans developed the cable and brought it ashore in Germany, the British would have been forbidden by international law from cutting it. The cable would have been property owned by a neutral, moreover a neutral whom the British took pains not to offend. After Gerard told this tory, early in the war, to president Ballin of the Hamburg-Amerika Line, the required permission quickly arrived from the Auswärtiges Amt. The only condition was that the line run directly to America, without it being brought ashore at any intermediary landing point.[160] But it was now wartime and the North Sea was full either of mines or the Royal Navy. The incident suggests that in the matter of submarine cables the German talent for strategic prescience had deserted them. If they had foreseen the capability of the British to cut their overseas cables, they would have welcomed the Western Union proposal.

The British, on the other hand, seem not to have considered a strategy of leaving the German cables intact and developing some means of tapping them. This approach would have given the British access to German messages. No such means of electric eavesdropping was available, and it appears that no great effort was made by either side to develop one.

As the war lengthened, the power of cable censorship and control became more extensive. In April 1917, the day after the USA entered the war, Britain adopted the policy of withdrawing "cable facilities from all banks in neutral countries which were proved to be actively assisting the enemy." This policy was applied to all banks in Europe and Latin America.

In mid-1917 Britain's chief cable censor was ordered to stop all commercial cablegrams to and from a neutral country.[161] The Admiralty had hurriedly improvised a code-breaking organization—the legendary Room 40 of Naval Intelligence—early in the war to break the German wireless transmissions. With the help of the French, who had "a preponderant cryptological superiority in 1914,"[162] and copies of the German diplomatic and naval code books furnished them from various colorful sources, Room 40 had by the end of 1914 developed the capacity to decipher the German wireless signals. Because the codes used could also be employed for German cable transmissions, that meant that the security of the enemy's cable transmission was also broken. As Room 40 began to develop the capacity to decipher the German codes, attention turned to the pile of uncoded diplomatic dispatches.

British sources are understandably reticent regarding the point in time at which the diplomatic cables crossing their territory were tapped. The biographer of Admiral Hall, who ran Room 40 for most of the war, is at pains to stress that the British government never entertained the idea of reading diplomatic cables in peacetime. But

> Now that we were at war we had to use every available weapon and it was no longer dishonourable to intercept and attempt to read all messages that passed through our cable offices or were sent by wireless.[163]

The British discovered, sometime in 1915, that the Swedes were permitting the Germans to use the Swedish diplomatic cable to transmit coded messages, which represented a double transgression against international law. The British informed the Swedish government that it knew that the Swedish diplomatic cable was being used for the transmission of German coded messages. The original route, called "the Swedish roundabout," seems to have been German cable Berlin-Stockholm, transfer to Swedes, then Swedish diplomatic cable Stockholm-Buenos Aires, then reversion to German diplomatic cable from Buenos Aires to Washington and other German embassies in the western hemisphere.[164] The British, however, did not make a diplomatic protest, so as not to call attention to the fact that they were monitoring diplomatic cables. Room 40 had decided to check the diplomatic dispatches passing through London and noticed the characteristic German code group in the middle of apparently innocuous Swedish diplomatic transmissions.

The Swedes apologized to the British and said that it would not happen again. They were true to their word—they used a new cable route. This time the British did not bring the matter to their attention, but simply monitored the German messages passing to and from the New World, gaining insight into the minds and the activity of their adversary. That explains why the British quickly extended their cable censorship beyond Austria-Hungary and Germany to all neutral countries in the form of a general information embargo. But the interception and decoding of enemy cables was a tactic opposite to censorship. It involved the free flow of information from one's enemy without his being the wiser.

Perhaps the greatest achievement of the cable intercepters and decoders was their part in bringing the United States into the war on the side of the Allies. In 1917 President Wilson, hopeful that he could bring an end to the war by acting as mediator, gave permission to the Germans to send messages in their own code through the American diplomatic cable. (The Swedish Roundabout took a week for a message and reply to pass between Berlin and the German Embassy in Washington.) Now Wilson was committed to the irregularity of using the American cable to transmit a belligerent's cable traffic in a code not known to us, in the interests of peace.

The Germans availed themselves of this arrangement to dispatch the 'Zimmermann' telegram to their legations in Washington and Mexico. This message was decoded by Room 40 in London as:

> We propose to begin on February 1 unrestricted submarine warfare. In doing this we shall however endeavour to keep America neutral . .(?) if we should not (?succeed in doing so) we propose (?Mexico) an alliance upon the following basis: (joint) conduct of war, (joint) conduct of peace . . . your Excellency should inform the President secretly (? that we expect) war with the U.S.A. (possibly). . . Japan and at the same time negotiate between us and Japan. . . please tell the President that. . . our submarines. . . will compel England to peace within a few months. Acknowledge receipt. Zimmermann.

Room 40 intercepted the message on all three of its routes. It passed through the American diplomatic cable on its way to Washington, it passed through the Swedish diplomatic cable on its way to Mexico City, and, according to some historians, it was also radioed from Germany to the German wireless station on Long Island.[165]

Hall in Room 40 quickly recognized he had in his hands the cable intercept of his career. Not only did it provide two weeks advance warning of a submarine blockade of the British isles, but succeeding days' work at decoding revealed that the Germans had promised the return of Arizona, New Mexico, and Texas to their hoped-for allies in return for their belligerency. The British, however, sat on their bombshell for several days in order to work out how the telegram could be used to precipitate the American entrance into the war without revealing to the Germans the secret of the cable intercept and decoding capability. The intercept arrived at Room 40 on January 17; it was formally handed to Page, the American ambassador, on February 23. The arrangement worked out was that an American would be wired the German code group in the American embassy in London, technically American soil, and would decode them with the aid of a British cryptanalyst and a German code book.[166]

Wilson had opened the American cable to the Germans to bring about peace. They had instead used it to suggest that a neighbor invade the United States. The news was published in the *New York Times* on March 1. Wilson, in response to questions, said that the message had been decoded on American soil, and sent a message to Congress that the text was authentic.* On March 3, to the amazement of all parties, Zimmermann admitted his authorship. This act freed the American authorities from an obligation to explain how they had obtained the telegram.[167] Slightly over a month later, when U-boat attacks had lowered the food reserves in Britain to a six-weeks supply, Wilson declared war on Germany.

This greatest intelligence coup of the war involved the technologies of cable, radio, cryptography, and cryptanalysis. Submarine cable became more important in the age of radio than it had even been before, precisely because of its characteristic security.

Domestic Censorship

After the Boer War, the British War Office had created a Special Intelligence Subdivision (SIS) in order to develop these capabilities under Lt. Col. George Cockerill. SIS did much of the pioneering British work in cables and cryptography. By 1904 it had prepared draft legislation for a war emergency that gave the government complete control of information.[168]

In an era of Liberal predominance in the House of Commons this proposal could not pass into law. It was later dusted off and became part of the Defense of the Realm Act, passed in 1914 on the fourth day of British participation in the First World War. Its prerogative was medieval; its powers foreshadowed the features of 20th-century totalitarianisms:

> *An Act to confer on His Majesty in Council power to make Regulations during the present War for the Defense of the Realm*. Be it enacted by the King's most excellent Majesty by and with advice and consent of the Lords Spirtual and Temporal, and Commons, in this present Parliament assembled, and by the authority of the same, as follows:
>
> 1. His Majesty in Council has power during the continuance of the present was to issue regulations as to the powers and duties of the Admiralty and War Council, and of the members of His Majesty's forces, and other persons acting on His behalf, for securing the public safety and the defense of the realm; and may by such regulations authorize the trial by courts martial and punishment of persons contravening any of the provisions of such regulations designed—
>
> (a) to prevent persons communicating with the enemy or of obtaining information for that purpose or any purpose calculated to jeopardize the success of the operations of any of His Majesty's forces or to assist the enemy; or
>
> (b) to secure the safety of any means of communication, or of railways, docks or harbours;
>
> in a like manner as if such persons were subject to military law and had on active service committed an offense under section five of the Army Act.[169]

The Parliament that passed the Defense of the Realm Act (or DORA, as it became known) believed, as did most contemporaries, that its disagreeable restrictions on traditional British liberties would have force only during a war that most believed would be over by Christmas. The legislation was drafted by the military at a time when they foresaw a possible war with the Russians and the French, in 1904. At that time, Colonel Davies of SIS was already negotiating with the Dominions on behalf of wartime Imperial cable censorship and a ban on coded cables. A speech was delivered in the House of Lords pointing to

> the dangers lurking in an unchecked dissemination of news and asked that, if legislation were required...it should be drafted in peace and kept ready for instant use in an emergency.

---

*Wilson saw the dispatch Saturday night, February 24. The next day he prepared an Armed Ship Bill for Congress. He addressed Congress that Monday, but did not refer to the Zimmermann telegram because a corroborating copy had not yet been located in the files of Western Union. The American copy had code numbers that differed from the original, necessary to keep the secret of the cable intercept in London. It was eventually found.

Cockerill of SIS on 23 May 1905 was permitted by the War Office to publish an article in the *Times* on "The Press as an Intelligence Agent in Time of War." It recounted a number of times when enemies of Britain had benefitted from information published in newspapers. If this was intended to prepare public opinion for information control in wartime, it backfired. The *Times* responded with a bristling first leader: "We are not accustomed to any form of Press censorship. The very name is abhorrent to our traditions." The *Times* suggested principles for the press during future conflicts: it should provide no disclosure of impending events; but it should be merciless in analyzing what had already occurred.[170]

By this time the question of information control had come to the attention of the Committee of Imperial Defense (CID), which had been set up in 1903 to advise the Prime Minister on questions of imperial strategy. The Prime Minister was its chairman, and the Secretary of State for War and the First Lord of the Admiralty always attended its meetings, together with other Cabinet Ministers as the subjects under discussion related to their departments. A digest of CID documents makes it plain that information control in war was a consistent concern of the imperial strategists, that that concern began early and rose in crescendo until the outbreak of war in 1914, and that submarine cables played a dominant part in the discussion.

Less than a year after CID first began its meetings, a paper was presented on "Intelligence from French colonial ports." (This occurred at a time of rising tension with France and Russia; the latter was preparing to send its Baltic Fleet around the world to the Far East, where it would be annihilated by Britain's ally, Japan, at Tsushima Straits.) In October 1904 the army presented a six-page report to CID on "Control of the press in time of war or threat of war," which led to a discussion on "control of naval and military information in cases of emergency."[171] Control of submarine cables must have figures largely in these papers and discussions. But CID's first consideration of information control was directed not at cables, but at radio, then only a few months old.[172]

Marconi, the developer of radiotelegraphy, had failed to interest the Italian authorities in his invention and had moved to London in the mid-1890s. By 1901 he had succeeded in transmitting a signal across the Atlantic and had the active support of the Admiralty, which was quick to see the naval importance of his device. Ships and naval stations quickly sprouted forests of antennas during the opening months of the new century. It was the awareness of this new means of communication as well as concern over cable censorship and codes that made information control in war such a consistent theme of CID discussions, in the context of anxieties raised by confrontations with Russia on the Northwest frontier of India and, after 1904, by the rapid increase of the Imperial German Navy. In July 1906 Asquith, the future Prime Minister, delivered a nine-page study of the "Ceylon Cable" at a time of border tension with Russia. In July 1906 the CID secretariat produced a further document on the control of the press during or in anticipation of war.[173]

As Britain is a group of islands, her government had an advantage denied to all other major European powers. Her press could be controlled, not through the onerous engine of direct censorship, but indirectly through control of the submarine cables. Thus British liberties could on the surface be maintained, while the government could exercise close control on information leaving or entering the country.

From 1906 to 1911 CID continued to study various strategic questions raised by the new medium,[174] and the related topics of espionage within Britain,[175] the treatment of enemy or neutral ships in war,[176] and the establishment of an Empire radio system.[177] The proposal of the Marconi company in 1911 to construct an Empire wireless system prompted CID to appoint a subcommittee to report on "Establishments of a chain of wireless stations throughout the Empire," and another committee to examine the question of "Submarine cable communications in time of war." The Agadir Crisis of 1911, which brought Britain to the brink of war with Germany in support of France, occurred while the submarine cable report was still being written.

The last months of peace were crowded with considerations related to the strategic value of the submarine cable. To cite the relevant documents in order:

Cab 20/3: "Submarine cable communication in time of war," 2 April 1912, Foreign Office and Government Post Office.

Cab 20/9: Minutes of 116th meeting, "Censorship of wireless in war...wireless system in Empire, submarine cable communications in time of war," 25 April 1912. [It was probably during this meeting that the decision was taken to cut the German overseas cable as the initial British act of war.]

Cab 21/21: "Submarine cables in time of war: Note of action taken on report of standing subcommittee," 1 June 1912, Colonial Office.

Cab 23/4: "Report and proceedings of the standing subcommittee: Enquiry regarding press and postal censorship in time of war," 22 January 1913.

Cab 23/8: "Report of the standing subcommittee: Maintenance of overseas commerce in time of war," 3 February 1913.

In early 1913, CID discussed the possibility of establishing direct radio contact with France and Russia. On 16 July 1913 the Overseas Defence Committee submitted a study on "Measures for safeguarding vulnerable points," which would have included landing places for cables; reports on means of press and postal censorship were received in December 1913,[178] though no action had been taken on setting up machinery for the latter by the outbreak of war in August 1914, when Britain cut the German cables. The Defence of the Realm Act with its censorship provisions was passed on the fourth day of the war.

Imperial cable censorship was brought "into full operation at once and without a hitch." The Special Intelligence Subdivision of the War Office, which had general responsibility for cable censorship and interception, entered the war with a total staff of eight. In the opening months of the conflict, its activities increased "a hundredfold." Director Cockerill was promoted to Brigadier-General in April 1915 and the Subdivision was transformed into the Directorate of Special Intelligence, the father of M.I.5, British counterintelligence.

Originally intended to prevent only "the leakage of naval and military information" to the enemy, Cockerill's group accidentally hit upon another military windfall for Britain accruing from British control of international cable traffic. Censorship of commercial cables often indicated the ultimate destination of goods. As the war deepened and it became evident that British dragoons would not enjoy their Christmas holidays in Berlin, a different pattern of war began to emerge, a long war of economic attrition. The traditional British weapon of blockade to commerce could be adapted to the international cable net.

Control of financial and commercial transactions through cable was not accepted as a weapon of war before 1914. But through judicious use of cable intercepts, Germany was cut off from the New York stock market and then, in 1915, from the practice of selling foreign stocks abroad to raise foreign currency. Then economic warfare by cable was extended to neutral nations. Firms trading with the enemy were put on a blacklist that brought a blockade of cable traffic. In this era, cutting off a company's access to cables was to threaten it with bankruptcy.

Cable interception also enabled the British to control securities and remittances to protect the British war effort. Profiteers intent on cornering the market in a particular commodity could be detected in the act and circumvented. Cockerill indicates that the British government was saved £600 million in one such operation alone.

Thus it was that under the impact of war Britain—the birthplace and bulwark of free speech and free trade—led the world in developing the modern means of information control. By making information infinitely more useful and more valuable, cables also had made information more dangerous; and the channels of information flow were also the system for its control. The Victorians, as we shall see, confidently believed that global communication would make war impossible; instead, it created Total War.

H. Oceanography

Since man has been fishing, sailing, exploring, and otherwise exploiting the oceans almost as far back as we can trace his development, there has obviously been a rich store of empirical observation and knowledge about oceans—but this knowledge was "lore," handed down or learned by individuals through experience, loosely codified if at all, conditioned and explained by the cultures in which it was embedded.

In the 17th century, Robert Boyle published his "Observations and Experiments on the Saltiness of the Sea," Newton contributed an explanation of tidal phenomena, and Robert Hooke lectured on methods for deep-sea research. In this early scientific period, most attempts to explain ocean-related phenomena were limited to what could be done in laboratories, or to the observations of naturalists about the coastal plants and animals. In 1776 Lavoisier analyzed the ingredients of sea water and Blagden related this information to the freezing point of salt water; and in 1820 Marcet established the identity or close similarity of ocean water around the globe. During the first three decades of the 19th century physicists studied and argued about the action of tides and currents.

During the 19th century many nations undertook the establishment of surveying bureaus and hydrographic offices, and in many cases cruises of exploration, as part of the burgeoning interest in commerce between countries and continents. The U.S. Coast Survey was established in 1807 (after some opposition to the effect that the Federal government had no constitutional authority to carry out scientific activities, it was justified under the commerce clause),[179] and the Depot of Charts and Instruments (U.S. Navy) was established in 1830. By the 1840s active investigation of currents, temperatures, salinity, and bottom deposits was under way. In 1838 the U.S. Exploring Expedition went out, poorly planned and organized; many of its reports were never published and almost none was published until ten years after the expedition. The head of the Depot of Charts and Instruments, Lt. Matthew Fontaine Maury, began in 1847 publishing his "Wind and Current Charts." He based his work on the earlier accomplishments of Redfield in the 1820s. Redfield had reconstructed storm paths by using thousands of ships' logs, establishing that "most storms...[traveled in] immense eddies or whirlwinds." Maury had been surprised to learn that navigators chose their routes on the basis of their individual experience and knowledge, which they jealously guarded. His new charts were based on special logs which he induced ship captains to fill out. In 1853 an International Maritime Conference was held and Maury's system was accepted by 10 nations.

After much effort and many delays, Maury had obtained his first research vessel in 1843. By 1848 Coast Survey Superintendent Bache was gathering samples of plankton fossils and remains. In 1853, the Brooke Patent Sounding Lead made it possible to bring up samples from the sea bottom and to provide much more accurate measurement of depths than had been previously possible.

A very different reason for studying the sea (as opposed to the commercialy motivated explorations and the samplings of dedicated naturalists) rose from the desire and ability to lay telegraph cables along the bottom of gulfs and channels. By midcentury more ambitious plans were being considered. More information than was intended or expected came from cable surveys, such as the discovery of deep-sea animals and planktonic oozes.

The cable companies needed and got help from government surveying offices to sound and sample feasible routes. In 1853 Cyrus Field had contacted Maury about the proposed transatlantic route. Maury answered, "The bottom of the sea between the two places is a plateau which seems to have been placed there for the purpose of holding the wires...." Later the Atlantic Telegraph Co. (organized in 1856) found the cable would have to span canyons and climb mountains, not lie on a plateau. The next year Maury published the first contour map of an ocean basin (the North Atlantic), on which he marked out the "telegraphic plateau." Based on 100 soundings, it gave the first indication of the Mid-Atlantic Range.

As each (cable) project was preceded by a sounding survey, a great deal was learned of the sea floor. Cable projects also led to many improvements in the design of sounding machines, which were in turn used by scientists who investigated the sea for less immediately practical purposes.[180]

In his book *The Physical Geography of the Sea*, published in 1855, Maury maintained staunchly that discovery and exploration were valuable for their own sake, any extension of human knowledge being a triumph of man over nature; but, he said,

> Already we are obtaining practical answers to the use of deep sea soundings for as soon as they were released to the public, they forthwith assumed a practical bearing in the minds of men with regard to the question of a submarine telegraph across the Atlantic.[181]

By midcentury the combination of interest, technical ability, and financial support necessary for large-scale scientific investigation was available. It was this combination that gave birth to the interdisciplinary science of oceanography. In the U.S. this took the path of concentration on exploring, charting, and studying the surface conditions of the ocean for the improvement of navigation; in Great Britain, the emphasis was on evolution of marine biology, which culminated in the *Challenger* expedition of 1872-1876. During the 1850s and 1860s, cable surveys gathered life specimens from several fathoms deep; but naturalists were also doing it independently.

In 1858 the British Admiralty assigned *Bulldog* to help Field, and in the course of 100 soundings starfish were brought up from 1260 fathoms. The naturalists refused to accept this find and insisted that the starfish had been caught as the sounding cable ascended. This was the time of great controversy over the ''azoic theory''—that life was impossible beyond a certain depth because of pressure, darkness, etc.

In 1860 a cable was pulled up from the bottom (near Sardinia) after three years and 15 animals were found cemented to the cable; that killed the azoic theory. In 1867 dredging of the straits of Florida, in the course of a cable survey, brought up other animals from great depths.

Cable surveys continued to stimulate deep-sea investigations that helped to resolve two controversies, over ''drowned continents'' and the ''continuity of chalk'' theory. The latter held that deep-sea oozes (of which samples were supplied by cable ships using Baille sounding machines and dredges) were identical to land deposits, which must have resulted from continental uplift. This theory, opposed by Agassiz and Kelvin, among others, was finally resolved affirmatively after the *Challenger* cruise.

By 1900 ships searching for cable routes had also established that greatest oceanic depths lie near land.

Cable-related research and surveys continued to stimulate oceanography: for example, in 1951 Heezen and Ewing studied the successive breaks of Grand Banks cables in 1929 and concluded that they resulted from turbidity currents down slope after the great earthquake of 1929; they measured the currents by the time and distance between cable breaks. This finding contributed to the discussion of whether turbidity currents could be the cause of canyons in the ocean floor and could account for observed patterns of sediment transported over long distances. The theory of Heezen and Ewings was further supported by study of cable failures after the Orleansville earthquake (Mediterranean basin) in 1954.

The laying of submarine cables thus had large benefits for the development of oceanography in stimulating interest in, and in some cases supporting, exploration and charting of the ocean floor; providing critical material (samples and data) on deep sea flora and fauna; stimulating and financially supporting the development of improved instruments for sounding, boring, and dredging; and providing some otherwise unavailable data sources contributing to knowledge of currents, turbidity, and seismic activity.

However, the relationship between oceanography and cable was a reciprocal one. In all the above areas scientific investigation had already begun for commercial or strictly scientific reasons; and particularly in charting of the ocean floor, the information already gathered contributed in some degree to the success of the cable project.

J. Time: Standards and Perceptions

Again and again, as Victorians discussed telegraphy and submarine cables, one phrase recurs in only slightly varying wording: ''science and energy have annihilated time and distance.''

So accustomed are we to this change, so much is it fundamental to our perception of the world, that we must exercise our imaginations to appreciate its drama for the men and women of the 19th century. Time is never, for humans, a constantly flowing river; a moment of fear or pain seems to stretch on and on, days of quiet happiness slip fleetingly by. But these impressions are very personal. In a more general sense, perception of time is somehow bound up with a sense of distance and with our ability and need to respond to events and stimuli. So long as news could travel no faster than man, or than horses or ships, only local happenings brought with them a sense of personal urgency in which minutes, or hours, or even days made a real difference. Events at a distance—both those of national

In the early conditions of the human race, when existence was free from the complications which civilization has led to; in the days when tribes followed pastoral pursuits, and each community was isolated from the other; when commerce was confined to a few cities, and intercommunications between distant countries rare and difficult; in those days there was no requirement for a common system of uniform time. No inconvenience was felt in each locality having its own separate and distinct reckoning. But the conditions under which we live are no longer the same. The application of science to the means of locomotion and to the instantaneous transmission of thought and speech have gradually contracted space and annihilated distance. The whole world is drawn into immediate neighborhood and near relationship, and we have now become sensible to inconvenience and to many disturbing influences in our reckoning of time utterly unknown and even unthought of a few generations back. It is also quite manifest that, as civilization advances, such evils must greatly increase rather than be lessened, and that the true remedy lies in changing our traditional usages in respect to the notions of days and hours, whatever shock it may give to old customs and the prejudices engendered by them. (Sir Sandford Fleming, International Meridian Conference, Washington, D.C. 1884.)

or global significance and those which were intensely personal, such as the death of a loved one, were over—beyond recall or control—before they were known; and though they might indeed change one's life, the span of time in which they could be absorbed rationally or emotionally, in which a response could be formulated and actions taken, must have seemed longer, slower, more even. Thus time was perceived in terms of seconds, minutes, or hours only for the few, immediate, personal events; in a larger sense it was measured in days, weeks, months, years—not a constant intimate harassing enemy but a distant and impersonal god ruling with an even hand.

Time, above all, was local—measured by the progress of the sun across the local horizon. Though one might journey like the American pioneers across a continent, the sun still rose and declined within the bounds of one's own sight, and the fact that others woke or went to bed by a different clock in the community where the journey began or in the region where it ended had no significance to the traveler.

When railroads were built, when the pace of work speeded up with the harnessing of steam, and daylight was extended by gas lighting, men marveled that Time itself moved faster. But only when communication was separated from physical movement did the striking change occur. When messages could be sent and received almost at the same moment, it was suddenly significant that that moment might be noon for the sender, and midnight for the receiver. When one can, potentially, affect the course of events although they are three hundred or three thousand miles distant—by sending an order, by supplying information, by setting into motion a corrective response—one may be compelled by duty, by responsibility, or by the hope of profit to do so. Thus rapid communication created new options, new duties, new responsibilities, and new anxieties.

The consequences of these changes were apparent in psychological, behavioral, and institutional responses, but only the institutional responses are simple to describe. Already, the transition from a rural agrarian society to an industrializing urban society had brought home to many countries the need to standardize both the measurement of time and behavior in response to time; factory workers could not appear for work or organize their tasks with the individual pace and flexibility of farm workers. The railroads created a new necessity for scheduling and metering time, and on larger geographical span. With telegraphy, this need became more urgent. Britain, for example, adopted a national standard time system in 1848. Sweden adopted a similar national standard in 1878.

In 1882 a committee of the American Society of Civil Engineers commented:

> The time system which we follow has been in use for centuries. It certainly answered all the purposes of mankind when there were no railways, no steamboats, and no telegraphs.[182]

The committee lamented society's having inherited a time system used for centuries in which

To apprehend that the question is one of importance, it is only necessary to glance at the existing condition of our time service. Mistakes in the hour of the day are frequent. In every State—in every city or town—discrepancies are met which produce great aggregate inconvenience. Thousands of engagements are broken. Innumerable disappointments and losses result. . . . These difficulties are not confined to this country. They are experienced in all civilized communities where lines of rapid communication have been established. . . . (American Society of Civil Engineers, 1882.)

"every spot of earth between the Atlantic and the Pacific is entitled to have its own local time."[183] Sir Sandford Fleming, the committee chairman, addressed the International Geographical Congress at Venice in 1881:

> If we take into view the whole earth, we have at the same instant in absolute time, noon, midnight, sunrise, sunset and all intermediate gradations of the day. The telegraph system, which is gradually spreading like a spider's web over the surface of the globe, is practically bringing this view of our sphere before all civilized communities. It leaves no interval of time between widely separated places proportionate to their distances apart. It brings points remote from one another, enjoying all different hours of daylight and darkness, into close contact. Under our present system of notations, confusion is developed, and all count of time is thrown into disorder.[184]

Fleming said that when rapid communications became universal, all times of different countries and localities must be in agreement or the necessity of making the cumbersome corrections in each transaction would "clog the ordinary business of the world."[185]

By the early 1880s standardized time in the commercial world was not just desired but essential. In November 1883 several Canadian and U.S. railway companies adopted a system of standards for time based on the Greenwich meridian. By 1884 the need for a world standard was sufficiently evident for 27 delegates to attend the International Meridian Conference in Washington, D.C., and adopt a world standard. Only the fear of surrendering local control of time and the question on which meridian to base the system remained. Submarine telegraphy partially assisted in the selection of the Greenwich meridian as the Prime Meridian, although an impressive case could be made for the choice regardless of the cable. Proposals to seek a more politically neutral area, such as the Azores, were faced with the fact that the Azores lacked adequate telegraph links with the world. Britain, however, the center of world communications, was ideally equipped to provide this world service.[186] It might be added that the inconvenient International Date Line slicing through the Pacific Ocean instead of the Atlantic Ocean reflects the domination by Western society of commerce, international politics, and science in the 19th century.

How new perceptions of time affected the Victorians psychologically can be told only anecdotally, and much will be recounted in their own words, in Chapter IV. Much as the temper of the times cherished Progress, science, and speed, they also recognized that these benefits brought with them anxiety and harassment. George M. Beard, a noted American physician, commented in 1887 that modern civilization differs "from the ancient civilizations mainly in these five elements: steam

. . . merchants were far less worried than now, and less business was transmitted in a given time; prices fluctuated far less rapidly, and the fluctuations which now are transmitted instaneously over the world were only known then by the slow communication of sailing vessels or steamships; hence we might wait for weeks or months for a cargo of tea from China, trusting for profit to prices that should follow their arrival; whereas, now, prices at each port are known at once all over the globe. This continual fluctuation of values, and the constant knowledge of these fluctuations in every port of the world, are the scourge of businessmen, the tyrants of the trade—every cut in prices in wholesale lines in the smallest of any of the Western cities, becomes known in less than an hour all over the Union; thus competition is both diffused and intensified. (Dr. George M. Beard, 1887.)

power, the periodical press, the telegraph, the sciences, and the mental activities of women. When civilization plus these five factors, invades any nation, *it must carry nervousness and nervous diseases along with it.*''[187]

Henry Thoreau, somewhat earlier, had already keenly observed the changes in psychology and behavior that were occurring:

> Have not men improved somewhat in punctuality since the railroad was invented? Do they not talk and think faster in the depot than they did in the stage-office?[188]

With telegraphy, this need for punctuality became for many a relentless urgency, as decision-making was bombarded by the ever swifter and more diverse flow of information. The dealer in the pit of the commodity market was only the most extreme example. The anxiety of the general struggling to coordinate the information from farflung armies, the diplomat monitoring a fast-breaking crisis across the world, were at least dimly mirrored in the uncertainties of ordinary citizens as they followed the excited dispatches from around the world in their daily papers or waited for the outcome of battles, accidents, or births about which they would have once known only long after they happened.

Chapter 4

HOPES AND EXPECTATIONS

CYRUS FIELD and his fellow-investors gambled large sums on the success of the transatlantic cable. They hoped, it is true, for excellent returns on their investment. But the historical records show that they also expected great benefits for their nations and for mankind in general. If they had any doubts on the latter point, the necessity of appealing to the investors' Victorian ideal of Progress and Civilization as well as the Victorian love of profits kept them from revealing such reservations. What, then, did other mid-19th century observers expect (or hope) would come from this new technology? Were the complex and far-reaching impacts described in the preceding chapter anticipated? And if so, by whom?

A. The Victorian View of the Future

The idea of progress. Mid-19th century was a period when the pace of change appeared enormously accelerated, when changes could be easily perceived within the span of one lifetime, and when people felt either buoyantly optimistic about the future or confused and threatened, according to their own conditions and disposition. It was, in a word, a period when people thought about the future as they perhaps do only in periods of accelerated change.

Bury, in his *Idea of Progress*, says that two conditions must be present for the idea of Progress to become an acceptable, accepted, and familiar idea: significant changes must occur and be given concrete manifestation, and there must be striking material evidence in the life of the common man of those changes.[1]

The 19th century was a period in which these conditions existed as seldom before. The harnessing of energy for production and transportation was one of the most significant breakpoints in human history. The age-old dependence on muscle power to perform all of the world's work (extended only slightly by mechnical assists and wind power) was broken. The steam engine brought about great increase in productivity in the early 19th century. The earth itself seemed to be made enormously more productive by new agricultural implements. Gas illumination had pushed back limitations on activity imposed by darkness. Such wonders seemed to give substance to the promise of a book title, *The Paradise Within Reach of All Men, Without Labour, by Powers of Nature and Machinery.*[2]

Improvements in carriages and roadways, in all the centuries since the Romans, had done relatively little to improve on the speed and endurance of horses as the most advanced form of land transportation. (It has been said that George Washington traveled from New York to Washington for his second inauguration at about the same speed at which Julius Caesar could have made the trip; Washington's equippage was better but the roads were poorer.) The improvement of wind-powered ships throughout history was greater, but the oceans remained an insuperable barrier to swift intercontinental communication.

It is therefore little wonder that the railroad and the telegraph were hailed as annihilating time and distance, and gave rise to great anticipation of international unity and cooperation. In 1829 South (in *Sir Thomas More*) related social and moral improvement to the invention of the steam engine. Tennyson (in *Locksley Hall*) spoke of this ''wondrous Mother-age'' and (misunderstanding how

We have girdled the oceans with our harbours and encircled the world with our cables, we have crossed the sea with our ships, we have harnessed the forces of nature and achieved over her victories which have only ceased to be wonderful because we are so familiar with their result. Chemistry has resolved matter into its elements, and seems to stand upon the very threshold of yet more stupendous conquests. The spectroscope has transported the analyst to the sun and stars. Photography has caught forms as they are. The phonograph has perpetuated transient sound. Steam and electricity have revolutionized the means of communication as machinery has revolutionized production. [A lead notice in *The London Daily Mail*, 1899, cited in Asa Briggs, Ed. *The Nineteenth Century: Contradictions in Progress* (1970), 337.]

railroad wheels fit rails) exhorted, "Let the great world spin forever down the ringing grooves of change!" The image of the railroad became an amazingly pervasive and powerful symbol in both literature and art.[3]

Between 1800 and 1830, then, technological change was highly visible. By the time of the Great Exhibition in London's Crystal Palace, in 1851, "progress" was a familiar idea. Faith in science, technology, and industry, and their transnational scope—as well as the benefits they implied for imperialism—had become firmly fixed in the English mind.

In America too, there was fertile ground for enthusiasm about the future and great expectations of material progress. As in Britain, minds had been well prepared by the thinkers of the Enlightenment. Early Americans had proved their desire for a different future by the very act of coming to this country. That initiative had been kept alive by the stream of immigrants, the unfolding of the frontier, the American revolution, the writing of a Constitution, and the establishment of new territories and states.

The displacement of the Indians, and later the arguments over Negro slavery, frequently brought forth the thesis that such primitive peoples were doomed by history or fate because they were "unprogressive"—unable to adjust to and exploit changing conditions.[4] The same thesis was used later against the Mexicans; nothing must stand in the way of Progress. War—against Indians, Mexicans, or possibly against Britain or France as the westward expansion continued—was both defended and denounced in the name of Progress: "War is necessary if the progress of the nation requires it;" or, alternately, "War is incompatible with progress."

By 1850, then, Americans and Britons alike generally believed that significant changes would come about within their lifetime, that their children's lives would be different from their own, and that it was very likely that the future would be somehow "better" than the past—in other words, they believed in progress.

Leo Marx in *The Machine in the Garden* portrays the way in which American thinkers and writers strained to incorporate the images of the machine and the industrial revolution into a view of the future which remained based on "the pastoral ideal," a Jeffersonian paradise of rural democrats and self-sufficient communities. (Jefferson himself, in the preceding century, had gloried in the promise of the steam engine, apparently thinking only of its incorporation into the rural economy and rejecting with horror the development of the English factory system.)

The Birth-year of the two Great Wonders of the Age—The Atlantic Telegraph, and the Non-explosive Gas Generating Cooking Ranges

The two Epochs of the Age—Electricity and Steam Ale

Placards carried in 1858 New York Parade, cited in C. T. McClenachan's *Detailed Report of the Proceedings Held in Commemoration of the Successful Laying of the Atlantic Cable* (1863), 104–105.

For the majority of Americans, who still lived and worked on farms, the rapid transformation of agriculture was highly visible and welcome—mowers, reapers, cornshellers, hay rakes, the cotton gin. A popular slogan went, "When Science speeds the plough, wealth and honor are awarded to the ploughman."[5]

Since technology was hailed as progressively freeing mankind from slavery to the soil, it came to be identified with increasing social democracy. But with the industrial revolution came the growth of industrial cities and new social problems, including the degradation and suffering of workers in the factories.

A debate began about whether the new working class as a whole benefited materially from the vast increase in productivity. Generally Americans were inclined to blame the social degradation which was a byproduct of the factory system in England on a lax government and a class-ridden society rather than on industry itself. In America, it was widely believed, the benefits of new technology would be more evenly distributed than in England. American writers extolled the wealth of productive factories and the growing importance of America in world trade.

Although factory-based poverty and suffering indeed developed in America, during the mid-19th century it was the benefits of the factory system, and their wide distribution, which most occupied the attention of the general populace. The domestic applications of technology had a profound effect on lifestyle. The sewing machine quickly became a fixture in middle-class homes. The rotary press led to a rapid increase in popular literature. Walt Whitman, as a reporter for the *Brooklyn Daily Times*, listed homely inventions and improvements that had been widely adopted between 1850 and 1857: baths, icehouses, ice coolers, fly traps, mosquito nets, housebells, marble stands, sliding tables, patent-ink stands, baby jumpers, street-sweeping machines—"Bless your star," he wrote, "that fate has cast your lot in the Year of our Lord 1857."

Another factor in changing expectations was the growth of public education, and the interest in self-culture and workingmen's education. Reformers put great faith in the efficacy of education as a means to both material and moral progress, especially in America. Through the schools the expectation of personal improvement and advancement was promulgated along with an admiration of science, technology, and industry.

A long process of geographical expansion—in Britain through colonization and in American through pushing back the frontier—contributed to the idea of progressive change and the promise of the future. From 1815 to the depression of 1837 and beyond, Americans came to glorify trade, expansion, and internal improvements as "Christian Republicanism." As noted, the taking of lands from Indians and Mexicans was justified on the grounds that such peoples were by nature unprogressive and doomed to oblivion. The idea of an American "mission" to spread democracy through "the peaceful working of our example" became transmuted into an ideal of political and material expansion for which violence may be necessary.[7]

After the failure of the European revolutions of 1848 there was in America a general turning away from Europe and the hope of democracy triumphing there, and a growing expectation that "the future"—a destiny of leadership for the New World—lay in the West. The railroad and the land telegraph lines were the ties that would unify all Americans into a great people. In Britain a spirit of complacency developed, but it was conditioned by general acceptance of gradual political reform and extension of suffrage, and enlivened by popular interest in the Empire and its glories.

Advances in science, although less widely understood and appreciated than technological applications, caused thinkers to speculate about a time when man could progressively subdue nature and overcome limitations of time, space, and even mortality. Darwinism did not emerge as a galvanizing debate until the 1870s, but as the science of biology evolved "from structure to process and systems,"[8] the groundwork was laid for the belief that progress (the future) was inherent even in the laws of nature. There were also forerunners of modern psychology—the realization that mental events too are subject to deterministic laws. In physics, including astronomy and optics, in medicine, and in other sciences, exploration of fundamental principles and their applications were productively interacting.

Morse, the inventor, supplied the germ;
Steers, the mechanic, supplied the Niagara;
Field, the businessman, completed the glorious work.

Americans exult! American invention, mechanical genius, and business energy and perseverance—furnished by Morse, Steers, and Field—do honor to America.

Parade Placard, New York City, 1858, as reported in McClenachan (1863), 39.

The minor chord. Although this dominant theme of optimism prevailed, there was also a minor but pervasive theme of regret for the loss of the "pastoral ideal"[9] and pessimism toward the future. Many writers and thinkers sounded a note of warning or had threaded through their works a note of loss and nostalgia for a simpler and better world. Conservatism marked some of the theologians and clergy, especially Catholic thinkers.[10]

"It is the philosopher alone who can conceive of the grandeur of Manchester and the immensity of its future," wrote Disraeli in 1844, but other observers recoiled. "Here civilization works its miracles and civilized man is turned back almost into a savage," said De Tocqueville.[11] The biting words of Marx and Engels a few years after the mid-century (although they looked to the future for retribution, justice, and hope) spoke for many others who saw the social costs of industry.

At mid-century the intrusions of the Industrial Revolution into both the "green and pleasant land" and the traditional land-based British class structure were hotly resented and resisted. The warnings of the historian Froude, although penned a generation later, are indicative of the fears of the future that some Englishmen felt even then:

A population given over to employments which must and will undermine the physical vigor of the race, generations of children growing under conditions which render health impossible. . . .[12]

In America such popular writers as Hawthorne, Melville, Cooper, and Simms often pictured material progress and technological development as equivalent to a loss of native virtue and simplicity, decay of human bonds, and moral degradation. They stressed the need for "natural aristocracy" to prevail over the rampant democracy based on materialism which they saw around them. Edward Everett, Charles Eliot Norton, and Francis Lieber called for a defense of property as a bulwark of freedom and responsibility.

In the South, which before 1830 had appeared ready to welcome industry, the threat to slavery caused Jeffersonian liberalism to give way. The defense of aristocratic ideals and "White civilization" took its place. The overwhelming number of Negro slaves was threatening and they were perceived as incapable of progress or civilization. The defiance of progress by resistance to any change in insititutions means, almost inevitably, a move from conservatism to reaction and a harking back to some mythical golden age.

Specific events reinforced the sense of erosion and loss of traditional values. The depression of 1837 seemed to signal the end of an inevitable increase in material wellbeing and more equal distribution of goods. The abortive European revolutions of 1848 ended the belief of many American

To show some of the advantages arising from rapid communication by means of the cable—the advantage of transmitting intelligence of important events—I am informed that but two weeks ago our government, or at least one of the heads of departments, would have given a million of dollars to have had an immediate communication with New Orleans. This shows the great need there is for immediate communication, and also how soon the transmission of important messages would pay the whole cost of the work. I will not detain you longer, as Mr. Field can better explain the whole subject, with which he is so thoroughly acquainted. [Peter Cooper, as reported in McClenachan (1863), 7.]

Scattered abroad over the whole world are Colonial possessions, dependent on the protection, and ministering to the high position, of these islands. In peace they are one of our chief sources of national wealth. In war they would probably form prominent objects of attack, and the foremost prizes of which an enemy would seek to possess himself. In the former case, their trade and importance would be advanced enormously by the newer connection which telegraphic communication with the mother country would confer upon them. In the latter case, it is hardly too much to say that direct correspondence with head-quarters would be pretty nearly equivalent to a doubling of their material means of defence. At present, not one of these possessions is in direct telegraphic *rapport* with England, although the means are ready at hand, and require only such an impetus as would set them in motion, either by Government itself undertaking the work, or by the grant of such a fair share of assistance on their part as would procure the capital elsewhere. [George Saward, *Deep Sea Telegraphs: Their Past and Future Progress* (1861), 44.]

thinkers that republics would triumph in Europe. The Texas and Mexican wars were setbacks for those who believed that peace and progress were inseparable. Warfare was moving toward more destructive weapons and larger scale, and looming over the American future was a growing likelihood of internecine conflict.

Expectations and predictions. As Americans faced the seventh decade of the 19th century they were transfixed by immediate prospects of a bloody civil war. But this preoccupation with a short-term crisis was only a break in the optimism with which most people in America and Britain viewed the future.

Mary Shelley, Hawthorne, Poe, Melville, J. D. Whelpley, Fitz-James O'Brien, and Dr. Silas Weir Mitchell filled the pages of popular journals with stories based on the marvels of science and prospects for future inventions.[13] *Godey's Lady's Book, the American Whig Review, Putnam's Monthly Magazine*, and the *Atlantic Monthly* carried many stories of the kind we now call science fiction, dealing with robots and automata, medical science and the shifting boundaries between life and death, interventions into the mind and altered states of consciousness, and space and time travel.

Elements of science and the supernatural were freely mixed. The more pessimistic writers such as Hawthorne and Melville frequently played on the theme of loss of innocence: to seek to inquire too far or to intervene too ambitiously into the workings of nature was to taste again the illicit apple in the

. . . While we write, a rapid interchange of communication with India would be of inestimable value. The power of instantaneously telegraphing the American outbreak to London, and thence to India and elsewhere, at its first commencement, might very possibly have made all the difference between plenty and starvation in thousands of Lancashire homes during next winter; and it is not too much to say, that, had the cables of the Red Sea and the Atlantic Companies existed in working order during the last six months, they would have paid for themselves twice over in value made and saved to this country at large, if not even to their owners also.

The amount of gold of which England has been drained since last November is over six millions, and the loss upon American securities, on this side the Atlantic, at the commencement of the secession movement, must have been enormous, considering that upwards of sixty millions of United States' securities are held in Great Britain alone. Had the Electric Telegraph then existed, these losses would have been very greatly mitigated; while an accurate knowledge, from day to day, of the state of exchanges at New York, would have materially checked the export of precious metals to that city, especially in large quantities, as it has frequently happened that the very gold sent out in expectation of securing, as a profit, the turn of exchange, has had the effect of so reducing the rate as to destroy all margin, and even convert the operation into a loss for the time being. [George Saward, *Deep Sea Telegraphs: Their Past and Future Progress* (1861), 47.]

Garden of Eden.

But more often writers such as Jules Verne reflected the Victorian admiration for exploration, adventure, and daring. Timothy Flint, in the *Western Monthly Review* (1827–1830) pictured for American westerners a world of peace and prosperity in which everyone was healthy (by virture of "5000 patent medicines") and balloons were flying from America west to China "as commonly as ships travel now."[14]

A playful note entered the popular discussions of the future. In France a set of playing cards ("Anticipations") bore on the back marvelous cartoons of 19th century ladies and gentlemen flapping through the air on mechanical wings, strolling decorously along the ocean floor in diving helmets, running a farm from front-porch rockers by means of electric wires, or producing elaborately curled coiffures by means of strange machines. In such cartoons and in the novels of Jules Verne and (in the 1880s) of Alfred Robida other strange new technologies were predicted: air travel, rocket ships, television and videophone, sound recording, chemical warfare, motorcycles, armored cavalry, elevators, dictaphones, permanent waves, processed food, motion pictures, news by loudspeaker, mobile homes, scuba diving, learning machines, hydroplanes, and mechanical floor polishers. Nor did the futurists confine themselves to physical technologies: the movement of women into professions, self-government in African colonies, and revolution in China are examples of other predictions.

Utopian literature stressed hopes of international brotherhood, and novels of future utopias often pictured them as coming about through an international language or some vague but miraculous new technology of instant transportation or communication across vast distances. It seems strange that the British with their experience of colonizing, and the Americans with their experience with the Indians, should have believed that to communicate is to understand and to understand is to love, but so it seemed to be.

> Now, who can see or foresee the result of all this? Cotton, which boasts of being "king," feels in this extended electricity new pledges of sovereignty. Rice is exultant. Tobacco even is not without joy. Gold, ever glittering, glitters yet more and yet wider in the sparkles of the wire. [James Brooks, journalist, speaking at the New York celebration; McClenachan (1863), 263.]

B. Expectations About the Cable

It is understandable that the successful laying of the transatlantic cable was a matter for public rejoicing; in America there were bonfires and dancing in the streets when the 1858 cable appeared to have been successful. The benefits of invention and innovation had been widely celebrated for a century, not only by men of finance and business but by the general populace whose standards of living and lifestyle had (for most classes) benefited enormously from the applications of science. The popular press had stimulated the enthusiasm for innovation, and first-of-a-kind technological breakthroughs against seemingly overwhelming odds were considered the stuff of romance and adventure, much as the man-on-the-moon saga of the 20th century captured the imagination of the public. Such exploits, moreover, were hailed in terms of national prestige and glory, and though the cable was British, the Americans celebrated the initiative of Cyrus Field and also assumed that—with their growing confidence in America's prospects as a manfacturing and trading nation in the future—the use of the cable would help them in mounting a challenge to Britain.

Even the pessimists who tended to equate technology with a loss of innocence, the radicals concerned with exploitation of workers, and the conservatives concerned with defense of property and class positions, felt little resistance to the cable beyond mild skepticism—the cable did not desecrate the wilderness, require the labor of women and children, or stimulate excessive "democracy." By 1866, despite the strains of the civil war, American attitudes toward Britain were not so much hostility as envy and competitiveness.

The telegraph will undoubtedly gradually take the place of postal communication, so far as a large proportion of correspondence is concerned, although the mere labor of rewriting or operating through the wires will always make the carriage of letters an important business. In financial affairs the telegraph will do away with the necessity of a considerable portion of the currency. Upon a system similar to that of the New York clearing house, large operations can be made through the telegraph without the medium of money, by depositing balances at commercial centers. Through the instantaneous action of the telegraph these balances can, for all practical purposes, be used as readily as if they were in the hands of their owners. The Bank of England has already proposed to pay the dividends of British consols, through the use of the telegraph, thus obviating the old, clumsy and troublesome system of requiring the personal presence of the bondholders or their attorneys at the place where the payments are made. [Article in a Buffalo newspaper, cited in *Journal of the Telegraph* (1868), 269.]

And—the critical point to the moralists who helped to form public opinion in the Victorian age—the cable appeared as a further and glorious step in providing bonds between the peoples of the earth and thus furthering the great ideal of peace and brotherhood. Most important of all, it held out the possibility of a good return on investment. Any innovation that incorporated the three great ideals of moral improvement, a dramatic triumph over nature, and return on capital, must have seemed to the men of the mid-19th century the wave of the future and thus a fit occasion for dancing in the streets.

Costs and benefits. The promoters of the cable had to convince potential investors that they would receive a good return on their investment. We know little of the arguments they used, since in general these negotiations were carried out person-to-person, among businessmen who already knew each other and whose prowess at making money was proven. After the British and American governments had committed themselves, in 1856 and 1857, to guaranteed usage of the cable, the promoters had a powerful argument at their disposal: the cable could hardly lose money.

But during the discussions which preceded governmental guarantees, the cost-effectiveness of the enterprise (to use a modern term) was debated. Senator Benjamin, for example, argued that the transatlantic cable, if it could prevent further wars with Great Britain, or "even defer a war for six months or twelve months, if one should ever become inevitable,. . . will in six or eight or nine months repay the expenditures of twenty-five years" by the U.S. government.[15]

. . . hence they are strongly tempted to order grain from those nearer ports with which they are in instant communication, in preference to ordering it from this side of the Atlantic, at the risk of so much delay, with other difficulties and dangers. If they order grain from this side, they must first write and ascertain what it can be got at. Then some weeks must elapse, a month at least, before they can get an answer, and then the order may come back about the time that prices change, and the merchant may not be in a condition to give a prompt order. Hence the temptation is to order at once at those places where they can ascertain with promptitude and certainty the price they will have to pay, and the profit upon which they can calculate with certainty. But when they order from this side, or when the merchant here ships to the European market, the prices ruling there when the grain arrives may all be changed, and the prices obtained may not pay the cost. If we were in immediate communication with Europe we should avoid sending our goods to unprofitable markets. Hence advantageous results would follow to the great agricultural interests of the country. The produce merchants of America are ready and willing and able to supply all parts of Europe with the necessary provisions. That is one great item in itself, and amply sufficient, if there were no other, to pay the cost of putting down and completing such a work. [Peter Cooper, on the occasion of the New York celebration; McClenachan (1863), 6–7.]

It will take the profits heretofore shared by the cotton speculators of Europe, and place them at the door of the cotton-planter; who at all times during his growing crop will have power to know what his crop is worth in the markets of Liverpool, which govern the markets of the entire world. [Senator Mallory (Fla.) speaking during the Senate debates on the telegraph bill as reported in the *Congressional Globe* (1857), 424.]

Senator Seward of New York drew an analogy from the domestic telegraph lines: the U.S. government had appropriated $40 000 to test the practicality of Morse's invention and "Now, there is no person on the face of the globe who can measure the price at which, if a reasonable man, he would be willing to strike from the world the use of the magnetic telegraph. . . ."[16] Senator Bayard of Delaware added, "I believe the government will obtain more service for the amount of money, than by any other contract that we have ever made, or now can make, for the transmission of intelligence."[17] And Senator Douglas of Illinois noted that Americans were ever ready to pay for speed: " . . . we pay twice as much to have mail delivered two days earlier."[18]

A few years after the debate in the U.S. Congress, Cyrus Fields wrote to Seward (by then Secretary of State) in regard to the *Trent* Affair: "I have no doubt that the English government has expended more money during the last thirty days in preparation for war with this country than the whole cost of manufacturing and laying a good cable between Newfoundland and Ireland."[19] Other Congressmen argued that assistance to the cable would be recovered through reduced diplomatic costs.[20]

On both sides of the ocean men argued that government assistance to, or subsidy of, a transoceanic cable would be a wise investment of public resources. George Saward argued in 1861 that a British government subsidy could be justified in terms of the savings in colonial defense alone, and that early notice of the secession of Southern states from the American union, by preventing losses from gold shipments and the holding of U.S. securities in Britain, would have paid for the cable.[21]

Not all political leaders were convinced by such arguments. Some argued that the cable would be a commercial failure because only the few extremely large merchant firms would be able to afford its services.[22] In the U.S. Senate, Toombs of Georgia attacked the promoters' promises.[23] Private investors, too, although promised "at least" a 10% return on investment, were often difficult to find; in America investment capital was small compared to that in Britain and there were many competing opportunities.

Commerce. But if individual businessmen hesitated to risk their capital on an enterprise which might well fail for technical reasons, there was yet little doubt that the long-run consequences of

Here I mark one thing, viz.; That while this wire, with those other co-related telegraphs on either side, will, in the first instance, work towards monopoly, in the second and main instance they will work towards diffusion and the common weal; for although commerce and politics, and the merchant class and the political class will in the first instance be the users, and so be benefited first by it, yet, in the main, the people will be the ones who will reap the benefit, for whatever thing brings now whole communities into circumstances of greater prosperity must needs, in the spirit and temper of our time, be distributive. If it were possible for knowledge to be confined to the minds of the few, if it were possible for monopolists to lock up the two ends of this wire, holding, on either side, the power of its intelligence, it might be disastrous to governments and to the people. But now it has a tendency to make knowledge co-extensive with the globe instantaneously. [Rev. Henry Ward Beecher, quoted in Charles F. Briggs and Maverick Augustus, *The Story of the Telegraph and a History of the Great Atlantic Cable*. New York: Rudd and Carleton (1888), 246.]

intercontinental communication would be an enormous increase in trade between Britain and America. The cable entrepreneurs constantly spoke of the advantages of "the commercial annihilation of distance" between North America and Liverpool or London in increasing both the total volume of trade and the American share of the profits to be made. Edwin Morgan, a wealthy New York financier, told a New York Chamber of Commerce meeting in 1863: "None can estimate or anticipate how largely our interests will be promoted. . . " and a committee reported at that same meeting that the cable would doubtless lead to "the improvement of the agricultural position of the country by extending to it the facilities, already enjoyed by England and France, of commanding the foreign grain markets. . . . "[24]

The promoters also spoke of other benefits of the cable to international trade: better management of shipping, reduction in fluctuations in exchange, abatement of the depreciation in public securities, less need for shipment of gold, and an end to financial crises rising from speculation in commodities.[25]

Political leaders in both countries also expected a great increase in commerce both between England and America and worldwide. American Congressmen recognized the special advantages for the USA. "We have an immeasurably greater interest in the foreign market than Great Britain or her subjects have in the markets of the United States," one Senator said.[26]

> Before the telegraph and the steamers were introduced, the merchants engaged in foreign commerce, as they decided upon their voyages and adventures, were only guided by the probabilities of the distant markets, and were obliged to calculate on a rate of profit which would cover all the risks. But many of our largest houses, while busily engaged in fitting out their ships an consignments, or on receiving their returns, had in the intervals time for reflection and rest, though often anxious to learn the result of shipments and adventures which it took months to complete. In those days they had packets for London or Liverpool twice a month, by means of which they corresponded with all parts of the world. . . . Comparatively they had an easy time.
>
> But now all this is changed, and there are doubts whether the telegraph has been so good a friend to the merchants as many have supposed. Now, sir, reports of the principal markets of the world are published every day, and our customers are continually posted by telegram. All the merchant can effect is a very small profit or commission. Instead of making a few large shipments in a year, he must keep up constant action, multiplying his business over and over again, using the telegraph to save time and capital . . . order the proceeds invested in commodities, the value of which is well understood, and which are again sold before their arrival. He is thus kept in continuous excitement, without time for quiet and rest . . . the poor merchant has no other way in which to work to secure a living for his family. He must use the telegraph. [W. E. Dodge, a merchant in 1868, as reported in the *Journal of the Telegraph* 2 (1868), 31–32.]

But the Congressmen argued lengthily about who would reap the benefits of increased trade. New York merchants and "Wall Street jobbers," said Senator Thomson of Kentucky.[27] Senator Pugh of Pennsylvania agreed.[28] British speculators, others thought. Senator Toucey expected that the country's cotton crop would be bought up by British speculators as soon as they got market reports by cable.[29] But other Southern Senators believed that rapid communications would allow Southern cotton planters for the first time to control the sale of their own crops and make bigger profits.

Clergymen, the opinion leaders of the day, preached sermons extolling the cable for its promise of new commercial worlds to conquer, for the Victorian religious leaders believed that Christianity and Commerce marched shoulder to shoulder and both signalled the advance of Progress and Civilization. The Reverend Henry Ward Beecher recognized that "markets will go up and fortunes will be made down in the depths of the sea."[30] But, he said,

> . . . although commerce and politics, and the merchant class and the political class will . . . be the first benefited by it, yet in the main, the people will be the ones who will reap the benefit, for whatever thing brings now whole communities into circumstances of greater prosperity must needs, in the spirit and temper of our time, be distributive. . . .[31]

. . . The nation sees and hails the future of this great port, when, connected fully with the Pacific by railroad and telegraph, as it is now with Europe by the Cable, it shall become the commercial centre of the world! Emphatically, with the Telegraph, there is no East, no West—only a centre. Time itself shall be no more! [Alexander Bache, head of the U.S. Coast Survey, at the New York celebration, McClenachan (1863), 271.]

As the great centre of commerce in America will be benefited in the greatest degree by the new and wonderful facility of inter-communication with Europe, so will her responsibility be increased to preserve that peace and good-will, to furnish to the world examples only of integrity and conscientious dealing. "To whom much is given, of him shall much be required." [Gov. Liston of South Carolina to New York City Commemoration Committee, cited in McClenachan (1863), 215.]

A thoughtful journalist, writing anonymously, also considered the question of whether the cable would benefit only the wealthy and powerful. So long as there was only one cable, owned by the corporation, he concluded that it would serve only the privileged, but that the same profit motive would create a remedy by creating competition, and the good of society would be served by "an equalization of the stocks and prices of commodities all the world over." Thus commerce would be restored to its rightful function, "from a game of hazard to the certain and moderate profits earned by placing goods where they are needed and wanted."[32]

The *Commercial and Financial Chronicle* (anticipating a major concern of Technology Assessment a hundred years later) provided a thoughful and prescient list of the parties who stood to lose from the cable's success: steamship lines with government subsidies for carrying mail, merchants trapped by sudden equalization of the markets, heavy shippers who would find their industry becoming increasingly competitive, bankers and brokers dealing with bills of exchange, dispatch agencies, and news vendors and news "inventors."[33]

Many predictions were made, in speeches, sermons, and newspaper columns, that America would become a world commercial center. The Commemoration Committee in New York (arranging for a city-wide celebration at the time when the 1858 cable appeared to be a success), noted that the most important influence of the cable, "so far as our own city is concerned, is necessarily that which it will have upon commerce: and upon the trade of New York depends, in a material measure, the prosperity and advancement of that of the Union."[34] The head of the U.S. Coast Survey agreed: New York "shall become the commercial center of the world!"[35]

Postal service. It was sometimes predicted that the cable and telegraph would "gradually take the place of postal communication, so far as a large proportion of correspondence is concerned." (This speculation appeared in a Buffalo newspaper.[36]) The *Southern Review* was more skeptical: "How long would it take to transmit by cable the letter mail alone of one of the Liverpool packets?" The *Review* concluded, sensibly, "The vessels that move over the waters between the two continents are, and must ever be, the great mail carriers."[37]

The Telegraph is only a transmitter of news, and news is only the occasion, not the effective cause of anything. Nor is it the principal news-transmitter even between Europe and America. The vessels that move over the waters between the two continents are, and must ever be, the great mail-carriers. The Telegraph is of course unequalled in celerity. But even on this point there may be some confusion of our ideas. How long would it take to transmit by the Cable the letter mail alone of one of the Liverpool packets? Nor can we see how the present enterprise can lead on to anything further, or more important of the same kind. (From an 1867 article in the *Southern Review*, 36.)

The telegraph is cosmopolitan. Not more British than American, it can neither be monopolized by government, nor stopped in its work of civilization by neglect. It is an institution for the people. Its office is, to diffuse intelligence; its effect, to allay differences. Men who talk together daily cannot hate or disown one another. (Dr. Ezra Gannett, First Church, Boston, 8 August 1858, as reported in an 1858 article in the *North American Review*, 543–544.)

Married, August, 1858, Europe to America. They were married by lightning, and "by thunder" they shall never be divorced

The Union of
ENGLAND AND AMERICA.
In place of the wedding
Ring, the Atlantic Cable

REJOICE ALL NATIONS
Married,
On Thursday, August 5, 1858,
In the Church of Progress,
At the altar of Commerce,
The Old to the New World.
May they never by divorced!

Placards in New York Parade, 1858, described in McClenachan (1863), 47–48.

Since the discovery of Columbus, nothing has been done in any degree comparable to the vast enlargement which has thus been given to the sphere of human activity. . . . The Atlantic Telegraph has half undone the declaration of 1776, and has gone far to make us once again, in spite of ourselves, one people. [London *Times*, cited in Carter's *Cyrus Field: Man of Two Worlds* (1968), 165.]

I confidently anticipate that when that terrible cloud which now hangs over the American continent is dispelled, and with it those fitful flashes of angry feeling which sometimes break through that cloud against this country, and when we are bound together by this intimate means of communication that natural affection will revive which ought to exist between the parent country and its children and cousins in America. [James Stuart Wortley's speech at the 7th annual meeting of the Atlantic Telegraph Company's Board of Directors (1864), 20.]

To lay the Ocean Cable
The Yankees proved quite able,
John Bull was not behind
Valentia Bay to find,
And thus the thing was done
By John and Jonathan.

The brothers need no bulwarks
No towers along the steep;
Our march is on the mountain wave,
Our lightning through the deep.

THE LAST NAVAL ENGAGEMENT BETWEEN
COLUMBIA AND BRITANNIA—A TIE

Parade placards, New York, 1858, as reported in McClenachan (1863), 39, 45, 191.

What saith the herald of the Lord?
The World's long strife is done;
Close wedded by that mystic cord,
Its continents are one.

[One stanza from John Greenleaf Whittier's cable hymn. Composed in 1866. Quoted from Carter (1968), 252.]

. . . promises fair to hasten the coming of that time when all men shall dwell together in unity, and wars and rumors of wars shall cease to be. [Introduction to the McClenachan Report (1863), ii.]

In this view, will not all nations of Christendom spontaneously unite in the declaration that it shall be forever neutral, and that its communications shall be held sacred in passing to their places of destination even in the midst of hostilities. [Henry Auricularius, a New York councilman, speaking at the celebration; cited by McClenachan (1863), 131.]

It is obvious that this "bond of peace" must be to an incalculable degree tightened and strengthened by the lightning-winged words which will daily cross and recross the ocean-path now opened to them. In addition to this, the ease with which accurate information can be obtained, the precedence of which we have spoken of fact before magnified rumor, the ease with which explanations may be made between governments, the facility for the settlement of incipient feuds before they have engendered anger, retaliatory speech, and vindictive purposes, must have a most momentous and blessed agency in hastening the time when men shall not "learn war any more." [From the *North American Review*, October 1858, 540–541.]

. . . the harbinger of union and brotherhood amongst nations and one of the chief agents in an enterprise which is destined to do more towards the realization of a millenium of love amongst men than the efforts of all the diplomatists and missionaries are ever likely to accomplish. [A comment on the cable in 1858 by a New York paper, cited in Isabella Judson's biography of Cyrus Field (1896), 90.]

Politics and war. Whatever their commercial interest in the cable, the Victorians continually reiterated one great expectation about submarine telegraphy: that it would mean brotherhood and peace, first between the great English-speaking nations who gave it to the world, and ultimately between all nations of the earth. In spite of their experience in subduing native peoples and establishing and managing colonies, the Victorians were inexperienced with impersonal, but interactive, electric communications: they had only their brief experience with telegraph lines in their own countries to guide them. Their simple, and seemingly universally shared, assumption was that from easy communication comes increased understanding, and that understanding necessarily produces sympathy, agreement, brotherhood, and peace.

From the first, Cyrus Field was determined that the transatlantic cable should be a joint Anglo-American enterprise, and the American investors who first joined him shared this commitment. The prospectus of 1857 said, "The Anglo-Saxon race which is in the van of all social progress, dwells on both sides of the Atlantic. Two thousand miles stretch between Englishmen and their own kith and kin. . . ."[38] Professor Samuel Morse was quoted: when "New York is a suburb of London and Washington the western half of Westminster, an American war will be about as likely a thing as Camberville organizing an attack on Camden Town or Peckham making a raid on Pimlico."[39]

American political leaders, facing the prospect of civil war in which British support was sought by both sides, were not quite so confident. "What security are we to have that in time of war we shall have use of the telegraph as well as the British Government?" asked Senator Hunter, acutely aware that both ends of the cable would lie in British territory.[40] And he warned, "We are contributing, and contributing largely, to place in [England's] possession the superiority . . . over us in time of war. . . ."[41] Other senators asked why the U.S. should support "a mail service under the surveillance of Great Britain,"[42] or aid Britain's relentless march for hegemony.[43]

Soon, like Orion's belt of fire,
  Its broad electric arm shall hold
With all a monarch's strong desire,
  The world and all its varied fold!
And from its tongue, through every sphere,
  Till time and earth together cease
Mankind, the glorious tale shall hear
  Of commerce, brotherhood, and peace.

[Stanza of poem, E. J. Reilly, presented in Charles Bright's *Submarine Telegraphy* (1898), 183.]

But a more optimistic view prevailed even in the U.S. Senate. Said Senator Seward:

My own hope is, that after the telegraphic wire is once laid, there will be no more war between the United States and Great Britain. I believe that whenever such a connection as this shall be made, we diminish the chances of war, and diminish them in such a degree, that it is not necessary to take them into consideration at the present moment.[44]

And other Senators agreed: Senator Thomson believed that the cable would "bind the two countries together literally with cords of iron that will hold us in the bonds of peace."[45]

Sermons and religious writings were full of rejoicing at the new bonds between English-speaking nations (in Protestant churches, at least, this was equated with "Christian nations"). "Our nation is clasped in the embrace of friendship by our former enemy. . . In this we see the power of Providence. . . and the two great branches of our race instruments in His hand," wrote the Mayor of Savannah to the Mayor of New York.[46] Clergymen were not content with the analogies of friendship or paternal affection; said the Reverend Dr. Ogilby: "England and America have been wedded by the Atlantic ring—a ring of love—a ring of peace—shall I not say, a ring of God? . . . . Those whom God hath joined together, let no man put asunder."[47]

Not only Anglo-Saxon unity but world brotherhood and peace were widely, and constantly, predicted. Cyrus Field's expectations were based on a practical hope that dangerous crises, such as the *Trent* affair, could be nipped in the bud before they developed an impetus of popular indignation, and many practical men of affairs agreed. "To avoid these misunderstandings would be one great step taken towards universal pace," wrote Peter Cooper.[48]

The moral sense of the world, rendered more delicate by the rapid circulation, along this great artery, of the religious sentiment of Christendom, will be shocked at the ruffian nation which shall rely upon powder and lead to enforce respect and deference for its rights. On the 5th of August, 1858, an era was established which divides the past from the future. The order was then given to all civilized nations to ground arms against Christian man, and that, henceforth, the engines of war for man's destruction should be employed only against the heathen and the barbarian. [Letter, Secretary of the Interior J. Thompson, cited in McClenachan (1863), 204.]

It is often amusing to see how sensible men, and sometimes entire communities, are led astray by words, especially figurative words. This telegraph wire is called a Cable, and the use of a Cable is to bind, and therefore the countries between which it lies, are, and will remain bound together in peace. This sort of running logic would not be tolerated any where but in prophecy. [From the *Southern Review* (1867), 45.]

Senator Toombs warned that it was dangerous to expend public monies on the pretext that the cable would bring perpetual peace and civilization when the "best insurance for peace is preparation for war."[49] But the feeling that the cable could "produce a new era in the world, a new feeling, a new sentiment among mankind,"[50] was almost universal. How could war exist, when communication was to establish "a common, a right, a happy understanding among all nations, tribes, climes, and interests . . ."?[51]

Again, a simple set of commonly accepted assumptions was operating: commerce and material wellbeing, Christianity, and western civilization are inherently bound together as part of Divine Providence, which must ultimately result in universal peace and brotherhood. "We are no longer to concede," said the Reverend Dr. Bellows, "that the business of the world is necessarily at war with the morality of the world, or to allow that a lasting internecine strife can go on among the various interests of man or the race."[52]

This hope was, of course, based on the realization that the transatlantic cable would be only the first of many which would connect all continents. " . . . [T]he extension of telegraphic communications in all parts of the world—thus annihilating space and drawing the nations together in close bonds of brotherhood—promises fair to hasten the coming of that time when all men shall dwell together in unity, and wars and rumors of wars shall cease to be."[53]

We have messages of peace and love for out brethren of the extreme West, but our hearts pant with burning words for the encouragement of our heroes in the East. If it be our mission to civilise the dark abodes of cruelty and anarchy in the East; if the empire of the East, as well as the Western Indies, is to be preserved for our rulers; if justice and peace are to succeed lawless rapine and demoniac strife, the electric cord which shall unite England with her Indian Empire cannot be longer neglected.

We desire, as the most ardent wish of our hearts—a wish which will find an echo in every Englishman's breast—that peace and good may be most speedily restored to our Indian Empire. We also hope that the sun will not run through another circle of the heavenly signs before the lightning words may flash through the Atlantic, Mediterranean, and Indian Oceans the angelic salutation of peace and good-will to the swarthy children of the sun; that Queen Victoria may have the intense gratification of sending with lightning speed the words of peace and pardon to her deluded children of the far East. [From the *London Daily News* (1865).]

Some influential observers, of course, were not ready to go quite as far as "all nations." U.S. Secretary of the Interior Thompson, for example, wrote that when on 5 August 1858 the first cable began its briefly successful operations, "The order was then given to all civilized nations . . . that henceforth, the engines of war for man's destruction should be employed only against the heathen and the barbarian."[54] The British in particular sometimes set their aspirations for universal peace in a context of their imperialistic drives, which had in them a strong component of missionary zeal and moralistic fervor. "If it be our mission to civilize the dark abodes of cruelty and anarchy in the East," the *London Daily News* editorialized, ". . . if justice and peace are to succeed lawless rapine and demoniac strife, the electric cord which shall unite England with her Indian Empire cannot be longer neglected."[55]

Not all Victorian observers were carried away by the optimism of the majority. The *Southern Review*, a thoughtful journal of opinion, mocked the unchecked sentiment which equated a telegraph cable with wedding rings and umbilical cords: "This sort of running logic would not be tolerated anywhere but in prophecy."[56] Rapid communications, the *Review* warned, might serve only to magnify political crises. "England and France have been telegraphically bound together for years in a much firmer connection [the Channel cable of 1837]; yet the international feeling is one of chronic distrust enlivened by acute paroxysms of terror. . . ."[57]

The British cabinet members who supported the cable foresaw that it would be of enormous

How can we bring before our minds, with sufficient distinctness, the consequence of a universal interchange of thought by the speediest method? The world, it has been said, will be made a great whispering-gallery; I would rather say, a great assembly, where every one will see and hear every one else. The press has for the last fifty years been the chief agent in educating society. It must now share that privilege with another instrument. The telegraph will anticipate the journal. Facts and opinions, the materials with which successive generations construct the road along which society advances with unequal, but never retrograde steps, will now be furnished to every one on the moment. [Dr. Ezra Gannet of Boston's First Church, reported in the *North American Review* (1858), 543–544.]

But I see, or think I see, it is the Printer who is to win the day, and through him, the universal Respublicas. The Electric Telegraph is but another way of printing. Its business is Letters. It lives on words, thoughts, ideas; and is a pure spirit—in nothing material. Through it and by it, our American thoughts, American principles, American precedents are to flash daily into the dusky rooms of St. James, the glittering halls of the Tuilleries, the majestic grandeur of the Escurial, the Austrian Schönbrunn, and the Roman Vatican. [James Brooks, a journalist, speaking at the New York Celebration; McClenachan (1863), 263.]

benefit in administering their far-flung colonies as well as in their defense. American observers agreed, although their outlook was somewhat different. Senator Pugh and others warned that the cable would help Britain maintain her control over Canada and end the lingering hopes of Canadian independence.

No observer specifically forecast (so far as we have found) that telegraphy might, by increasing the political sophistication of colonized peoples, some day contribute to the breakup of empires. But many Victorians went almost that far—they perceived the cable as a great force for education and civilization, and education itself as an agent for the spread of liberalism and social democracy. The press, said the eminent divine Dr. Ezra Gannet, "has been for the last fifty years the chief agent in educating society." The telegraph, he expected, would be an equally powerful force. "Facts and opinions, the materials with which successive generations construct the road along which society advances . . . will be furnished to everyone on the moment . . . and improvement will go on steadily, as well as rapidly."[58]

His fellow minister, Henry Ward Beecher, added: ". . . If it were possible for monopolists to lock up the two ends of this wire, holding on either side the power of its intelligence, it might be disastrous to governments and to the people. But . . . it has a tendency to make knowledge co-extensive."

Victorians in both England and America had a strong belief that knowledge brought with it virtue, a sense of responsibility, and a love of liberty—this indeed might have been their definition of civilization.

> The miracle of this age of miracles had been wrought! What mind can conceive the mighty effects to flow from this practical demonstration of the power of man over the subtle element? Who can foretell the changes it is destined to work in the polity of the world? And, will it not mark the commencement of an era of the most serious revolutions—revolutions that through its means will, in all probability, be as bloodless as they will be important and startling by their rapidity and extent! [Introduction to the McClenachan Report (1863), 7.]

If civilization, then, is to be brought by the cables to the darkest corners of the world, mighty forces are unleashed. "Hence then comes a mighty conflict of antagonistic principles—an equalization of thought, the fraternity of mankind," wrote James Brooks, a journalist. "Something is to topple over—but whose that something is, no Cassandra . . . can predict."[59] And the New York Commemoration Committee in 1858 asked, "Who can foretell the changes it is destined to work in the polity of the world?"

Some 19th century observers did step down from these heights of optimism and warning to comment on more specific implications of the cables for international relations. U.S. Congressmen predicted that the cable would reduce the costs of maintaining a large diplomatic service, although others argued that it would be impossible to use the cable for diplomatic purposes because of the lack of security and privacy.[60]

Diplomats themselves argued these points. The Select Committee on Diplomatic Service of the British Parliament held discussions on the subject in 1861. Distinguished career diplomats testified that the cable connections would be useful to diplomats, but not an unmixed blessing, and on the whole they were inclined to think that the impact on the service would not be great. Relying on telegrams, they had already found, meant relying on scanty information that often proved at variance with information arriving in dispatches.[61] All subtlety was lost, and there might be many mistakes in transmission. Increase in the speed and volume of messages, diplomats correctly foresaw, would place a tremendous burden of work on their missions, which they had not the staff to handle.[62]

Though outside observers, including Congressmen, might expect that the diplomatic corps could be abolished and Heads of State communicate directly with one another, diplomats had no such fears. There might be, eventually, some diminution in their authority, some recognized, but as one Foreign Officer noted, ministers exist to prevent the necessity for written communications.[63]

The Atlantic Telegraph: An artery through which the living blood of Liberty shall be infused through all nations over the sea. [New York parade placard, 1858, cited in McClenachan (1863), 46.]

THE ATLANTIC CABLE—A RAY FROM THE SUN OF LIBERTY, DESTINED TO ILLUMINATE THE GLOOMIEST RECESSES OF THE WORLD. [New York parade placard, cited by McClenachan (1863), 101.]

William Russell in the *Fortnightly Review* pointed out that direct communication between those ultimately responsible for action was not much to be desired:

Mr. Adams very recently expressed his gratification at the absence of an Atlantic cable, during the civil war in the United States. Perhaps he feared a Sewardism so pungent that it might any day be taken as a declaration of war.[64]

News. It was obvious to everyone that the news media would make immediate and great use of the cables. Paul Julius Reuter, whose news service had been built around the early landline telegraphs, told Field in 1857 (when Reuter accompanied him on a visit to the Prime Minister to request support for the cable) that one cable would not be enough—Reuters alone, he predicted, would guarantee £5000 business a year.

But newsmen, though they competed to use the cable, were not entirely sanguine about its effects. Mowbray Morris of the London *Times* predicted that it would be the most frivolous, irrelevant, and trivial news reporting which would benefit most from speed.[65] (To be sure, the position of *The Times* as arbiter of international news was unchallenged until the cable spawned international news services.) Some others agreed: Thoreau predicted that the news provided by the cable would, all too often be that "Princess Adelaide has the whooping cough."[66] We have already pointed out, however, that on the broader level of "intelligence" the Victorians had a strong sense of the political forces that might be unleashed by broad and rapid access to worldwide information, and they explicitly connected this development with the spread of democracy.

Religion. The clergy of America and Britain preached untold numbers of sermons about the cable. Many of their comments have been quoted elsewhere: they expected the cable to carry education, civilization, progress, and ultimately democracy to all nations. In part the cable's beneficent influence would work through expanding trade relations which would spread material wellbeing and prosperity. But they also expected that telegraphy would be a direct harbinger of the gospel, carrying the messages of religion (Christian, Protestant religion, of course) into lands "where the living missionary dare not . . . utter the same, except at the immediate peril of his life."[67]

Thus the Reverend Dr. Ogilvy predicted that ". . . man shall speak to man, from the one end of the world to the other, of the gospel of the Savior and of the glory of the Lamb. . . ."[68] And Dr. Nott of Union Theological Seminary prayed that through the cable ". . . the way be prepared for the diffusion of the knowledge of the Gospel of peace and the introduction of the reign of the Son of God."[69]

Science. Science, and "progress"—with which science was almost synonymous to the Victorians—would be one of the greatest beneficiaries of the new technology. The spread of information and knowledge through use of the cable would stimulate scientific thinking everywhere. The cable itself was hailed as another triumphant demonstration of the applicability of scientific

. . . The Queen, the Czar, the Emperor, must use it as they use the air all breathe. Hence then comes a mighty conflict of antagonistic principles—an equalization of thought, the fraternity of mankind. Something is to topple over—but whose that *something* is, no Cassandra at a dinner-table like this can predict. [Journalist James Brooks in McClenachan (1863), 273.]

But it is to be observed that newspapers and diplomatists are rather afraid of telegraphs, and the latter much mistrust an agency which gives a meaning to words and takes so much gilding off political gingerbread. Mr. Adams very recently expressed his gratification at the absence of an Atlantic cable, during the civil war in the United States. Perhaps he feared a Sewardism so pungent that it might any day be taken as a declaration of war. Lord Lyons certainly reciprocated the feeling of the American minister; and it is not to be denied that ambassadors and ministers would all cut their wires in secret if they could,—as surely as post-admirals or naval officers would so the same, for they have had no rest since the Admiralty got hold of electricity. [An article by William Russell entitled "De Profundis" in *Fortnightly Review* (1865), 330.]

It is supposed that every Cabinet of human government has a special department which is not honest, but which comes under the head of "Intriguing Diplomacy." Your Cable will put an end to this special department, and it is to be hoped that no cabinet minister, no private individual, no missionary, no merchant, on either side of the Atlantic, will be so profane as to tax Heaven's lightning, which is to abide in your Cable, with the message of a lie. . . . The tranquillity of its repose will be almost sufficient to forbid the transmission of angry messages, or, if transmitted, it will mollify the asperity of temper in which they were dictated. And thus it will silently suggest to thinking men that they should rise above the moral atmosphere in which human passions exercise their violent and destructive influence. [Henry Auricularius, speaking at the New York Celebration; McClenachan (1863), 135–136.]

We are in great haste to construct a magnetic telegraph from Maine to Texas; but Maine and Texas, it may be, have nothing important to communicate. Either is in such a predicament as the man who was earnest to be introduced to a distinguished deaf woman, but when he was presented, and one end of her ear trumpet was put into his hand, had nothing to say. As if the main object were to talk fast and not to talk sensibly. We are eager to tunnel under the Atlantic and bring the old world some weeks nearer to the new; but perchance the first news that will leak through into the broad flapping American ear will be that Princess Adelaide has the whooping cough. [Henry David Thoreau, *Walden* (1961), 67.]

Christianity is the basis of civilization. The nations that have rejected it from the beginning remain in *statu quo*. They have not advanced. Other nations that have expelled after having once received it, have invariably relapsed into their primitive stupidity. The electric Cable may be made, in a certain sense, the harbinger of Christianity to all peoples. It never can be the messenger proper. But it will announce or insinuate in a thousand ways the truth of revelation, in lands and amid communities where the living missionary dare not, in the first instance, utter the same, except at the immediate peril of his life. [New York Alderman James Davis, speaking at the New York Commemoration, cited in McClenachan (1863), 134.]

. . . May it be instrumental in bearing only the messages of peace, extending the glad tidings of salvation, the gospel of thy dear Son, and hastening the day when from every corner of the earth shall rise that blessed song, "Peace on earth, good-will towards men. [Prayer at New York City Celebration, 1858, cited in McClenachan (1863), 70.]

Let the joy of the world in rich harmony rise.
   Let the sword keep its sheath, and the cannon its thunder;
Now intellect reigns from the earth to the skies,
   And science links nations, that war shall not sunder!
Where the mermaids still weep, and the pearls lie asleep,
Thought flashes in fire through the fathomless deep;
Now mind reigns triumphant where slaughter has been!
Oh, God bless our President! God save the Queen!

[Hymn sung at New York Celebration, cited in McClenachan (1863), 10.]

. . . the celebration of an event which must so quicken the pulsations of the world of mind, and so connects the nervous texture of the great family of man that it must ultimately, and at a period not very remote, unite in one common brotherhood all the civilized nations of earth, and rapidly hasten the day when the arts and sciences shall be cultivated throughout the habitable globe. . . . [Letter from Gov. Stewart of Missouri to New York City in commemoration of the laying of the cable; cited in McClenachan (1863), 212.]

. . . We may well ask what class of men, what departments of busy life, what interests of literature, or science, or religion, what nation, however remote, what point of time in the future of the world, are not to be directly or indirectly affected by it? We are lost in the calculation of its results to the present and the future! It seems to us, to be *the benefaction of science to the world and all ages of the world*. [Henry Auricularius, New York alderman, in McClenachan (1863), 127.]

Among the important results of the oceanic telegraph we must name the easy detection of fugitives from justice, and the added difficulty and hazard thus placed in the way of crime. The tendency of all modern science has been in this direction. Chemistry, by its unerring tests, may detect vestiges of the murder in the decomposed body of its victim, and proofs of the murderer's guilt in the slightest discoloration of a garment or a blade. Photography may furnish transcripts of the features of a thief or a swindler for the picture-gallery of every police-office in Europe and America. And now he who would shun exposure and infamy may "take the wings of the morning," yet shall hardly have lost sight of his native shore before Transatlantic ministers of the law are preparing for him an ungenial welcome. [From an article in the *North American Review* (1858), 539–540.]

principles to practical affairs. It was described as "the triumph of science and energy over time and space."[70]

The Governor of Missouri predicted that the cable would "rapidly hasten the day when the arts and science shall be cultivated throughout the habitable globe. . . ."[71] And the Governor of Massachusetts said:

> . . . All the great inventions and discoveries of the last three centuries have contributed incalculably to the elevation of industrial pursuits. They have enlarged the resources of men and enabled them to supply their constantly increasing wants. Humanity keeps pace with material prosperity, and the development of great truths in the material world seems to contribute as directly to the equality of men and the welfare of the human family as the highest progression in institutions of government. As it has been in other cases it will be in this, the greatest of all.[72]

It was left to Lord Napier of Britain to deliver a solemn warning, which somehow sounds more like the mid-20th than the mid-19th century:

> . . .we shall do well to remember that the magnetic telegraph forms no exception to the category of inventions which, however apt and proper, and willing to be the vehicles of benevolent designs, are also the unresisting tools of every blind or intemperate impulse in our nature.[73]

In general. In the more sober journals of opinion, there were sometimes forecasts of specific uses of the cable that sound remarkably practical, and accurate, after the sweeping promises of most commentators. The best of these articles appeared in the *North American Review* in the fall of 1858. It forecast in sober terms the growth of responsible news services and market reports, very likely under "competent authority" to assure their veracity and general access; equalization of world commodity prices; and international cooperative organizations for the scientific detection and apprehension of criminals.[74]

The potential users of the cable, especially the businessmen—who were also the potential investors—tended to be most skeptical about its promised benefits, at least in America. Some of them had already discovered that landline telegraph, together with the railroads, was transforming their way of doing business, exacting costs along with benefits—demands for speed, efficiency, control, and more complex management structure. Political leaders also were cautious, scenting risks as well as opportunity in anything which could upset delicate balances between nations, regions, interests, or social classes. The general public, on the other hand, was wildly enthusiastic (in America) or smugly proud (in Britain), their view of the cable as a triumph of Anglo-Saxon genius and an unsullied promise for mankind was indeed fostered and fed by the excessive exultation of the clergy, the press, and the opinion leaders of the day.

They were not concerned with practical detail, or accurate forecasting. Perhaps they grasped the essential fact about the new technology—that it was unique and awesome because it fulfilled a dream mankind had dreamed for countless generations—to be able to talk back and forth across the barriers of time and space. For whatever mundane, even selfish, purposes that capability might be used by practical men, its potentialities were so limitless that they could fittingly be described only in grand and mystical images.

For this the Victorians—especially the clergymen who were the social commentators of the age—ransacked their vocabularies:

> . . . we now live on a globe . . . not as scattered members, but as connected limbs of one organic body . . . vitalized by a spinal cord animating all its limbs, harmonizing all its movements, and giving coherency, energy, and unity to its whole being.[75]
>
> . . . the edict of Babel is reversed—the dispersion of the nations is to be undone, in God's time and in God's way, by bringing them together as one in Him . . .[76]
>
> . . . the renewal and repetition of the Pentecostal wonder . . .[77]
>
> . . . these traveled lines of thought to be stretched over continents and islands and beneath the waters of the sea. . . .[78]
>
> . . . A consecrated lightning, consecrating the very ocean through which it travels, consecrating this glorious, blessed day . . . .[79]
>
> . . . God allowed us to call the lightnings of Heaven and employ them as our messengers . . . .[80]
>
> . . . that silent cord lying in the sea, perfectly noiseless, perfectly undisturbed . . . sublime, itself impassive, quiet, still, moving either hemisphere at its extremities by the tidings that are to issue out from it.[81]
>
> . . . an influence more important and beneficial than the armies and fleets of a hundred kingdoms . . . .[82]
>
> . . . a network of electrical nerves that shall enwrap the world.[83]

Whether the images are mechanical, anatomical, or celestial, they express wonder, awe, and pride at what man has done. And for Americans there appears a strong sense of the beginning of a new role for America in world affairs, born of pride in American industry and science.

Ballyragget, Ireland
September 1, 1858—3½ o'clock, P.M.

To Wm. Eglington, 140 Fulton Street, New York

Faith, Bill, I'm very glad to hear that you are well, and able
To send a graphic sketch to say so through the cable.
No cows here have got the "murrain," no sheep have got the rot;
Potatoes they do flourish now, throughout the bright green spot.
I like to hear from all of ye, but I'd like to see you better;
Send back a toast by "cablegraph," and the liquor in a letter.
But won't it be a glorious thing, when science will be able,
To let us hear your own sweet "brogue" by electrographic cable.
Dam vivimus bibamus—[While we live let us drink.]

CHAS. COLTER

Answer

For your telegraph dispatch I paid cheerfully $20.

W.E.

[A "transparency" used in the public celebration, in New York, cited in McClenachan (1863), 40.]

Does it seem, I say, all but a miracle of art, that the thoughts of living men—the thoughts that we think up here on the earth's surface, in the cheerful light of day—about the markets and the exchanges, and the seasons, and the elections, and the treaties, and the wars, and all the fond nothings of daily life, should clothe themselves with elemental sparks, and shoot with fiery speed, in a moment, in the twinkling of an eye, from hemisphere to hemisphere, far down among the uncouth monsters that wallow in the nether seas, along the wreck-paved floor, through the cozy dungeons of the rayless deep; that the latest intelligence of the crops, whose dancing tassels will, in a few months, be coquetting with the west wind on those boundless prairies, should go flashing along the slimy decks of old sunken galleons, which have been rotting for ages; that messages of friendship and love, from warm, living bosoms, should burn over the cold, green bones of men and women, whose hearts, once as warm as ours, burst as the eternal gulfs closed and roared over them centuries ago! [Edward Everett, a trustee of the Boston Public Library, opening of the Dudley Observatory in Albany in 1857. Reported in H. M. Field's *Story of the Atlantic Cable* (1866), 373.]

Yet the (speed of transmission) has a very curious interest, a showing plainly how our notion of time is not absolute but only an idea we get from comparing one length of it or another of it with one whirl or one swing of the globe. [From an article in *Galaxy* magazine (1966), 89.]

I thought all the way in riding down here tonight how strange it will seem to have that silent cord lying in the sea, perfectly noiseless, perfectly undisturbed by war or by storm, by the paddles of steamers, by the thunders of navies above it, far down beyond all anchors' reach, beyond all plumbing interference. . . . To me the functions of that wire seem, in some sense, sublime: itself impassive, quiet, still, moving either hemisphere at its extremities by the tidings that are to issue out from it. [Sermon by Reverend Henry Ward Beecher. Quoted in Carter (1968), 166.]

. . .And true to His great plan, God allowed us to call the lightnings of heaven, and employ them as our messengers. This wondrous agent was known to Him from the beginning, and like the gold was held in reserve for the men and the times. Did he not speak of it to Job, when He challenged him out of the whirlwind: "Canst thou send lightnings that they may go, and say unto thee, Here we are?" [Rev. Anderson, in a sermon in San Francisco (1858), 7.]

. . . A consecrated lightning, consecrating the very ocean through which it traverses; consecrating this glorious, blessed day; consecrating anew that time-honored flag of England—the banner of a thousand fights; consecrating the stars that glitter on the flag of freedom, which, in less than a century, has won for this nation its place among the ancient empires of the world, and which, wherever the rights of men are to be asserted, forever floats and blazes in the van. Consecrating, shall I not say, beloved friends, consecrating anew our hearts to the love of man and to the glory of the living God. [A sermon cited in McClenachan (1863), 72.]

C. Prediction and Forecasting, Mid-19th Century

Prediction is an assertion about, or description of, future events and trends. When tied to a specific chronological period in the future, predictions are usually called forecasts ("a forecast of energy supply and demand, 1975–1990"), but the two terms are often used interchangeably.

Prediction seems to have been a part of nearly all cultures although the idea of "progress" over time is peculiarly modern.[84] In modern times predictions are generally based on "observed regularities among consecutive events of the past,"[85] that is, on a developing pattern of events—and usually on an explanation of the causal factors in such a pattern. In earlier or less "scientific" cultures, however, prediction was often based on either the attempt to create a sefl-fulfilling prophecy by influencing the mysterious forces which controlled human destiny, or the attempt to give guidance (policy advice) to decisionmakers—either consciously or unconsciously disguised as divine revelation or the outcome of a religious ritual.

This traditional, mystical, and religious form of prediction perhaps persists in modern times in the form of revelations by seers and cult leaders. Such revelations may be sanctified by appeal to religion or mysticism or by an appeal to extrasensory (metascientific) sources of information. All such prediction rests not on interpretation of collected data or on a rational deduction but on belief that the predictor has a source of knowledge that is direct and personal and must be accepted on faith.

A third line of development of prediction lies between the "prescientific" revelation and the "scientific" extrapolation of the past.

Such predictions consist of interpretations of history—rational arguments propounded by thinkers such as Malthus, Smith, Marx, and Toynbee. They are based, at least formally, on observation and explanation of alleged patterns or evolutionary trends, but not necessarily on systematic data collection and extrapolation. The prediction rests ultimately on the persuasive power of the explanatory treatment, rather than on empirical data, which are usually treated in highly generalized or abstract terms.

Modern prediction and forecasting "had to wait the development of empirically oriented research and statistical/mathematical tools."[86] The words "statistics" and "statistical analysis" were introduced into English about 1790 from the German, and are derived from the word for "state."[87] Sir John Sinclair in his Statistical Account of Scotland (1798) defined his techniques as "an inquiry into the state of a country for the purpose of ascertaining the quantum of happiness enjoyed by its inhabitants and the means of its future improvement. The word "future" is important here, although for some time statistical analysis was viewed more generally as a possible way of "improving" the future (otherwise assumed to be much like the past and present) than as a technique for predicting the future. Sir John, for example, developed a system of "County Reports" for Britain as a way of increasing agricultural production.

Adolphe Quelelet, a Belgian professor of mathematics, organized the census of the Netherlands in 1829. From this experience and his collection of statistical information on Belgium he developed a deep interest in "moral statistics" or attempts to quantify and record psychological and sociological facts, to describe the "average man," and to develop "social physics" or "laws" of marriage, suicide, crime, etc. Such "laws," of course, in the first instance descriptive, could also be used in a predictive mode—the "experience tables" used by modern sociologists to "predict" (in the sense of assigning probabilities) individual behaviors.

The U.S. government had necessarily been involved in collecting "statistics" from the beginning because of the Constitutional requirement adopted in 1787 for allocating seats in Congress on the basis of a decennial census—the first regular periodic census ever legislated.[88] However, census figures were collected in terms of families (rather than individuals) by untrained census takers, whose procedures varied considerably from place to place, and the census was a somewhat hit-or-miss head count until the census of 1850. In 1850, organized by James D. B. De Bow and Lemuel Shattuck, data collection was systematic, categories were rationally defined, information was gathered by individuals rather than about families, and many new kinds of information were recorded. These data allowed the insurance companies, which burgeoned between 1840 and 1870, to construct actuarial tables (a form of prediction) and allowed new kinds of analysis of, for example, health, life span, and income. Hinton R. Helper used the census data of 1850 to write *The Impending Crisis in the South* (1857) around the argument that slavery was gradually impoverishing small farmers and the South in general. This approach was perhaps somewhere between the "grand thesis" type of prediction and empirical extrapolation.

> **We hold the ball of the earth in our hand, and wind upon it a network of living and thinking wire, till the whole is held together and bound with the same wishes, projects and interests. [Senator Latham, as reported in the Communication of Hon. William H. Seward, Secretary of State, on the subject of an intercontinental telegraph connecting the Eastern and Western Hemispheres by way of Bering Strait (1884), 44.]**

In the discipline of economics prediction is often dated from Clément Juglar's work on business cycles.[89] In 1856 the (French) Academy of Moral and Political Sciences offered a prize for an essay on commercial crises. Juglar, a physician who had become interested in the new activity of collecting statistics on disease and health, won the contest in 1862. Before Juglar crises had been explained individually on the basis of, for example, bad harvests, wars, or excessive issuance of paper money. Juglar constructed a history of commercial crises by putting together time-series data on consumer prices, interest rates, gold prices, central-bank balances, etc., one of the first examples of attacking a problem of prediction by beginning with empirical, systematic collection of data. He observed periodicity in these time-series and ultimately explained crises in the context of larger cycles of economic activity. Juglar concluded that such crises were natural, unavoidable, and desirable—"natural purgings." All later work on business cycles and much later economic prediction stem from Juglar both in terms of theory and in terms of statistical methodology, but these methods developed slowly.[90] Ernst Engel, another pioneer, developed cross-section data techniques on workmen's wages, prices, etc. (1883); still later (about 1914) economics began to borrow more sophisticated techniques of statistical inference from physics and biology.

Sociological forecasting developed even more slowly. Sociology did not emerge as a distinct discipline in Europe until the last part of the 19th century, still later in America. However, social research had begun (as "political arithmetic") in the early part of the 19th century with the growing interest, described above, in collection and quantification of social data relevant to governmental decisionmaking. Reiss[91] relates the new interest in social statistics research to the differentiation of civil from political sectors and the disassociation of religious/moral thinking from social thinking beginning with Hobbes, Rousseau, and Locke. Collection of statistics soon allowed the construction of "experience tables" in which individual experiences or behavior are categorized in terms of race, age, education, income, family status, or other factors, which allow one to assign probabilities to other individual outcomes—i.e., predict behavior. Durkheim is usually mentioned as the father of modern sociology both in abstract conceptualization and in acceptance and development of methods of empirical research, from which came modern predictive techniques. In modern sociology, Scheussler points out, "predictions ideally are to be drawn from theory [that is, from understanding or postulation of causal relationships] but for the most part they have been little more than statistical projections based on compilations of empirical data within categories of little theoretical significance."[92] Such projections, however, may give rise to new theoretical explanations. Thus Ogburn, in 1922, based his hypothesis of cultural lag on (empirical) growth curves of inventions.

Thus the techniques most often used in technology assessment (economic forecasting, social forecasting, technology forecasting) were only beginning to develop in the early and mid-19th century; the relevant disciplines had just disassociated themselves from philosophy and religious and theological thought; and systematic data collection for governmental purposes was also recent. This was true not only on a societal scale but to some extent on an individual scale. Boorstin points out in another connection that "until well after the Civil War it was the rare merchant who knew the precise nature of his income"[93] (or his inventory) and such indispensable aides to gathering and analyzing data as adding machines were not invented until the 1880s.

National income figures became available after 1900. Obviously records were kept—the commercial empires of earlier centuries were not possible without record keeping—but they were not often compiled in comparable categories or through compatible techniques and were not easily gathered, compiled, or compared for any but the most immediate, direct purposes.

Moreover, since both transportation and communication were slow and expensive, information (and experience) was probably not nearly so widely shared; hence perception of recurring patterns of experience—for example, pollution from industrial growth, employment patterns, crime rates, family disruptions, etc.—may have been difficult. (On the other hand, rapid and profound changes were occurring—such as concentrated industrialization in the midst of an agricultural society—which may have been for that reason more highly visible than they are today when society as a while is more diverse, fluid, and interdependent.)

The impacts of the cable on marketing, shipping, arbitrage, and news gathering were well within the daily experience of large numbers of people (merchants, traders, diplomats, writers) but only on the individual, nonsystematic and highly generalized basis we have described earlier in this chapter. More subtle, third- and fourth-order social effects (how people, or political processes, change their behavior when they have more immediate and colorful knowledge about events in other countries) are perhaps only slightly more predictable today, since social science is still fairly rudimentary. Some areas of knowledge which have progressed fairly rapidly since the mid-19th century—for example, second- and third-order ecological effects—were not critical in the case of submarine telegraphy.

Although the present research team attempted to test directly some hypotheses that might have been formulated had a technology assessment been undertaken in the 1850s—for example, forecasting the reduction in fluctuation of stock and commodity prices—we found that necessary data either were not collected, or were not preserved (for example, daily price fluctuations before and after the cable). That is natural enough; since the question was not asked, the data to supply the answer were not collected. This is a clear example of the chicken-and-egg problem: data suggest new hypotheses (questions), which require collection of more data for testing.

We therefore tentatively conclude that a formal exercise in forecasting, as part of a technology assessment in the 1850s, would have been frustrated by lack of time-series data and adequate information on long-range trends. Yet because the technology of cable communications was so unique, daring, and highly publicized, it provoked a large volume of social commentary, in which most of the significant impacts of the new technology were "predicted" in general, thoroughly value-laden terms by at least some of the many observers. The extent to which a more systematic, quantitative, comparative, and integrated assessment—had such been possible—either would have or "should" have changed the outcome of public decisions and policy formulation, is considered in our final chapter.

D. The Hypothetical Assessment: An Experiment with Gaming and Delphi

The expectations of Victorian observers of submarine telegraphy hardly constituted social forecasting or technology assessment. They were generally polemical and hortatory, unorganized, unsystematic, and as far as we know not brought together and integrated for the use of decisionmakers (although they were sometimes intended to influence decisionmaking). We therefore asked ourselves, had a "Technology Assessment team" been assembled and put to work in 1861, would they—limited to information available at that time—have anticipated additional or different impacts of submarine telegraphy? That question can never be answered, but two experiments—a gaming workshop and a Delphi—provided intriguing and suggestive insights.

A group of twenty-two experts not previously connected with the study met in January 1976 in response to an invitation purportedly issued in January 1861 by Her Majesty's Joint Committee to Inquire into the Construction of Submarine Cables. They were asked to play the roles of scholars and specialists like themselves in 1861, assembled in the first planning session of a Technology Assessment to be conducted to assist Her Majesty's Government in evaluating the new technology. The team was made up of thirteen persons chosen for their expertise in various disciplines and their knowledge of the historical development of those disciplines, five historians, and four individuals (also from varied disciplines) who were experienced in performing or using present-day Technology Assessments.*

Several weeks earlier another gaming workshop had been held with fourteen graduate students enrolled in a course in Technology Assessment in the Science, Technology, and Public Policy Program of The George Washington University.

*Disciplines and fields represented included business administration, chemistry, economics, education, engineering, foreign service, management sciences, physics, political science, public administration, security intelligence, sociology, and telecommunications. The historians specialized in social history, history of science, maritime history, history of international relations, and American studies.

THE JOINT COMMITTEE
TO INQUIRE INTO THE CONSTRUCTION
OF SUBMARINE TELEGRAPH CABLES

London, January 1, 1861

I have the honour to request your presence to testify and provide evidence before the Joint Committee appointed by the Lords of the Committee of Privy Council for Trade and the Atlantic Telegraph Committee, from 5:30 to 10:00 p.m. on Wednesday, January 14, 1861.*

The Joint Committee has been appointed to inquire on behalf of Her Majesty's Government into the proposed construction of submarine telegraph cables. A technical report on their feasibility, entitled "Report of the Joint Committee to Inquire into the Construction of Submarine Telegraph Cables," has already been prepared. With your able assistance and that of other distinguished scholars and men of affairs, we now seek to explore the broader social implications and possible societal impacts of further construction of submarine telegraph cables and especially of the proposed laying of a cable between Valentia Island, Ireland, and Heart's Content, Newfoundland. We will ask you to address yourselves to the possible profound consequences for the Empire, for Her Majesty's Government, and for Her Majesty's colonies. We should also like you to address the effects in North America on relations with the United States, whose Government and merchants, we anticipate, will be important users of such a cable.

Respectfully yours,
Douglas Galton
Chairman, Joint Committee

*The meeting will be held in Room 730, The George Washington University Library, 2130 H Street NW, Washington, D.C., January 14, 1976, at the time given above. A buffet supper will be provided from 5:30 to 6:30 p.m.

In both workshops, participants were given some background information about cable technology, the plans of the promoters of the transatlantic cable, and the problems which had been encountered up to 1861. They were then asked to generate, through informal and free-flowing discussion (and argument), a list of potential "impact areas" that should be investigated. It was stressed that they should use only as much information as they believed their peers in 1861 would have had at their disposal, and that they should propose any impact, or sequence of developments, which seemed possible—however remote or dubious.

The research team fully realized that it was unlikely that the participants could play out their 1861 roles with any accuracy or realism. The gaming had three purposes. The least critical was to test our own highly tentative ideas as to what questions could have been asked, or possibilities postulated, in 1851 against those of the participants. The second purpose was to glean from the knowledgeable participants substantive information about changes that occurred in the 19th century, particularly in their own areas of expertise. Most important, the workshops provided an opportunity to check against the imagination and insights of others the list of possible impact areas that the research team itself had generated for investigation. The results of the simulations agreed well with the work done by the research team; no significant impacts were suggested that had not already been investigated or listed for investigation. Some participants provided helpful information about additional experts or sources who should be consulted.

The gaming workshops were followed up by an extensive Delphi* questionnaire that went to all

*Delphi is a technique for gathering the informed opinions of experts about a question which cannot be answered from empirical data (e.g., how long will it be before organ transplanting is sufficiently safe and efficacious for widespread use?).In a classical Delphi exercise, divergent opinions and the supporting arguments are fed back to participants, a second round of questioning follows, and the exercise may be repeated. This tends to produce convergence of responses which is considered more likely to be accurate than that of any one expert. Our questionnaire, sent only once, was technically not a Delphi, but was a useful device for systematic questioning of the participants and for getting them to quantify individual judgments.

workshop participants, members of the project's oversight committee, and a few other historians and Technology Assessment specialists.

The questionnaire listed 74 potential impacts (the list generated from the workshops, suitably winnowed and condensed), grouped in five categories: economic, military, political, social, and cultural. (See the Delphi Questions for list of impacts.) Participants were asked to rate each impact, on a scale of 1–10, in relation to five questions. They were to respond to the first three questions "as an informed observer might have responded in 1861": What do you think is the probability of this impact occurring as a result of the cable, or with that as a contributing causal factor? How desirable does the occurrence of this impact appear (from the British perspective, from the American perspective)? How important do you expect this impact will be if it occurs? The last two questions were to be answered from the vantagepoint of 1976: *If you believe that this impact did occur,* to what extent was the cable a causative factor? and how important was the impact in making the contemporary world as we know it?

Participants were thus asked to make six decisions about each of 74 impacts, 444 in all. The fact that 25 questionnaires were returned out of 40 distributed shows the degree of interest and involvement felt by workshop participants and oversight committee, or their commitment to be of service.

As expected, the anticipated probability of an impact occurring as a result of the cable from an 1861 perspective was related on the average to the assessed extent that the cable was a causative factor from a 1976 point of view. This general concurrence of opinion on these two questions could either reflect the inseparability of contemporary and retrospective views or the implicit focus on cable causation in suggesting these 74 impacts during the gaming and workshop sessions. In contrast, there was little or no relationship between the assessed importance of an impact and the belief that the cable caused it.* Since degrees of importance remain a matter of opinion and surveys are a valuable tool for getting at opinions, the independence of assessed causality from relative importance offers hope that Delphi techniques will not necessarily fall prey to "wishful thinking" when used as an assessment tool.

In the workshops, participants had been asked only to generate a list of all possible consequences of the cable rather than to rate them for their likelihood, importance, or desirability. Therefore "TA" or assessment was done unsystematically. However, the Delphi results seemed generally to be in agreement with the tone of discussions during the simulation. The impacts rated most clearly "expectable" in 1861 were: (1) increased domination of trade by the British, (2) an increase in the volume of world trade, (3) greatly improved scheduling of shipping, (4) the growth of large-scale newsgathering organizations, (5) utilization of the cable for military surveillance and intelligence, and (6) increased unity between English-speaking peoples. These impacts were in fact widely predicted in 1857–1866, as we have seen. The second, third, fourth, and fifth did clearly occur, the first occurred for a short time at least, and the sixth is at least defendable. Thus we cannot say whether the Delphi participants were extremely good at putting themselves into the role of observers in 1861, or whether they knew from their own reading of history (they had no access to the project research) some of the consequences of submarine telegraphy and were unable to divorce their game playing from this knowledge.

Impacts rated most *important* from a 1976 perspective were: (1) increased concentration in industry and commerce and growth of monopolies, (2) increase in total volume of trade, (3) strengthening of U.S. position in world trade, (4) increased capability for self-government by colonies and demands for independence, (5) international cooperation in the control of communicable diseases and epidemics, (6) advances in basic sciences (physics, chemistry, oceanography, etc.), and (7) demand for technical, scientific education. Only one impact of the 74 was both among those considered most important and most expected—the increase in total volume of trade. Also considered highly likely and quite important were: (1) growth of large-scale news-gathering organizations (press networks) and (2) improved relationship between Britain and America. Overall these valuations were not surprising to the research team.

---

*Only the six historians showed a bias toward evaluating the more likely impacts as more important.

DELPHI QUESTIONS

ECONOMICS

E1. Increased concentration in industry and commerce, growth of monopolies.
E2. Development of international businesses (''multinationals'').
E3. Development of international cartels.
E4. Ability to manage or suppress depressions before they become worldwide.
E5. More rapid propagation of financial crises.
E6. Increased stability of currency (especially the pound sterling).
E7. Instability of currency (especially the pound sterling).
E8. Increased British domination of trade and commerce.
E9. More competition for world trade.
E10. Substitution of communication for transport.
E11. Increase in total volume of international trade.
E12. Creation of large individual fortunes.
E13. Greatly improved scheduling of shipping.
E14. Stimulation of development of faster ships.
E15. Faster, more efficient grain transactions, increased food supply, lower costs.
E16. Manipulation and control of world's food supply, higher prices.
E17. Strengthening of U.S. position in world trade.
E18. New forms of speculation, especially arbitrage.
E19. Increased competition between firms, better position for small firms.
E20. Stimulated growth of large industrial cities.
E21. Long-term domination of cable manufacture and operation by Britain.
E22. Growth of large-scale newsgathering organizations (press networks).
E23. Creation of futures markets, changes in buying and selling.
E24. Recruitment of labor from labor-surplus to labor-shortage countries.

MILITARY

M1. Value of British naval supremacy reduced.
M2. Utilization of cable for military surveillance, monitoring, intelligence.
M3. Increased direction of military strategy from the capital (Britain, America).
M4. Decreased impact of rumors in military operations.
M5. Less likelihood of warfare, because of increased international understanding.
M6. Increased military interventions (because of more immediate news of disputes.)
M7. Increased military shipbuilding to protect cable.

POLITICAL

P1. Government (Britain, America) intervention in communications; censorship, control.
P2. Government (Britain, America) intervention to protect currency and regulate markets.
P3. Tighter control of colonies by Britain, strengthened imperialism.
P4. Increased capability for self-government by colonies, demands for independence.
P5. Concentration of management of imperial (colonial) affairs in Home Office.
P6. Increased popular demand for suffrage and political participation stemming from the news coverage of events.
P7. Increased power for the ''Jewish cabal'' of international financiers and bankers.
P8. Decreased power for the ''Jewish cabal'' of international financiers and bankers.
P9. Increased international understanding, less likelihood of wars.
P10. Improved relationship between Britain and America.

SOCIAL

S1. More realistic expectations on the part of arriving immigrants (America).
S2. Increased immigration (to America) resulting from maintenance of closer ties between immigrants and homeland.

S3. New types of criminal behavior: theft of information, purveying false information.
S4. Development of codes and cryptography.
S5. International cooperation in detection and apprehension of criminals.
S6. International cooperation in control of communicable diseases and epidemics.
S7. International cooperation in control of famines and natural disasters.
S8. Rerouting of immigration because of knowledge of epidemics, local conditions.
S9. Selective immigration of educated and technically trained (brain drain).
S10. Interference by papal hierarchy in national affairs.
S11. Advances in basic sciences (physics, chemistry, oceanography, etc.).
S12. Increased political radicalism and subversive philosophies (stemming from media coverage).
S13. Increased violence, stemming from media coverage.
S14. Growth of knowledge about other countries.
S15. Changes in news coverage to emphasize immediate events rather than background.
S16. Changes in religious institutions, more emphasis on missionary work.
S17. General increase in bureaucracy reflecting value of information exchange.

CULTURAL

C1. Changes in substance of communication (some ideas more easily expressed in brief form than others).
C2. Under-communication, ambiguous communication, poor communication.
C3. Changes in language structure and word usage, jargon.
C4. Reduced civility resulting from abbreviated and compressed messages.
C5. Less emphasis on, and less prestige for, classical education.
C6. More relevance and modernity in education, more emphasis on current affairs.
C7. Demand for technical, scientific education.
C8. Altered image of America in Europe—recognition as commercial nation.
C9. Altered image of America in Europe—more realistic view of American life.
C10. Increased interest in Europe on the part of Americans.
C11. Increased travel abroad by Americans.
C12. More prestige for commerce and trade (America and Britain).
C13. Increase in self-confidence and national pride (both British and American).
C14. Increased prestige for English language; English becomes international language.
C15. Increased unity between English-speaking peoples.
C16. Increased acceptance of ideal of international brotherhood and peace.

The last major finding of the Delphi survey related to the differences in perceptions of relatively direct impacts of the cable and of the indirect or ''higher-order'' impacts. One would expect a balanced list of possible impacts to include few immediate or direct impacts along with numerous possible indirect contributions to more remote events or impacts. When the 74 impacts were classified according to their degree of immediacy or, conversely, remoteness, many more did appear in the remote categories than in the immediate ones.

For the Delphi participants, expectations of a causative connection between the cable and its possible consequences diminished on the average for relatively more remote impacts. In contrast, there was no evidence for remote impacts being considered any more or less important than the immediate ones.

Therefore, according to those who generated the impact list during the gaming sessions and evaluated the impacts in the Delphi, the submarine cable was viewed as contributing to an ever-expanding set of possible impacts, but participants had ever-diminishing confidence in the cable's role as a cause. Since the more numerous remote impacts were assessed as equally important as the more immediate, their combined influence could have been greater than that of the immediate ones, even with a greatly reduced role of the cable in their causation. Unfortunately, the quantification produced by the Delphi was only relative and could not be aggregated to address this point.

Yet the essential motivation behind Technology Assessment as a discipline is revealed in this result: The consequences of a technology were viewed to be simultaneously far reaching, significant, and increasingly uncertain. Technology Assessment strives to enumerate the far-reaching consequences, to determine their relative importance, and to reduce the uncertainties accompanying them. The Victorians either held technology as inherently beneficent or allayed their concerns with future uncertainties by trusting the judgments of more knowledgeable decisionmakers.

Analysis of the Delphi

Owing to the small sample size and the single survey that formed the database from the retrospective Delphi, several analytical tests were performed on the survey responses to verify their utility in drawing conclusions. These tests are presented in Appendix III; their results are summarized below:

- Interpretation of the impact statements revealed some variation from respondent to respondent, but these differences were shown to be unimportant.
- The consistency of individual responses was verified to hold well within the bounds necessary to substantiate the overall conclusions given above.
- The consistency of the responses among the three participant groups—the six historians, staff historians Finn and Jaras, and the total—supported the overall conclusions.
- Comparison of the 1861 perspective with that of 1976 reveals a high correlation and supports the expectation that Delphi participants could not divorce themselves from their contemporary views and experience.
- Comparison of the importance of impacts with their likelihood shows no significant correlation.
- Enumeration of the most likely and most important impacts shows that all three participant groups are mutually consistent and that likelihood is independent of importance according to the group's perceptions.

The following subsection describes in detail the principal finding of the Delphi.

Comparison of Direct and Indirect Impacts Relevant to Technology Assessment in General

Immanuel Kant argued in his *Critique of Pure Reason* (2nd edition, 1787) that certain concepts are implicit a priori to all rational thought. Space, time, and causality were among these a priori concepts. Classical physics, underlying the rapid growth of Victorian technology, strove to separate the world of phenomena into universal theories describing the way causal influences propagated and initial conditions that determined which of all possible causal sequences would be observed in a particular situation. Given a comprehensive theory, a complete specification of the initial conditions, and the intellect sufficient to work through all of the implications, in principle all future events were believed predictable.*

As we all know, modern physical theories no longer hold dogmatically to the expectation of predicting specific discrete events; rather, they determine a probability distribution for all possible outcomes. Quantum mechanics, the uncertainty principle notwithstanding, is as deterministic in its philosophical presuppositions as was classical physics. Theory is still expected to *determine* the complete set of future event probabilities, given past ones.

It would be a natural extension of the concept of causality to attempt to describe the events spawned by submarine telegraphy as an unfolding causal sequence of interactions. Of course, lack

---

*The world system is very complicated and it is clearly impossible for the human mind to understand it completely. Man has therefore devised an artifice which permits him to abstract a domain in which simple laws can be found. The complications of the world system are regarded as accidental initial conditions, the regularities as laws of nature. Unnatural as such a division of the world structure may appear to be, and probable though it is that such a division has its limits, the underlying abstraction is one of the most fruitful the human mind has made. It has made the natural sciences possible. (E. P. Wigner, *Symmetries and Reflections*, Bloomington: Indiana University Press, 1967).

of adequate data, research time and support, and a host of other "practicalities" would act as a hedge against failure to complete such an ambitious task. A more fundamental consideration is that the intellect performing the analysis is viewed by most contemporary scientific thought as a part of the same world of events that is undergoing analysis. This view is contrary to the requirement for a limitless intellect to execute the analysis, since the whole (the world of events to be forecast) is always more complex than the part (the intellect performing the analysis). Nevertheless, "impact trees" and terms such as "primary, secondary, tertiary, and higher-order impacts" appear frequently in the Technology Assessment literature. They reveal the underlying desire to work through the causal sequence paradigm for impact identification and evaluation to at least an approximate degree.

The responses to the Delphi provided an opportunity to identify quantitatively some of the features of this conceptual orientation implicit in the thought processes of the respondents—and perhaps in all of Technology Assessment. Events or impacts that are more remote from the initial causative factor are intuitively considered less certain that those more immediate or direct. This ordering explains the use of terms such as "higher-order impacts," which are often thought to be overlooked. Remoteness can be simply interpreted as further away in distance, later in time, or separated from the initial causative factor by a long chain of sequential events. Figure 4–1 illustrates an hypothetical causal sequence constructed from the impacts listed in our Delphi questionnaire.

Conceptually, at least three sources of uncertainty enter into long causal chains. To illustrate them, first consider a fully articulated impact tree which for simplicity shows each impact as spawing two more distinct impacts (Fig. 4–2a). Uncertainty can enter into this ideal construction in the following ways.

- The rapid proliferation of discrete impacts soon defies detailed description and requires a reduction in specificity and a more efficient representation. Aggregation of individual impacts is accomplished by the coining of new terms and concepts with broader, yet more diffuse, interpretations (Fig. 4–2b). For example, an immediate, specific impact of the cable was "faster, more efficient grain transactions," whereas a more remote impact was broadly hypothesized as "the development of international cartels" without specifying how.
- Different causal chains can overlap producing conflicting influences leading toward a given single event (Fig. 4–2c). The cable could have contributed to both a "concentration of management of imperial (colonial) affairs in the Home Office" and "increased capability for self-government by colonies and demands for independence."
- The initial causal factor of central interest to the Technology Assessment is not the only prime-mover producing change even in a limited analysis. Several original causal events can have subsequent indirect impacts that interact to produce observed consequences. For example, steam-powered steel ships augmented the growth of shipping that the cable helped coordinate, and accelerated the growth of international trade and the spread of Western influence to colonial markets. One probably cannot distinguish that portion of the growth caused by the cable from that caused by steam-powered steel ships. More likely their combined effect was much larger than the sum of their individual effects might have been.

Perfect understanding of the processes by which these causal influences combine to produce results would, in principle, remove the necessity for believing that the more remote impacts were "less certain." Unfortunately there are uncertainties in each step of these causal sequences. Once an error or uncertainty is introduced, all subsequent implications are distorted. Uncertainties grow cumulatively with each successive step—at frighteningly swift pace for those engaged in Technology Assessment studies.

Major technological innovations appear to set in motion a characteristic sequence of events. At first the new technology simply performs a familiar task better: it substitutes for a previous method. As the availability of the technology persists, new uses are discovered and user groups modify their activity patterns to take advantage of the opportunities it presents. As applications

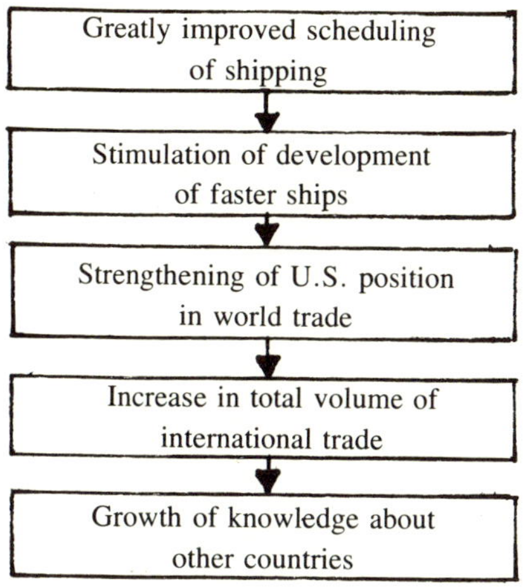

FIG. 4–1.—Possible causal sequence, constructed from Delphic hypothetical impacts.

continue to spread, the technology helps to sire a new way of life. Attitudes toward former patterns of activities are modified, different relationships among people are established, and eventually an altered view of the world develops.

This causal sequence can be paralleled by five successive categories of impacts from the Delphi survey; together these categories form a progressive scale of remoteness starting from the initial causal event of laying the transatlantic submarine cable. In rank order from most immediate to most remote these are:

1. *Options made available to user and promoter groups*. Upon completion of the cable, it could immediately substitute for previous slower, presumably less effective means. Included are services that are rendered technically feasible or in immediate demand, with consideration given only to applications requiring little or no structural change within the existing user and provider institutions. Illustrations: greatly improved scheduling of shipping; utilization of cable for military surveillance, monitoring, intelligence.

2. *Structural changes in user groups*. As the use of the cable progresses, its users can adapt to its presence. They can alter their organizational structure or format of activities to utilize the cable's advantages better. User groups may be defined narrowly to encompass only one institution, but usually the term is used to refer to an entire industry or activity sector. Illustrations: growth of large-scale newsgathering associations; creation of futures markets; changes in buying and selling.

3. *Changes in British-American relations* (with one another or with respect to the world at large). Since the early submarine cables were built under patent rights restricted to British industry, since the capital to sponsor cable construction was British, and since the British were the dominant sea power, they were the closer to the focus of the new technology's impacts. America was to be connected by a fast communications link to Britain and Europe. Not only was there a shared culture and language, but both nations were still in the process of rapid economic and territorial expansion. Many of the impacts of submarine telegraphy on the world at large could be expected to emanate from these two nations. Illustrations: increased British domination of world trade; increased prestige for the English language—English becomes an international language.

4. *Changes in worldwide patterns of activities*. Submarine telegraphy was to be the first of a series of technologies to unite the world through rapid communications. Time and distance were no longer locked together for communications. Now cities with telegraphs and cable access were in

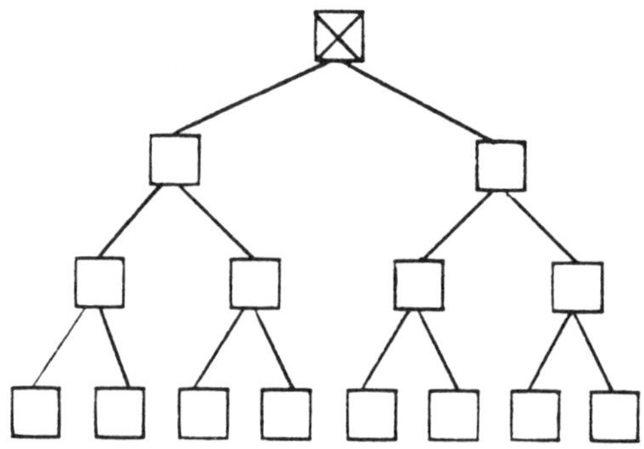

(a) Fully articulated impact tree.

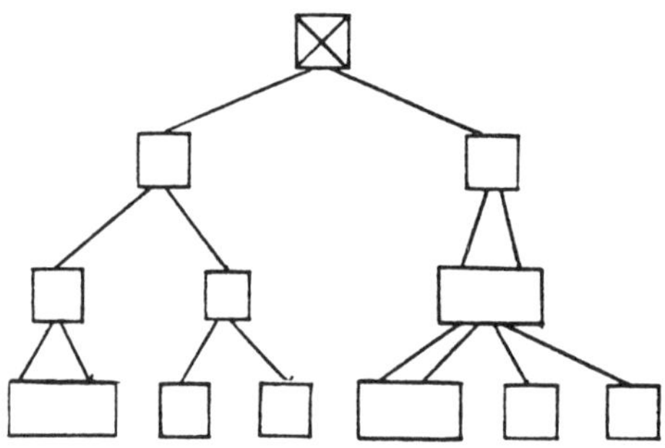

(b) Reduced specificity.

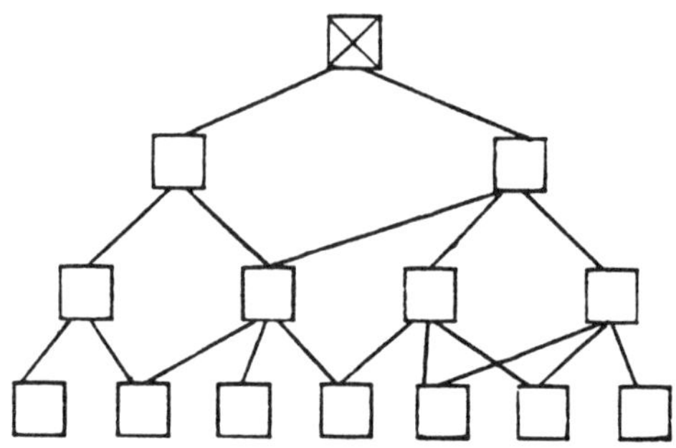

(c) Cross links.

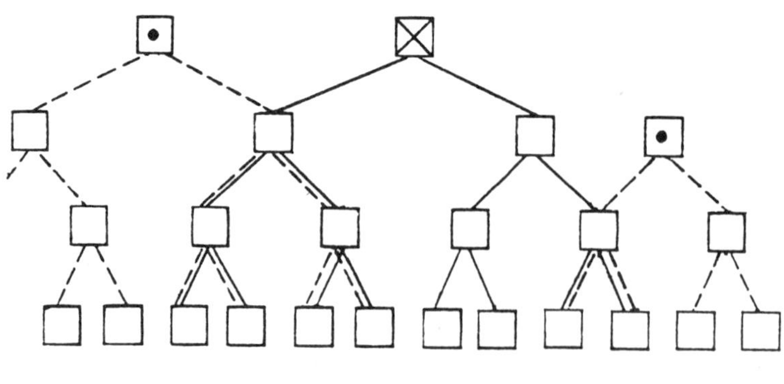

(d) Multisource causation.

FIG. 4–2.—Graphical illustration of sources of uncertainty in causal sequences of impacts or events. (□ impact or event; ☒ primary causal factor; ▣ secondary causal factor. Impacts spread downward in each diagram.)

closer touch with London, Europe, and New York than with their hinterlands or less developed neighbors. Such a redrawing of the map for communications would be expected to have profound implications for each of the impact sectors considered in the survey. Again, these impacts are broadly felt and the exact specification of the chain of events (or chains of events) producing them must be left somewhat open to interpretation. Otherwise, the organic interdependencies of many of the components of worldwide activities would need to be understood better than is deemed practical, even without the limitations of research effort. Illustrations: increase in total volume of international trade; international cooperation in control of communicable diseases and epidemics.

5. *Subjective changes in values and attitudes*. As the influence of rapid communication overseas by submarine telegraphy contributed to the alteration of major institutions and the reshaping of the relationships among nations and peoples, attitudes and values could be expected to change. Not only would the pattern of human interrelationships shift, but so would the perception of those interrelationships. The question about degree to which submarine telegraphy contributed to these shifts was asked of the survey respondents in a variety of hypothetical impacts. Illustrations: demand for technical, scientific education; increased popular demand for suffrage and political participation stemming from the news coverage of (current) events.

Table 4–1 shows the number of impacts that fall in each remoteness category under each subject area heading. As expected, the number of impacts per category increases toward the more remote impact categories. Also, economic impacts appear to be considered as somewhat more immediate than those in other subject areas.

The average responses from the total group in their estimate of the relative likelihood of the impact occurring, assuming an 1861 perspective, are shown in Fig. 4–3. (There were some ambiguities in assignments to the remoteness categories considered during the analysis. These ambiguities did not affect the results presented here and were not pursued further.) The black square indicates the average of the impacts in each category according to the scale of values listed along the top of the figure. The systematic tendency for the more immediate impacts to have higher likelihood scores is highly significant; it would occur by chance in only one out of 50 trials (by a chi-square test applied to the hypothesis that all the category averages were equal).

Analysis of the historians' and Finn/Jaras's consensus evaluating the degree to which the cable caused impacts seen from a 1976 point of view yielded nearly identical results. The category averages and their statistical uncertainties are shown in Fig. 4–4. Again the same systematic trend appears. These trends are significant at the 2.5% confidence level for Finn and Jaras—at the 0.5% level for the historians.*

TAB. 4–1.—Number of impacts falling in each remoteness category according to subject areas. (Economics is less remote at the 2% level of confidence.)

	SUBJECT AREAS					
	Economic	Military	Political	Social	Cultural	Total
Options made available	3	2		3		8
User group structural change	6	2		1	2	11
British-U.S. position change	4	1	4	1	4	14
Worldwide patterns change	10		2	6		19
Values and attitudes shift		3	3	6	10	22
Totals	24	7	10	17	16	74

*The 1861 perspective gives the same systematic trend toward reduced estimated likelihood with increased remoteness. This trend holds even with each category having nearly one unit higher average value for the 1861 expected probabilities.

Mean responses for each impact item on Delphi question, "In 1861, what do you think is the probability of this impact occurring as a result of the cable, or with that as a causal factor?" (Scale from 1 to 10).

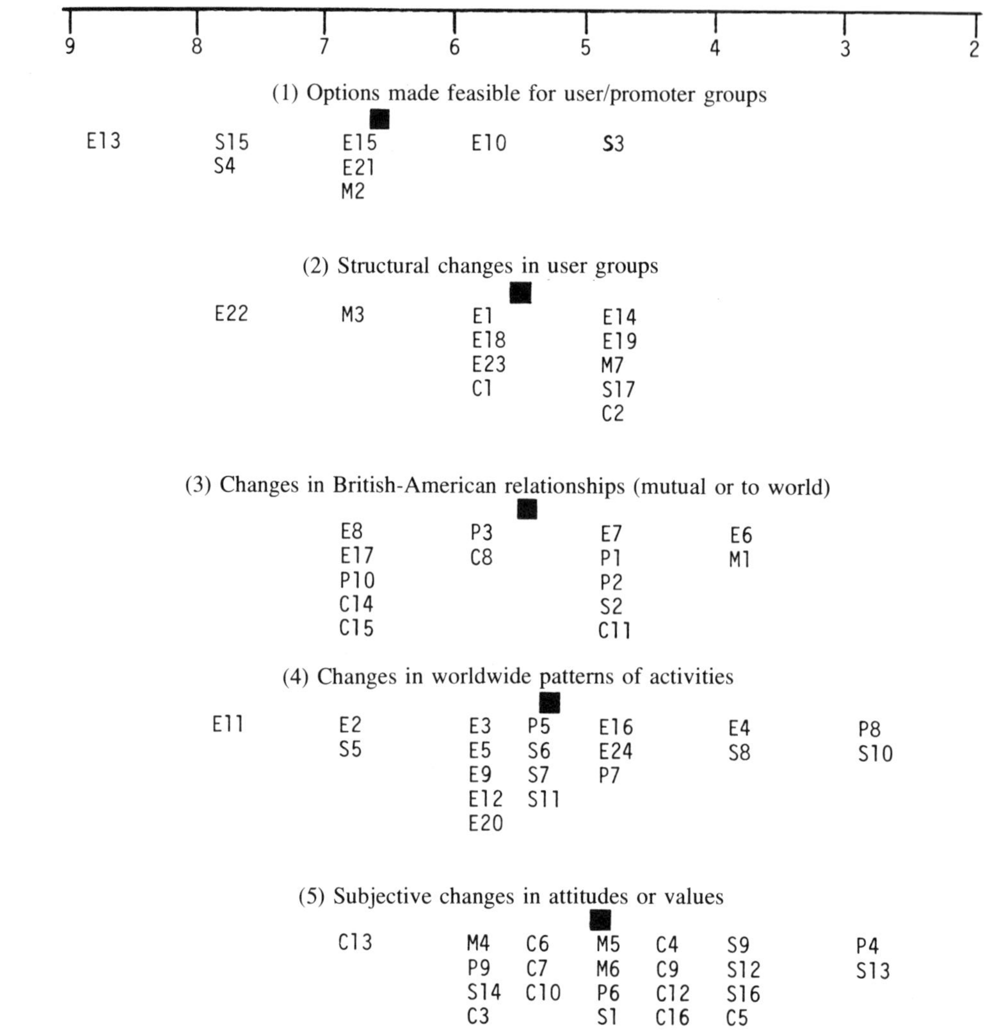

FIG. 4–3.—Reduced perceived probability of cable causation with increasing remoteness. (*E*, economic; *M*, military; *P*, political; *S*, social; *C*, cultural; ■ mean of cluster of responses in each category.)

In sharp contrast, the identical test applied to the category mean responses for question "E" (how important was this impact in shaping the modern, 1976, world?) gave no evidence for a systematic trend.

The average value assigned to each impact according to the total group is shown in Fig. 4–5 binned along a ten-point scale. Only the 43 relatively more likely impacts were analyzed, yet no trend was evident from the mean values (indicated by the black squares). The degree of scatter of these means about a common average would occur in 25% to 50% of all random trials. Similarly, the means and error ranges for the remoteness categories are given for both Finn/Jaras's and the historians' evaluations in Fig. 4–4. All impacts were considered. Therefore, in both cases, roughly 50% of all random trials would exhibit this degree of scatter. There is no evidence for a systematic trend connecting the importance of impacts to the degree of remoteness they have from the initial technological cause.

Taken together, these findings imply that:

• The hypothetical impacts identified during the gaming and workshop sessions increased in number with ever-greater remoteness from the submarine cable's direct causation. This result is expected from the general systems principle that branching causal chains will spread with each step forward in time.

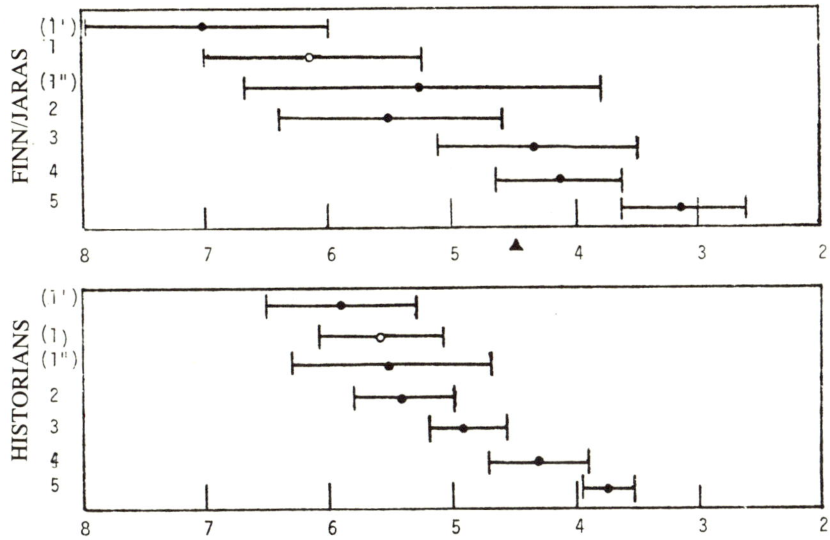

(a) Cable *causation* on modern world.

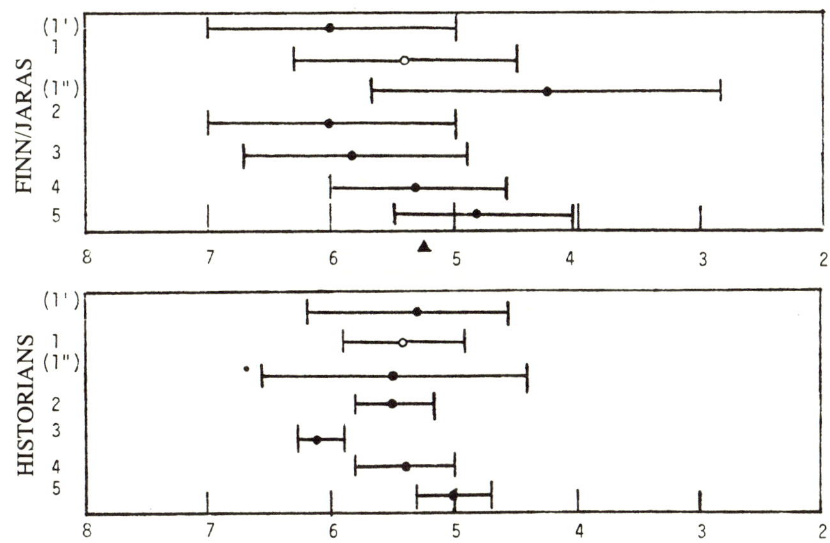

(b) Impact importance to modern world

FIG. 4–4

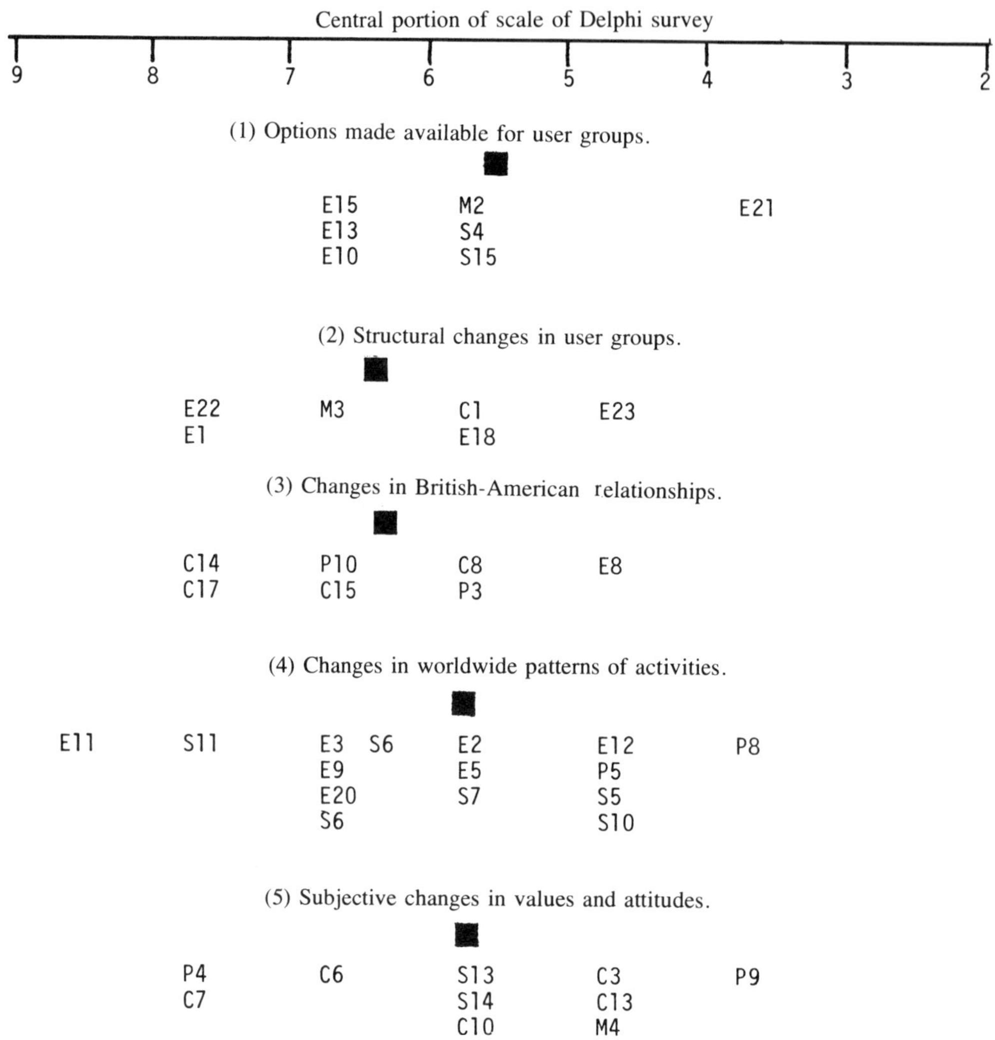

FIG. 4–5.—Responses to question E on impact importance, classified according to the remoteness scale categories. (■ mean of cluster of responses in each category.)

- Isolated consideration of each impact's likelihood still revealed a subjective assessment that results independently judged to be more remote or indirect were less predictable than those judged to be more immediate. Attitudes reflect what one would expect from the general systems principle that uncertainties accumulate along indirect causal chains (unless an error-suppression strategy is possible, as in satellite and computer systems).
- There is no diminution of estimated importance of impacts even though they are more remote from the cable's direct causation.
- Those charged with the responsibility for initiating and shaping the development of a new technology are therefore pictured as setting in motion sequences of events with spreading impact to ever more numerous possibilities and ever diminishing perception of their precise implications—truly opening Pandora's box.

Since the above argument is derived from very basic systems principles and revealed in the subjective evaluations of the Delphi respondents, one might expect that it was held at the time of the introduction of submarine telegraphy. Optimism in the future benefits of successful technologies may have overriden the anxiety of producing new uncertainties.

Today this optimism no longer protects us from concern about the dimly perceived possible consequences of technology. Experience has demonstrated many pitfalls in that optimism. Technology Assessment as a specialized discipline addresses this very point. In a complex world of many specialized disciplines and professions no one can expect those charged with decisions to have both a comprehensive and a detailed view of the implications of their decisions. Conversely, the techniques of science and the judgment of experts, both brought to bear on the task of identifying the implications and estimating the significance of the consequences of introducing a new technology—the tools of Technology Assessment—do not necessarily guarantee that the inherent uncertainties of decision will be reduced to an "acceptable" level.

Neither success nor failure to reduce future uncertainties through Technology Assessment studies based on scientific approaches will remove or alter the fundamental reality that ever-unfolding possibilities of technology at some point in the future may introduce significant yet poorly understood consequences into the course of events. Methods for monitoring the sequence of events as they unfold, options for shaping the consequences as corrective actions, and the assessment of risks or assignment of delegated responsibilities are supplementary tools of Technology Assessment intended to reduce the anxiety caused by inherent uncertainties. Yet even if these strategies fail, one is still confronted with need to act, the need to know, and inherent sources of uncertainty—Technology Assessment must persist by one title or another—as long as technology is recognized as producing important changes.

Chapter 5

THE CABLE INDUSTRY AFTER 1866

BEFORE 1866, long cable projects had proved as financially disastrous as the shorter cables had been profitable. Of a total of 77 cables laid, only 17 had been failures; yet these cables represented 7300 miles of the 15 000 miles laid—nearly half.[1] In 1866 the first enduring transatlantic cable was put down. Possibly of even greater importance was the successful recovery of the lost end of the 1865 transatlantic cable and its repair and completion. From this point on long cables could be designed for ease of repair, which greatly reduced the risk of loss to investors. The submarine telegraph industry thereafter expanded rapidly to satisfy the increasing demand for worldwide communications.

The long service life of most cables after 1865 is illustrated in Fig. 5-1. Longer cables were most persistent owing to the cost advantages of repair over replacement.

The long service life of submarine telegraph cables necessarily implies a low rate of technological innovation for the cable design itself—a stranded copper core surrounded by successive layers of gutta percha and "Chatterton's compound," wrapped with impregnated jute and finally iron wire. As recently as 1973 machinery dating from about 1929 (and virtually identical to that of the 1850s) was being used to wrap jute and iron wire on submarine telegraph cables. A cable seaman of a century ago would feel quite comfortable on a telephone cable ship today, as soon as he became used to the modifications in machinery introduced to accommodate the bulky electronic repeaters.

Reluctance to innovate can be explained by several factors: the risk of losing a large investment by altering proven design; the lack of competition from an alternative communications technology; the incremental advantage of repairing long cables rather than replacing them; and possibly the excessive cost of building new manufacturing, laying, and maintenance equipment to serve a modified cable design. Despite this reluctance, changes did come—both from without and within the cable telegraph industry.

Transatlantic cables that carried the heaviest traffic for the distance they spanned exhibited the most rapid rate of technological advance as measured by the increase in signaling speed. Between 1866 and 1965 there was a 68-fold improvement in signaling rate (Table 5-1), of which only a 6- to 9-fold increase could be attributed to improved cable designs. Figure 5-1 also shows that the service life of cables was cut short by the events surrounding World War II and later those of the mid-1960s. Electronic technology at first improved the performance of submarine telegraphs, but eventually it led to practical submarine *telephone* cables of many times the capacity of telegraph cables. By the mid-1960s satellite broadcast relays emerged as practical. The older cables were supplanted on all the world's major communications routes—relegated to serve only the out-of-the-way areas in which reinvestment in improved communications was not yet warranted.

The story of the cable industry after 1866 subdivides naturally into three topics. First is the technological evolution, best exemplified in the transatlantic routes. Second is the development of international telegraphy, beginning with the rise of Britain to dominance and followed by the competition from the telegraphy of other nations and from radio. Third is the impact of electronics and international telephony by cable and satellite. Finally, a comparison of the competitive and complementary patterns of interaction between cable and broadcast options for international communications reveals several parallels and affords contrasts between telegraphy and telephony.

TABLE 5–1.—Technological innovations on transatlantic telegraph cables.

Year	Technological Innovation	Performance (words/min)	Improvement	
			Individual	Cumulative
1857–58	Mirror galvanometer	3–7	—	—
1870	Condensers	12	1.7	1.7
1872	Siphon recorder	17	1.4	2.4
1879	Duplex	24	1.4	3.4
1890	Automatic transmission and curbing	*	1.4	4.8
1894	Larger diameter**	72–90	3.0	10.2
1915–20	Brown drum repeater and Heurtley magnifier	100	1.4	14.3
1923–28	Magnetically loaded lines**	300/320	3.1	44.3
1928–32	Electronic signal shaping amplifiers and time division multiplexing	480	1.5	68.6
1950	Repeaters on the continental shelf	100 up to 300***	3.6	(51.4)
1956	Repeatered telephone cables**	21 600†	45	3085

*Bright (*Submarine Telegraphs*, p. 674) gives capacity increase at best performance (whether used or not) as a 40% improvement.

**Improvements in the cable itself; all others are improvements in terminal technology.

***Primarily brought lower-capacity cables up to comparable rates of the loaded lines. Loaded lines could have been improved, but were not.

†432 channels × 50 words/channel (Kelly and Rodley, 1956).

A. The Evolving Technology

Transatlantic telegraphy has always attracted the most advanced signaling technology available at the time.[2] The more than 2000 miles separating Britain and North America requires a large investment in each cable. For shorter connections (those under 500 miles), it was often more economical to lay additional cables than to introduce sophisticated devices at the terminals. Despite the distance and the costs of laying cables and of improved terminal technology, the high demand for international communications across the North Atlantic has encouraged the construction of 24 telegraph cables and nine telephone cables between 1866 and 1976 (Table 5-2).

Table 5–1 lists the 10 major innovations to submarine telegraphy before polyethylene and repeaters revolutionized the industry in 1956. Overall, the performance of telegraph cables was improved 68-fold according to the highest-capacity individual cable. Telephony enjoyed a 45-fold increase in signaling capacity the moment it was introduced; however, this increase takes into account that each telephone channel is easily equivalent to nine telegraph channels in information capacity. Since the first telephone cable, there has been an 83-fold increase in single-cable capacity as of 1976, as will be discussed in more detail later.

The ten cable telegraph innovations were as follows.

The *mirror galvanometer*[3] was introduced by William Thomson (Lord Kelvin) during tests on the 1858 Atlantic cable (Fig. 5-2). Weak signal pulse currents from the cable produced a magnetic field in a stationary solenoidal coil. A mirror with magnets mounted on its back was suspended by a thread pivot inside the solenoid. As the magnetic field from the solenoid changed, the torque on the mirror induced it to rotate and to deflect a beam of light reflected from it. Operators distinguished the successive deflections of the light beam and could achieve a signaling rate of seven words per minute.

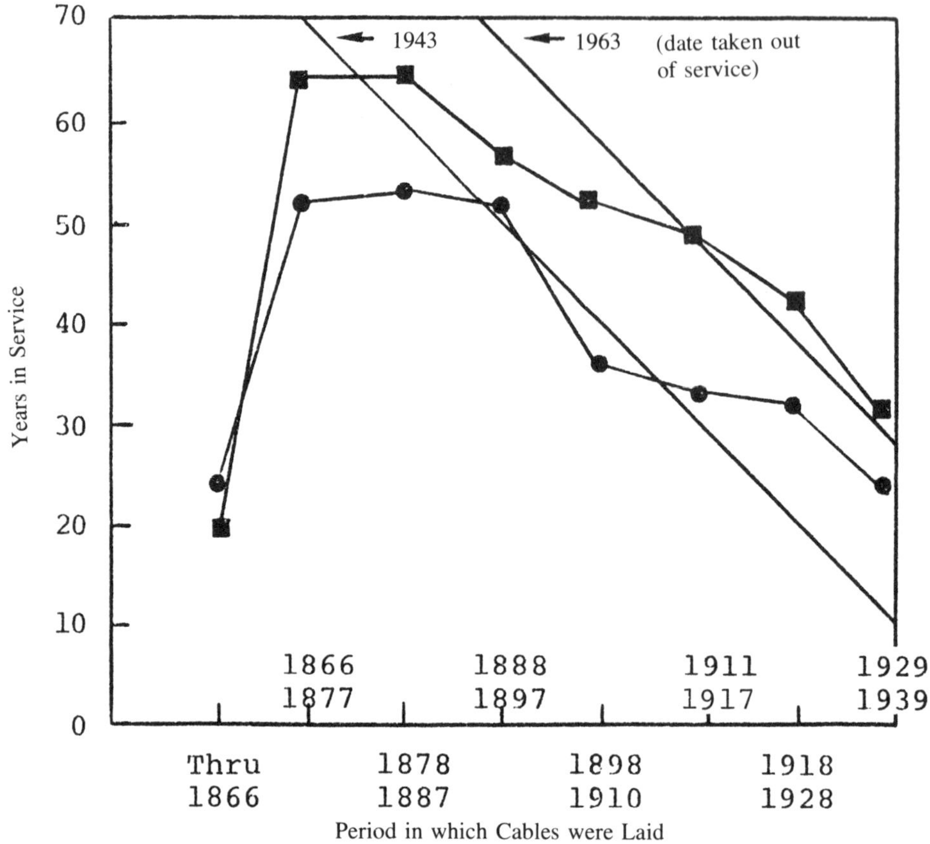

FIG. 5–1.—Longevity of stranded-copper gutta percha cables, based on average years of service, by period in which cables were laid. (■ cables over 500 miles long, ● cables 5–500 miles long.)

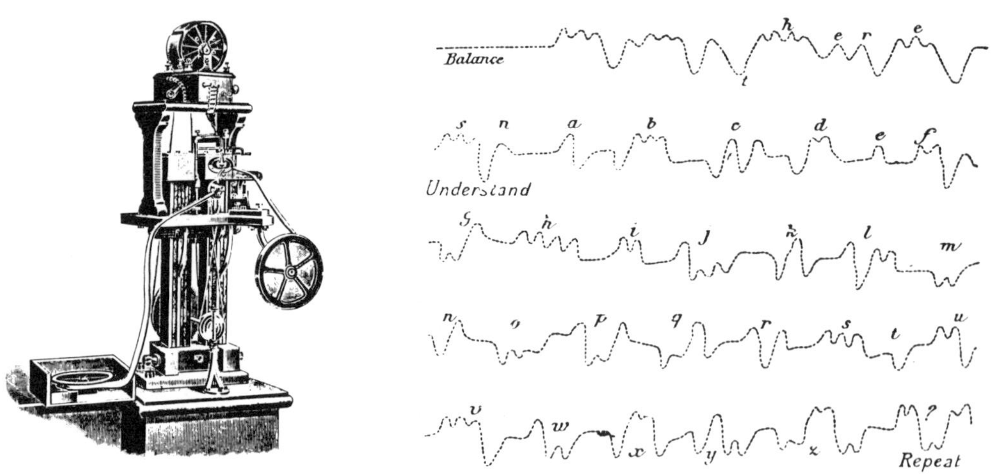

(a) (b)

Fig.5–2.—(a) Thompson siphon recorder with permanent magnet; (b) signals received with siphon recorder (from Ref. 3).

Condensers[4] were used to isolate the cable electrically from the ground in 1870. Rather than flowing into or out of the ground, signal currents simply charged or discharged condensers at each end of the cable. This arrangement maintained the electrical neutrality of the cable as a whole and produced a stronger reverse current to discharge the cable for the next pulse; signaling rates increased to 15-17 words per minute.

The *siphon recorder* produced a permanent paper-tape written record of the cable signals.[5] This innovation improved the efficiency of the cable operation more than the rate of signaling itself. As modification of the mirror galvanometer concept, a siphon tube to an ink reservoir (rather than a mirror) was attached by strings to a coil driven by the cable signals. Either electrostatic charge on the ink induced flow without the siphon touching the paper, or vibration of the paper tape served to reduce friction.

Duplex (Chap. 2) nearly doubled the capacity of cables in 1879 by enabling signals to be transmitted in both directions simultaneously.[6] Duplex was based on electrically "balancing" the cable against an artificial line (an electrical circuit that mimics the response of the true cable as seen from the sending end). The sensitive receiving instrument bridges the true cable and the artificial line. Powerful outgoing electrical pulses are balanced across the receiving instrument and produce no effect; incoming signals upset the balance and are detected by the receiver. Duplex increased the rate to 24 words per minute soon after its first application.

Automatic transmission and curbing[7] had the potential of increasing sending rates by about 1.4; however, the innovation was not adopted until similar improvements in receiving efficiencies were accomplished. Curbing is a type of signal shaping. Instead of simply transmitting a single pulse per symbol, a series of carefully controlled and proportioned pulses is sent. The sum of this series of pulses provides a sharper output signal by more rapidly discharging the line through the action of counter pulses. Punched paper tape was used to activate the automatic transmitters.

Cables of larger diameter were introduced in 1894; they permitted signal rates of up to 90 words per minute.[8] Larger cables increased the signal rate by reducing the total electrical resistance (more copper) and line capacitance (more gutta percha insulation). The signaling rate theoretically increases with the size of the core, but the marginal costs of the cable itself do not improve with core size.

Brown drum and Heurtley magnifier combined to provide a practical means of amplifying incoming signals so that they could be automatically relayed. The Heurtley magnifier used a fine-resistance wire as a replacement for the swinging siphon of the siphon recorder. The battery current flowing in the fine-resistance wire increased whenever the incoming cable signal pulse caused the resistance wire to swing into the breeze from a small fan.[9] (Chilling a hot resistance wire reduces its resistance, so that more current flows for a given voltage.) The increased current from the magnifier was sufficient to drive the Brown drum repeater.

The coil of a Brown repeater moved a wire resting on a rotating drum, shifting it from one conducting band to another, as the sliding motion produced by the drum's rotation reduced friction. The current passing between wire and bands acted as signals to the next relay station. This arrangement greatly reduced the need for manpower to read and retransmit messages. It did not appear to reduce the manpower requirements for operating the telegraph stations, possibly because of the need for constant repair and maintenance of this electromechanical equipment. Signaling rates would have been increased 1.4 times by these innovations alone. In conjunction with automatic transmission and curbing, the actual rates of the transatlantic cables appear to have doubled.

Magnetic loading was the single most important innovation in the design of cables themselves. Oliver Heaviside had argued the value of magnetic loading to help push currents through electrically sluggish cables by means of magnetic induction. Michael I. Pupin and George A. Campbell were rival claimants to patents awarded in 1900 for loading telephone land lines with regularly spaced magnetic coils.[10] However, loading of long submarine telegraph cables had to wait for the development of permalloy and mu-metal materials of high magnetic induction. These metals were immediately applied as wrappings of cables in the mid-1920s to three cables across the Atlantic and one

TABLE 5–2.—Listing of all transatlantic cables, 1858 to 1976.

Key: Upper dash designates laying, *lower* designates termination of its use. Numbers represent signal rates in words per minute; parentheses designate an inferred rate.

NATIONALITY DESIGNATION	GB "58"	GB "66"	GB "65"	Fr	GB* 1VA	GB* 2VA	Dir	Fr PQ	GB* 3VA	GB* 1PZ	GB* 2PZ	US Com 1	US Com 2	US Com 3	GB* 4VA	Fr Dir	Ger
YEAR LAID	1858	1866	"1865"	1869	1873	1874	1875	1879	1880	1881	1882	1884	1884	1894	1894	1898	1900
	═ 3																
1860	a	— 7 12	— 7 12	—													
1870		17 — bc	17 —	(11)	— (12)	— 2x12	—	—	—								
1880			bc	11 — d	2x12 e	e	2x16 e	(2x12)	(2x12)	— 13 s (2x12)	— 18 s (2x12)	— (2x12)	— (2x12)				
1890														— 2x36 f	— 2x45 f	— (2x10)	
1900																	— (2x40)
1910					(2x50)	(2x50)	(2x50)	(2x50)	(50)	(50)	(2x50)	(2x50)	(2x50)	(2x50)	(2x50)	(2x40)	(2x50)
1920					r	r	r	r	r	r	r	r	r	r	r	r	r
1930						2x48 j											
1940					3x60 x 2x50	3x60 x 2x50			60 60 p —	60 60 kpq	60 60 pq	—		2x57 p			
1950					5x60 pm 6x60 w	5x60 pqm 6x60 w				5x60 m	5x6C m		2x50 t	2x50 t	6x60 m		
1960					—	—	(—)	(—)		—	—		—	—	—	(—)	(—)
1970																	
1980																	

Table 5–2 (cont'd)

	US Com 4 1901	Ger 1904	US Com 5 1907	GB 3PZ 1910	US Com 6 1923	US 1HO 1924	US 4PZ 1926	US 2HO 1928	TAT-1 1956	TAT-2 1959	CAN/ TAT 1961	ICE CAN 1962	TAT-3 1963	TAT-4 1965	TAT-5 1970	CAN/ TAT-2 1974	TAT-6 1976
1860																	
1870																	
1880																	
1890																	
1900	(2x40)	(2x40)	(2x40)														
1910	(2x50) r	(2x50) r	(2x50) r	(2x40) (2x50) r													
1920					(2x50) r	5x60 g	4x80										
1930							8x60 g	5x50 h									
1940	2x50 x		2x72 x	2x60 x	4x70 x	5x65 pq	8x60 x 8x50 q	4x62 pq									
1950				6x60 m	6x50 i		8x60 m		48x y	48x y							
1960											160x y	24x y	138x y	128x			
1970														138x y	720x 845x y	1840x y	4000x
1980																	

Sources

a. Ref. 3, p. 51
b. Heart's Content cable station diaries.
c. Ref. 3, pp. 105, 539–540.
d. *Ibid.*, p. 556.
e. *Ibid.*, p. 123.
f. *Ibid.*, p. 574.
g. Ref. 11, p. 489.
h. Haglund, ref. 10, pp. 108–110.
i. C. H. Cramer, *El. Engg* 66:765–767, 1947.
j. Haglund, ref. 9, p. 96.
k. Cramer (1951), ref. 12, p. 90.
m. Cramer (1956), ref. 12, p. 176.
n. WU files in Smithsonian, 1 January 1946.
p. WU report to FCC, August 1950.
q. WU report to FCC, December 1950.
r. Millnor, ref. 9.
s. Scott, ref. 13; Bright, ref. 3, p. 573.
t. American Cable Co. report to FCC, August 1950.
w. WU report to FCC, updated to January 1951.
x. Report to FCC, mid-1940s.
y. H. H. Schenck, *The World's Submarine Telephone Cable Systems*, U.S. Department of Commerce, Office of Telecommunications contractor report 75–2.

under the Pacific. Unfortunately the Depression and the growth of radio stifled the demand for more loaded cables. Competition resulting from threefold increase in capacity was not enough to cause any of the existing cables to be abandoned.

Electronic signal shaping, electronic amplifiers, and, time-division multiplexing were all introduced in the late 1920s and early 1930s.[11] Together, they improved signaling rates by about 1.5 over the previous electromechanical technologies. Their greater reliability was probably the cause of a significant drop in manpower at cable stations such as Heart's Content in Newfoundland. Early electronics improvements were based on the rapidly evolving vacuum-tube technology promoted mostly by a demand for radio systems and other electronic devices. In time-division multiplexing, several parallel signal streams were created by synchronization of the rotation of disk switches (commutators) at each end of the cable.

Repeaters on the continental shelf grew out of the communications surge of the Second World War. Underwater repeaters were impractical in deep-sea portions of cables, but such amplification devices could be placed at the edge of the continental shelf. The repeaters improved the ratio of signals to noise, which deteriorates especially over the last third of the distance. Repeaters were placed in the nonloaded Western Union cables during 1950-1951 and successfully brought their signal rates up to nearly those of loaded lines.[12]

Telephone cables were made practical by improvements in vacuum-tube reliability so that they could be expected to last 20 years without servicing. In 1956 an entirely new type of transatlantic cable was laid through cooperation between AT&T and the British Post Office. It was coaxial, used polyethylene rather than gutta percha insulation, and had repeaters submerged with the cable at twenty-mile intervals.[14] It could carry 36 telephone conversations (later increased to 48 by a reduction in the information capacity assigned to each). Routing was no longer constrained by cable length; telephone cables landed in Rhode Island or New Jersey rather than Newfoundland—Europe or England rather than Ireland. The telephone cable made all existing telegraph cables obsolete, since it could carry the equivalent of 45 times the message rate of the best telegraph cable and 5.6 times that of all previous transatlantic telegraph cables combined.

The relative importance of technological innovation to simply constructing more cables can be assessed by a comparison of the fraction of total capacity deriving from innovation and that from laying cables. Table 5–3 shows the growth in the capacity of the best cable of each period, the growth in numbers of cables, and the growth in total transatlantic message capacity within successive time periods. In each time frame, technology contributed more than numbers of cables to the increase in total capacity. The highest growth rates of all three measures occurred in the earliest period, except

TABLE 5–3.—Rate of transatlantic cable development.

Time interval	Best technology: Capacity (words/min)	Best technology: Rate (%)	Cables: Number	Cables: Rate (%)	Total capacity: Capacity (words/min)	Total capacity: Rate (%)	Type of signal
1866–1879	7 to 24	9.9	1 to 5	7.3	14 to 99	6.2	Telegraphy
1879–1900	to 90	5.4	to 13	4.7	to 480	7.6	Telegraphy
1900–1924	to 320	6.1	to 19	1.6	to 1980	6.0	Telegraphy
1924–1950	to 480	1.8	to 18*	−0.2	to 3853	2.6	Telegraphy
1950–1976	to 1.8×10^6	36	to 0**	−1.2	to 21600	28	Telegraphy to telephony
1956–1976	48 to 4000 (channels***)	25	1 to 9	11.6	48 to 7248 (channels)	28	Telephony

*Representing a net loss of one cable, from 19 to 18.

**Representing the phasing out of telegraph cables; fully 18 were abandoned between 1960 and 1965.

***Representing the transition from telegraphy in 1950 to mature telephony in 1976.

for the small increase in technology's contribution from 1900 to 1924. Total capacity grew generally somewhat more slowly than the sum of the contributions from technology and numbers combined because the older cables could not improve as rapidly as the newer state-of-the-art ones. Yet the introduction of continental-shelf repeaters to unloaded cables brought up total capacity faster than that of the best loaded cables.

During the first 20 years of repeatered telephone cables, the capacity of each new design generation had increased at an average annual rate of approximately 25%; submarine telegraphs showed technological improvements of 9.9% per year from 1866 to 1879. Electronic technology produced 2.5 times the rate of innovation found in the fastest growth period for 19th century telegraphy (derived from the Table 5-3). Increases in transatlantic communications capacity derived from the laying of more cables were comparable for the two technologies—telegraphy grew at 7.3%, telephony at 11.6% annually. For both technologies, technological innovation contributed more to the growth in transatlantic message capacity than increases in numbers of cables. This effect is especially pronounced for telephony: total capacity has grown at 28% per year; that of the largest cables, at 25% per year.

Table 5-1 reveals pauses in an otherwise regular progression of cable layings from 1884 to1894, 1910 to 1923, and 1928 to 1956, as summarized in Table 5-4. Each pause is statistically significant (in terms of the previous long-term averages) at the 5%, 2%, and 1% levels, respectively. Moreover, the rates of laying cables are significantly above average during the periods of early expansion, loaded cables, and telephone cables. In each of these cases there has been significant innovation in comparison with the years immediately preceding.

Several hypotheses could explain these pauses. Telegram prices declined at 15% per year during the first period of rapid cable expansion. The first pause, 1884 to 1894, coincided with the bottoming out of this rapid price decline (Fig. 3–8); prices may have fallen to the real costs of providing service and further investment was not profitable until demand again rose because of continued national development of the communicating parties. The third and fourth periods of rapid expansion—1923 to 1928 and 1956 to 1976—occurred during post-World War economic booms. Each war produced a large demand for cable capacity, but economic resources were devoted to more critical activities. After the wars, demand remained higher than before, if diplomatic message volumes are any indicators of institutional reactions. Thus, renewed interest in cable construction would naturally

TABLE 5–4.—Periods of rapid expansion and pauses.

Periods of rapid expansion	Pauses	Long-term averages
Technological refinement, 1857–1873 5 cables, 3.2 years	Interruption of attempts, 1858–1866 Rate increase, 2.3 times*	
Early expansion, 1873–1884 9 cables, 1.22 years	First pause, 1884–1894 Rate increase, 3.0 times*	Early expansion and pause, 1873–1894 2.3 years
Secondary expansion, 1894–1910 8 cables, 2.0 years	Second pause, 1910–1923 Rate increase, 3.1 times*	Through second pause, 1873–1923 2.35 years
Loaded cables, 1923–1928 4 cables, 1.25 years	Third pause, 1928–1956 Rate increase, 45 times*	Through third pause, 1873–1956 4.95 years
Telephone cables, 1956–1976 9 cables, 2.22 years		Overall average 1858–1976 4.4 years

*Represents technological advance of cables laid soon after pause relative to those laid during the previous period of expansion.

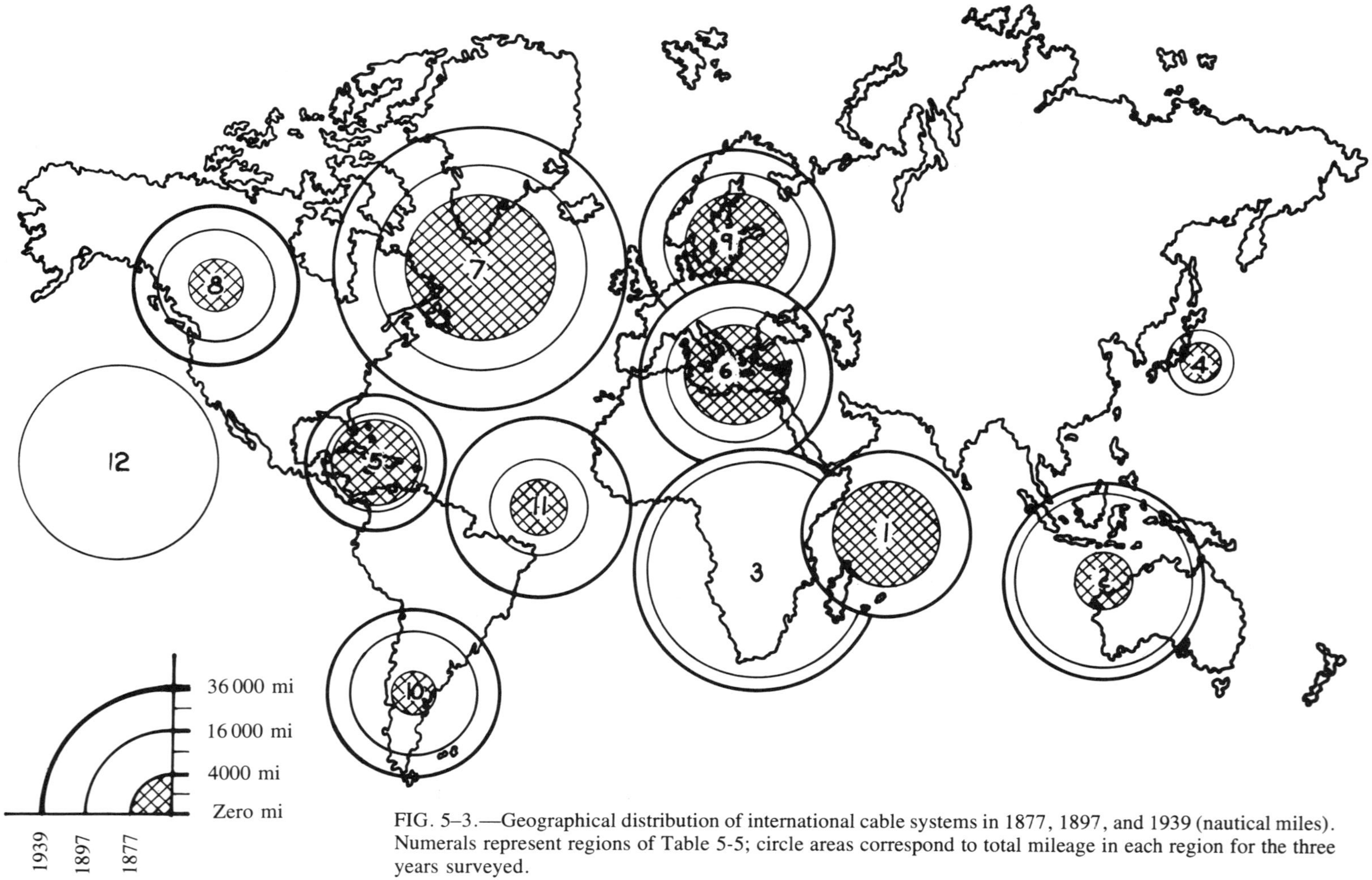

FIG. 5–3.—Geographical distribution of international cable systems in 1877, 1897, and 1939 (nautical miles). Numerals represent regions of Table 5-5; circle areas correspond to total mileage in each region for the three years surveyed.

follow. The improvements in technology were by themselves probably not responsible for renewed cable construction. For example, the first cable built after the 1910-to-1923 pause was of essentially of the same design as those before the pause. However, the loaded cables laid in 1924, 1926, and 1928 followed upon materials technology breakthroughs. The improved cable designs following pauses were therefore more likely the result of improved understanding gained in the intervening years than the other way around—cable construction induced by technological advance.

B. The International Telegraph Industry

Figure 5–3 shows how cables reached out to embrace the globe. Figure 5-4 depicts the growth and relative mileage of domestic and international submarine cables. By 1871 the major land masses of the world were joined together. But that was only the beginning. Multiple cable routes were necessary—sometimes for the volume of traffic, always for reliability. It proved profitable to connect even smaller units together. Isolated islands with relatively few inhabitants were suddenly thrust into immediate contact with the rest of the world.

As with other communications systems (postal service, landline telegraphy) internal cables across rivers and harbors, and to and between off-shore islands, were developed by governments. But most international lines were financed and managed by private business (Fig. 5–5).[15] The major exception was the British Pacific cable at the turn of the century, which the government undertook only reluctantly after private firms refused to accept the risk. One advantage of private ownership was the lack of any need for intergovernmental agreements. When the British nationalized their cable industry after the Second World War, they purchased all the stock and left the form of the private organization intact. Existing agreements with other countries thus could be left untouched and negotiations could be conducted as if by a private corporation.

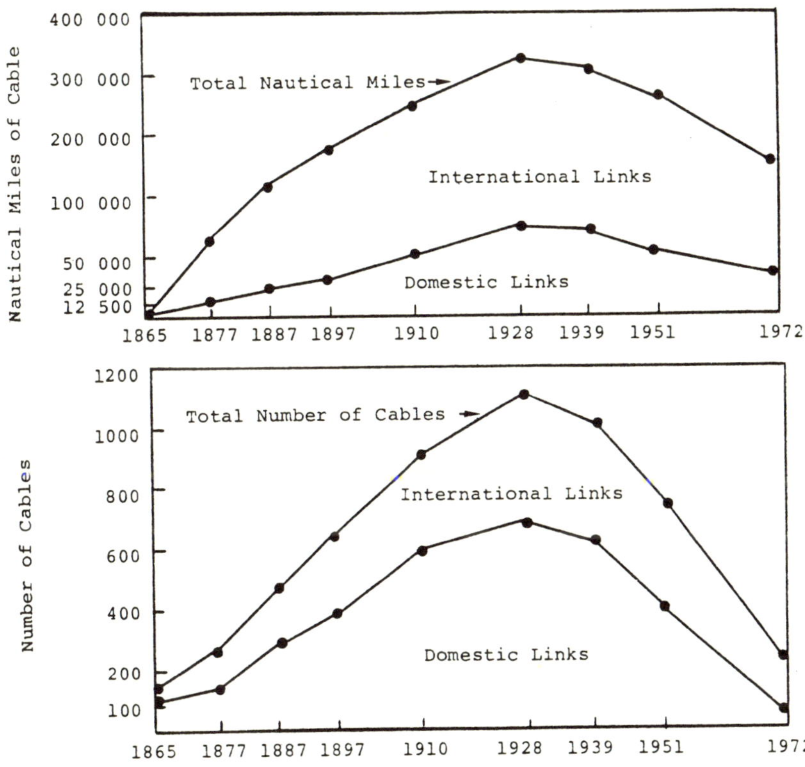

FIG. 5–4.—Length and number of all cables more than 5 miles long, from 1865 to 1972.

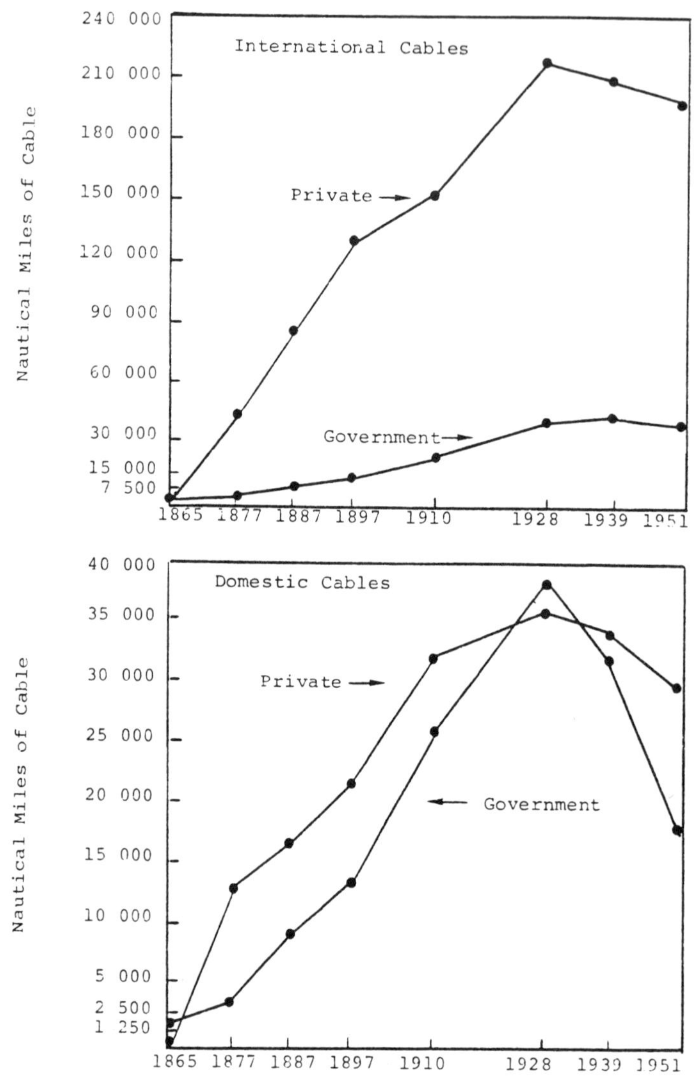

FIG. 5–5.—Length of international and domestic cables under government and under private ownership, 1865 to 1951 (after which joint ownership makes distinctions impractical).

Exploring for Factors Correlated with the Spread of Cables

The most common use of correlation analysis is to verify the importance of a factor once it is identified, or to justify an hypothesis. Technology assessment is an exploratory discipline, as much concerned with what might happen, what the options are, what the relative importance of alternatives are, or what might have been overlooked as it is with formulating and proving scientific hypotheses. It is in this exploratory sense that rank correlation analysis is used in this section to sift among the more obvious factors associated with the spread of submarine telegraphy on an international scale to assess their relative importance and range. Rank correlations are easier to treat than full factor analysis, especially when the factors involve a fair degree of judgment on the part of the analyst before a common framework of comparison can be achieved. Factors with strong associations then guide a more detailed examination of many related sources of evidence. In retrospective technology assessment, such evidence comes from the historical record.

The number of miles of cable laid in 1877, 1897, and 1939 (Fig. 5–3) provides a global pattern of the spread of international submarine telegraphy. Certainly one would expect that the total miles of cable that were laid in any one region of the world would be an important factor in determining the early pattern of network development. Total mileage is listed in the first column of Table 5–5; the rank order of mileage from the area of least effort (3000 miles interconnecting Japan and the Far East) to the area of greatest effort (45 000 miles for the North Atlantic routes) is given in the second column.

The relative level of effort in each region, measured against that necessary to create a mature cable network for that region, can be obtained by ranking the *percentage* of the total which was completed at both intermediate dates—1877 and 1897, respectively. The percentages are given in the 4th and 14th columns; their ranks (from greatest effort ranked first to least last), in the 5th and 15th columns.

In addition to the total mileage laid, the importance of the connection to the growing British Empire would be expected to be a major factor in the sequence of cable construction. Britain's colonial interests and naval superiority would suggest that submarine telegraph connections from England through the Mediterranean and the Middle East to India and from there on to the Far East or down to Australia would be the backbone of their network. Connections to British possessions in North America, the Carribean, and Africa would tie the remaining pieces together. The growing technical and mercantile interests of the still underdeveloped USA would add increasing importance to the North Atlantic, American Coastal, and South American routes. Finally, the transpacific routes would tie the USA to her newly acquired island territories and provide Britain with an alternative route for crucial messages to the Far East and Australia.

We may estimate how important the connection was to the British Empire from the military strength in the region or the amount of trade with the region. Military strength has been selected because the data are readily available. These troop concentrations and naval-deployment patterns are listed in Table 5–5 (columns 6, 7, and 8 for 1877; and 16 and 17, for 1897). U.S. interests were considered minor in 1877 but increased somewhat by 1897 for the parts of the world where having an influence was most crucial. Some judgment had to be exercised in order to provide a match between the regions used to describe cable spread and the location of British military bases; these considerations are described in the footnotes to Table 5–5.

In 1877 British influence and the number of miles of cable needed to connect the region were unrelated; in fact, there is a slight anticorrelation (Table 5–6). The total mileage contributed only somewhat less to an understanding of where cables were to be found than did the degree of British influence—correlations of 0.52 and 0.60, respectively. Therefore, a combined measure was constructed by the addition of the rank of British influence to the rank order of mileage; the resulting numbers were again ranked. This rude composite measure is strongly correlated (0.81) with the actual ranking of the percentages of cable mileage laid by 1877. As expected, the combined measure is correlated with British influence and total miles in roughly the proportions used in its construction. The residual variance is well within the degree of judgmental freedom used to construct the British influence factor; that is, an average difference in ranks of only two units per each region.

By 1897 the total mileage was no longer a significant factor in determining which regions were to develop first; its rank correlation had fallen from 0.52 for 1877 to only 0.20—not even significant at the 20% level. British influence remained as significant as before, 0.6l compared with 0.60. The pattern of development in 1897 was only somewhat correlated with that of 1877: 0.49 occurs at random in one out of ten cases.

Close examination of the differences in ranks between the pattern of cable laying in 1897 and British influence reveals a large deviation (5.5) for Northern Europe and the Mediterranean—that is, the regions surrounding Europe. Somewhat smaller deviations (4.5 and 4) occur for South America and the Far East and Japan. Only the latter seem unrelated and a product of the simplicity of the analysis. (One should note that trade with South America was important to Britain even though her military presence does not reflect that importance.)

TABLE 5–5.—Ranks for the rate of cable network spread throughout twelve regions of the world. (Columns 1-2: completed system, 1939; 3-12: early system, 1877; 13-21: 1897.)

Column number	1	2	3	4	5	6	7	8	9	10
Region	Miles of cable (thousands)	Rank	Miles of cable (thousands)	% of 1939	Rank	Troops (thousands)	Ships	Sailors	U.S.	Rank[a]
1. Middle East,Asia	15	3.5	6.6	44	1	63	22*	2111*		2
2. Philippines and Australia	23	10	7.6	33	4	6.3	20*	2091*		3.5
3. Africa	32	11	0	0	11.5	3.5¶	11	1199		5.5
4. Far East and Japan	3	1	1.2	40	2*	3.3	11*	1387*		7.5
5. Caribbean	11	2	4.3	39	3	4.1	8	848**	U.S.	5.5
6. Mediterranean	21	7	6	29	5	9.8***¶	16	2035		3.5
7. North Atlantic (transatlantic)	46	12	12	26	7	1.8†	8	848**	U.S.	7.5
8. North American Coast	15	3.5	2	13	8				U.S.	10
9. Northern Europe	22	9	6	27	6	96	59	639		1
10. South America	16	5	1	6	10		5‡	313‡	U.S.	9
11. South Atlantic	19	6	2	10	9		‡	‡		12
12. Transpacific	21	8	0	0	11.5		3*	389*		11

*Pacific fleet divided one-third to region 12 and two-ninths each to Regions 1, 2, and 4.
**Split of North American fleet.
***Includes half of recently removed Ionian garrison.
†Canadian forces reduced from 8230 in 1851 to 4268 in 1961.
‡Derivative need to connect to South America.
¶Half the Gibraltar garrison.

A combined measure was constructed by beginning with British influence and reducing the relative rate of cable development around Europe prior to 1897. This combined measure significantly improved the rank correlation with the pattern of cable laying. What then could justify a relative reduction in the rate of construction surrounding Europe? Several hypotheses seem plausible:

- Continued economic growth of the Western nations, Europe in particular, outstripped that of the rest of the world and therefore demanded more cables than other areas from 1897 to 1939.
- Britain's domination of the seas inhibited other nations from developing a cable system of comparable extent until the 20th century.
- International communication by radio permitted other nations to circumvent the need to build long-distance international cables, so continued cable construction concentrated on Europe itself where spectral congestion, secrecy, and access to alternatives gave cables their greatest advantage.
- The overseas cables may have had more than enough capacity to carry the increasing message traffic. They may have been laid to create a connection and to provide alternative routes. However, growing traffic surrounding Europe may have required increased capacity.

Clearly, the inhibiting force of technical difficulty influenced the first decade of successful long international submarine telegraphs. But by the end of the century, after 40 years of development, the physical barriers were no longer important. Relative to other regions, Europe underwent a burst of cable construction between 1897 and 1939. This analysis guides the examination of the historical record to follow, from early British dominance to the rise of international competition for cable and radio connections.

Table 5–5. (Cont'd.)

11	12[b]	13	14	15	16	17	18	19[c]	20	21[d]
Sum (combined)	Rank	Miles of cable (thousands)	% of 1939	Rank	Troops (thousands)	Ships Sailors	U.S.	Rank	Diff. (combined)	Rank
5.5	1	16	106	1	73	Singapore		2	1	1
13.5	6.5	18	78	4	Colonial	Singapore		4	0	2.5
16.5	9	26	81	2	14.4	Capetown		4	2	2.5
8.5	3	2.9	80	3	8.2	Singapore		7	4	6
7.5	2	6	54	6.5	5.3		U.S.	7	0.5	6
10.5	5	11	52	9.5	17.9	Malta, Gibraltar		4	5.5	9
19.5	12	24	52	9.5	1.8		U.S.	7	2.5	6
13.5	6.5	8	53	8			U.S.	9.5	1.5	9
10	4	12	54	6.5	132	Gibraltar, Dover			5.5	4
14	8	9	56	5			U.S.	9.5	4.5	9
18	10	5.4	28	11				12	1	12
19	11	0	0	12			U.S.	11	1	11

Combination Rules: a. Average British troops and naval strength; subtract one unit for U.S. interests; rank.
b. Sum = Rank (British Influence) + Rank (1939 Mileage)
c. From troops and fleet subtract 3 for U.S. interest; carry over Australia from 1877 to reflect colonial troops.
d. Ranks for British influence were increased by 5.5 for Northern Europe and the Mediterranean; the resulting sums were reranked to produce the combined measure.

TABLE 5–6.—Rank correlations and their statistical significance for the model of international submarine telegraph development; top number is rank correlation, bottom number is significance of correlation (after M. W. Tate and R. C. Clelland, *Nonparametric and Shortcut Statistics*, 1957), where

$$\text{Rank correlation} = 1 - 6 \sum_i^N [(R_i - R_i')^2] / [N(N^2 - 1)]$$

		1887				1897			
		British	Total mi	Combined	% Laid	British	Total mi	Combined	% Laid
1887	British	•	−0.15 Nil	0.56 5%	0.60 5%	0.98 <0.1%			
	Total mi		•	0.66 5%	0.52 10%		1.0 Perfect		
	Combined			•	0.81 1%			0.49 10%	
	% Laid				•				0.44 10%
1897	British					•	−0.22 Nil	0.85 1%	0.61 5%
	Total mi						•	−0.16 Nil	0.20 Nil
	Combined							•	0.83 1%
	% Laid								•

Early British Dominance

From the first, Britain completely dominated the manufacture, laying, and financing of submarine cables. The early years of effort produced a group of British-based financiers, manufacturers, scientists, and engineers with the necessary experience and organization to duplicate their 1866 success elsewhere.

An important member of this group was John Pender (Chap. 1), a director of both Atlantic Telegraph Co. and Anglo-American Co. and chairman of Telegraph Construction and Maintenance Co. (TC&M). TC&M, then the largest and most experienced firm in the field, soon controlled much of the world's supply of gutta percha, the most satisfactory insulating material for submarine cables until the late 1930s.

Within two years of the Atlantic success, Pender and other associates of the Atlantic venture began to establish the more important cable routes of the Empire: cable links with the British Mediterranean possessions, India, the Far East, Australia, and South America. The rapid development of these routes reflected the eagerness of the private British investor to support cable ventures, as well as the continued technical and financial success of the Atlantic cables in the first few years of operation, the availability for reinvestment of £8 million returned to the shareholders in the U.K. when domestic landlines were nationalized in 1868–1870, and the resulting indirect promotion of confidence in telegraph enterprises.

Submarine telegraphy permitted Britain to free herself from dependence on European land lines. A Malta-Alexandria cable was laid in 1868 by Anglo-Mediterranean Telegraph Co. and within two years the Falmouth, Gibraltar and Malta Telegraph Co. linked Malta to the United Kingdom. In 1870 Egypt and India were connected via the Red Sea route by British India Submarine Telegraph Co. A related venture was introduced by Marseilles, Algiers and Malta Telegraph Co. to funnel French traffic over the developing cable network to India.

In 1872, under Pender's leadership, these four cable-operating companies were amalgamated to form Eastern Telegraph Co., which provided a single system of submarine cables from England to India controlled and operated by private British interests. During the next two decades the company continued to expand and extend its routes, particularly in the Mediterranean, where over 80 cables were laid.

In 1869 British-India Extension Co. had established cable communications between Madras and Singapore via Penang. By 1871 Pender's China Submarine Telegraph Co. laid a Singapore-Hong Kong cable via Saigon and eventually extended the system to Japan. The company shared the Shanghai route with Great Northern Telegraph China and Japan Extension Co.—an arrangement preferable to competing with this Danish firm's developing landline system to the Far East. The same year, 1872, Pender's British Australia Telegraph Co. linked Australia with the outside world.

In 1873 John Pender amalgamated the three companies beyond India to form the Eastern Extension Australasia and China Telegraph Co. and actively set about filling the gaps in the Far East. Australia and New Zealand were linked in 1876, a second Penang-Singapore cable was laid in 1879, and Hong Kong and Manila were connected in 1880. Numerous other new links and duplication of existing routes were carried out through the 1890s.

In South America, beginning in 1873, two Pender-controlled companies established a Brazil-Europe cable route and linked the major ports along the Brazilian coast by cable. In 1879 TC&M laid cable down the east coast of Africa (Aden-Zanzibar-Mozambique-Delagon Bay-Durban) to South Africa for the newly formed Eastern and South Africa Telegraph Co. Again in time the route was duplicated in sections and further developed by connection to Mauritius off the coast of Madagascar in 1893. On the west coast of Africa, beginning in 1886 a number of cables were laid, principally to serve French and Portuguese territories. Within three years all major cable facilities serving both coasts were controlled by the "Eastern" group.

Pender's cable empire grew in the Americas as well. In the West Indies, beginning in the late 1860s, West Indies and Panama Telegraph Co. and Cuba Submarine Telegraph Co. had established a number of local cable links. Three decades later, in the 1890s, a Bermuda-Halifax (Nova Scotia) cable was laid and a few years later Jamaica was linked to this route at Bermuda.

The Boer War stimulated an "Eastern"-laid cable from Cape Town to St. Vincent via St. Helena and the Ascension Islands, as well as other cable projects that improved the reliability and speed of communications to South Africa. A similar development, the improvement of British Strategic Communications to China, occurred at approximately the same time as a result of the Boxer Rebellion.

When John Pender died in 1896, his cable empire, known as the Eastern and Associated Telegraph Companies, operated over 50 000 nautical miles of cable and was the giant of world telecommunications; it had the influence of a monopoly and a future closely linked with that of the British Empire. The achievements of the Eastern and Associated companies reflected the domination of cable technology by British firms, the financial advantage of operating as a conglomerate, and the positive support of the British government for a system of privately owned cables.

Nationalism and Cables

Beginning in the 1880s a non-British cable industry started to take form as the industrialized western nations realized that a national capability to manufacture, lay, and repair cables was essential if dependence on Britain was to be avoided. In France the Minister of Posts and Telegraphs in 1887 supported a scheme for linking French colonial possessions in the Western Hemisphere. By 1890 a French company, Société Générale des Téléphones, was capable of manufacturing both a cable core and the outer covering, and in 1895 Compagnie Française des Cables Télégraphiques was formed from the merger of the existing two French operating companies. French cables were then laid on the transatlantic route, in the Western Hemisphere, in New Caledonia, and elsewhere in the Far East.

The Italian cable industry developed at approximately the same time. The Pirelli company of Milan carried out limited cable projects in the late 1860s, and by 1886 was able to manufacture a complete submarine cable. Some cables were laid in the 1890s, notably a Spain-Balearic Islands cable for the Spanish government. Not until the 1920s did Italy launch a major effort in cable expansion, and then foreign manufacturers accounted for three-fourths of the output.

In Germany the firm of Felton and Guilleaume built and laid a short (less than one mile) cable in 1854, but as late as 1896 the first German North Atlantic cable was too great an undertaking for the firm; consequently, TC&M carried out most of the project. Two years later Germany built a new submarine cable factory and acquired a cable ship. The next year a new company, Norddeutsche Seekabelwerke, was created by Felton and Guilleaume and by German Atlantic Telegraph Co. A second Atlantic cable was laid in 1903-04 and several new routes were established to German colonies in the Pacific. Between 1908 and 1911 a German cable to South America was placed in operation, followed by cables to possessions in Africa and elsewhere. The German cable industry, although slow to develop an overseas capability, expanded vigorously once established.

In contrast with Germany, American companies operated a number of cable systems at the turn of the century but the local cable manufacturing capability was less developed. Simplex Electric Co., one of the few American firms able to manufacture submarine cable, faced the same problem as most other firms: a shortage of gutta percha, which was controlled by British manufacturers.

Thus a handful of nations were capable of manufacturing and laying submarine cable by 1900, but approximately 70% of cable was manufactured and laid by TC&M, and 75% of the international cable routes were probably financed by British capital.

Although only 17% of the world's cables by length were government-owned in 1900, most cable systems had some form of state support, ranging from direct subsidy to political support in the international arena. The distribution of cable systems, by country, for cables in existence and under construction in 1902 is shown in Table 5–7 and Fig. 5–6.

Until 1883, the transatlantic cable routes were controlled by the "pool" monopoly agreements among British, French, and American owners. There was no cable owned and operated solely by Americans. John W. Mackay, who made his fortune at the Comstock silver lode, and Gordon Bennett, publisher of the New York *Herald*, felt that rates were too high and there was opportunity in constructing a competing system. They founded the Commercial Cable Company and proceeded to

TABLE 5–7.—Distribution of cable ownership in 1902.

Country*	Length (Nautical Miles)
Great Britain	140 332
United States	39 495
France	19 494
Germany	12 362
Denmark	7 303
Japan	2 150
Spain	1 740
The Netherlands	1 240
Italy	1 072
All Others	1 709
Total	226 897

*The assignment of cables by country is a distortion in one sense. Most cables were privately owned and therefore the assignment of a national category reflects where the firm is registered rather than where the cable system is located. For example, over half of the United States total reflects the North Atlantic holdings of two firms, most of whose cable was between Canada and Great Britain.

lay two transatlantic cables. (The exclusive landing rights of the Anglo-American Company had expired.) Rates were cut and a price war with Jay Gould's cables ensued. With the construction of their third cable in 1894 Commercial controlled a third of all transatlantic telegraph capacity. By the turn of the century this share had fallen to 25%; but in 1912 Western Union leased all Anglo-American's cables, which gave the U.S. firms control of 83% of all capacity. The transition was complete by 1928. Britain retained only her portion of the German cables obtained after World War I; France had an equal share of these cables and one of her own; and Commercial sold out to International Telephone and Telegraph. The U.S. firms now controlled 90% of the total transatlantic capacity. This shift of ownership reflected the growth of American capital more than nationalistic control of the communication routes.

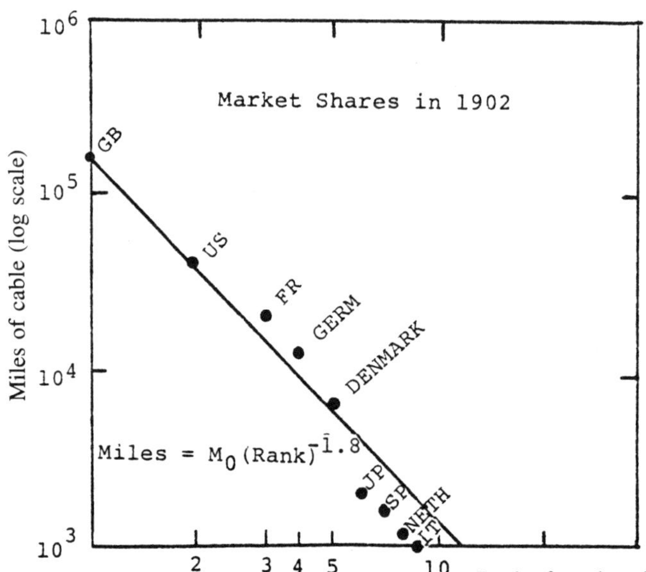

FIG. 5–6.—Highly concentrated cable ownership when ranked in decreasing order follows approximately the Pareto distribution form with an exceptionally high concentration; decreasing with a −1.8 power (−1.0 is typical).

The early 20th century brought a continued emphasis on national systems in worldwide telecommunications. Great Britain's system grew from the revenues of commerce rather than national prestige in empire building or the desire to exert naval authority over the seas. But commercial incentives were insufficient inducement for the Eastern and Associated Telegraph companies to construct a transpacific cable. Pressure from New Zealand, Australia, and Canada led Great Britain to enter into a cooperative agreement to lay the world's longest cable section as a part of the Pacific link from Canada to Australia, thereby providing an alternative strategic route to the Far East. The German and French government-supported cable projects continued until 1914.

In the USA John Mackay's desire to extend a competing network around the globe to challenge Western Union's domination brought private investment support for an American Pacific cable from the West Coast through Hawaii and Midway on to the Phillipines. The USA never did nationalize its cable network.

At the outbreak of World War I, all German cables were cut or diverted for Allied use. After the war, these cables were distributed among the Japanese, French, British, and Americans. Germany retained its cable-manufacturing capability and eventually laid a number of new cables.

France also manufactured and laid cables between the World Wars. The French government supported a number of new cables in the Mediterranean but the major private firm, Compagnie Française des Cables Télégraphiques, sold off its Caribbean cables and its cable ships. After World War II, the government assumed the responsibility for maintaining the firm's remaining cables, but the company continued to operate them. Eventually, all French radio and cable comapnies were nationalized and combined into one orgainzation.

Radio: A Competing Technology

Radio offered serious competition for the first time in the 1920s. Research by the Marconi company on beam transmissions increased the range and substantially reduced power requirements. The result was a highly competitive system, which the British Post Office immediately adopted in 1926-27 for radio communications with Canada, Australia, South Africa, and India.

The effects of the system on cable revenues were almost immediately apparent. Members of the Commonwealth requested the British government to evaluate this turn of events. In 1928 an Imperial Wireless and Cable Conference recommended a merger of radio and cable facilities of the Empire into a single comapny, Imperial and International Communications, Ltd. Several important considerations entered into this decision: concern over comparative costs, national security, dependability of service, and the distribution of services. The capital cost of beam radio installations was much lower and the capacity of the system much greater than existing cable systems. By 1928 there were unmistakable signs that the strong financial position of the existing cable companies would be eroded. If unrestricted competition were permitted without some form of support the private cable networks would in time have to liquidate their assets or sell out to foreign interests.

Foreign ownership was not an acceptable solution for reasons of national security and because radio did not fully duplicate existing cable services. Radio transmissions were subject to fading and occasional prolonged interruptions caused by atmospheric conditions — a handicap to both commercial and political users. Moreover, maintaining a back-up means of communications was essential for security. Furthermore, an adequate number of radio facilities would not be available for some time to provide service to many areas currently linked by cable.

The problem was how to preserve the economically vulnerable cable networks and still benefit from the new radio technology. Possible solutions included: subsidy, minimum revenue guarantee, and a pooling scheme or fusion of all cable and radio interests. Fusion or amalgamation was chosen as the most permanent and lasting solution consistent with national goals.

Imperial and International Communications, Ltd., acquired the cable interest shares of the Eastern and Associated Telegraph companies and the international communications portion of the Marconi company. It also acquired the Post Office's Imperial Atlantic cables and new beam radio stations. The final result was a government-guaranteed private company to provide communications through two media — radio and cable.

In 1934 the company's name was changed to Cable and Wireless, Ltd. To consolidate imperial communications and contend with a world depression, several cable systems were liquidated. With the outbreak of World War II, traffic on the imperial communications system increased, eventually to 400 per cent, but a considerable number of the company's cables were interrupted, especially in the Mediterranean and the Far East.

Cable and Wireless emerged from the war intact but a growing concern for national control of telecommunications led representatives of Britain, Canada, New Zealand, Australia, South Africa, Southern Rhodesia, and India to meet in July 1945 in a Commonwealth Telecommunications Conference. The Conference recommended government purchase of all shares of Cable and Wireless and of all private interests in the overseas telecommunications services of Britain, the Dominions, and India. In November 1946 Cable and Wireless was nationalized. The European pattern of development seemed to lead inexorably toward nationalization.

On the North Atlantic route, where private American companies owned most of the cable systems, radio posed less of an economic threat and national security considerations were minimal. Consequently, the role of the U.S. government in the regulation of international telecommunications carriers was minimal.

In spite of the large concentration of traffic over the North Atlantic cables, no new transatlantic cables were laid between 1928 and 1956, largely because of the growing competition from transatlantic commercial radiotelegraphy, radiotelephony, and international airmail services. Figure 5–7 and Table 5–8 illustrate the growing volume of radio traffic handled by U.S. carriers. Cable traffic ceased to increase, attracting a decreasing share of the total market. In short, competitive radio technology between the wars had an important adverse effect on transoceanic cable systems and on the cable industry as a whole.

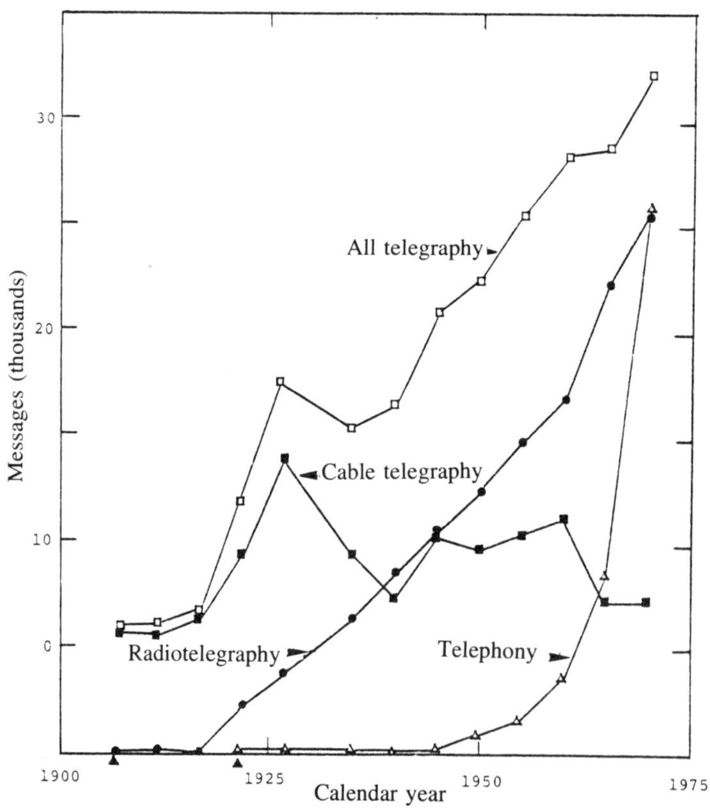

FIG. 5–7.—Growth in international message volume (from Table 5–8); telephony has grown exponentially since 1922. (From the U.S. Department of Commerce *Historical Statistics of the United States*.)

TABLE 5–8.—Shift of traffic from cable to radio.

Year	Message volume				Total	Telephone traffic growth (messages)
	Telegraph cable	%	Radio	%		
1907	5 869	97	155	2.6	6 024	
1912	5 841	95	280	4.5	6 121	
1917	6 451	98	122	1.9	6 573	
1922	9 603	80	2 365	20	11 968	10
1927	13 987	79	3 778	21	17 765	12–23
1935	9 050	58	6 619	42	15 669	28
1940	7 667	46	8 952	54	16 619	70
1945	10 531	50	10 516	50	21 047	360
1950	9 969	44	12 609	56	22 578	1 000
1955	10 671	42	14 971	58	25 642	1 742
1960	11 186	40	17 092	60	28 278	3 713
1965	6 487	22	22 363	78	28 830	8 108
1970	6 548	20	25 693	80	32 241	25 813
Annual growth rate					2.7%	19.1%

The reaction by the British Empire and the United States to radio as a competing technology differed considerably. Britain relied on a cable-based telecommunications network, with the United Kingdom, Commonwealth members, and colonies involved to varying degrees in the ownership and operation of imperial and domestic cable systems and other means of communication. In contrast, fewer U.S. international telecommunications links were critical for national unity, nor had the state an active role in the ownership or operation of telecommunications facilities. The U.S. Navy's early dominant position in radio communications ended in the early 1920s and competition was limited to the private sector. In one sense radio was a liberating technology for the USA because it provided a worldwide telecommunications capability not dependent on British cables and a European-centered network.

Cable dominance of international telegraphy ended by 1928; from that time onward, all growth in telegraph message volumes was absorbed by the radio network. The message volume by cable remained constant as it provided a complementary means of communication for the occasionally undependable radio links. While telegraph traffic continued to grow at 2.7% per year, a rate comparable to growth in real GNP, radiotelephone traffic grew at a steady pace of 19% per year—over seven times faster. It took nearly 40 years for international radiotelegraphy to rise from a 20% share of the cable telegraph market to a 60% share. If the relative growth rates of international telegraphic and telephonic communications continue, telephony will have gone from 20% of the message traffic to 60% in only 12 years. (Telephony has risen from 22% of total message traffic in 1965 to 44% by 1970 and was estimated to have reached the 60% mark by 1977.) Telephony, as a service, appears inherently to grow faster than telegraphy, even when both were using the same technology—radio transmission.

C. International Telephone: Cables and Satellites

Electronics has brought a technological revolution to international communications. Beginning in 1956 with the laying of the first transatlantic telephone (TAT-1) cable, growth has continued unabated. TAT-1 had 48 voice channels, equivalent to 432 telegraph channels in information capacity or "bandwidth." By 1976 TAT-6 was laid with 4000 voice channels, an 83-fold increase in only 20 years.

Earlybird, the first transatlantic communications satellite, went into service in 1965. It had 240 voice circuits. Recent satellites have up to 10 000 voice circuits, a 42-fold increase in only 12 years.

The rate of increase in technological capability is thus increasing rather than abating. This development reflects the rapid refinement of solid-state electronics, which displaced the former vacuum-tube circuits. From 1956 to 1965 the submarine cables were increasing in capacity from one

TABLE 5–9

(a) Shift of transatlantic telephone traffic from cable to satellite.

Year	Cable			Satellite		
	Designation	Circuits	% of total	Designation	Circuits	% of total
1956	TAT-1	48	100			
1959	TAT-2	48	100			
1961	CAN-TAT	160	100			
1962	ICE-CAN	24	100			
1963	TAT-3	138	100			
1965	TAT-4	128–138	69	EARLYBIRD	240	31
1966			53	INTELSAT II	240	47
1968			24	INTELSAT III	1200	76
1970	TAT-5	720–845	45			55
1971			15	INTELSAT IV	5/6000	85
1974	CAN-TAT-2	1840	29			71
1975			15	INTELSAT IVA	10 000	85
1976	TAT-6	4000	29			71

(b) Annual growth rate (per cent) in capacity.

	Cable		Satellite	
	Best cable	Total capacity	Best technology	Total capacity
Growth rate, 1965–1976	27	36	48	40
Growth rate, 1956–1976	28	24		

generation to the next at an average annual rate of 12.5% per year. From 1965 to 1976 they improved at a rate of 36% per year, and satellites improved at 40% per year (see Table 5–9 for a summary). Such rapid development is unprecedented.

As with the competition between cable and radiotelegraphy, satellites—in a single decade—have overtaken cables in supplying the majority of circuits.

Cable Telephony

Since 1950 over 50 000 miles of submarine telephone cable have been laid. Over 150 cables are in operation throughout the world and approximately 50 more are scheduled to be laid by 1980.

Today's cable industry, which includes the manufacture of the cable, the submerged repeaters and equalizers, and the terminal equipment, represents the wedding of traditional submarine-cable manufacturers with the telephone industry. Standard Telephones and Cables, a major firm in the industry today, had its origins in telephone instruments and at one time was part of Western Electric Co. in Britain. In 1925 the company became part of International Telephone and Telegraph Corp. (ITT) and moved into submarine cable telephony. American Telephone and Telegraph Co. (AT&T) entered the field in the 1920s in the development of coaxial cable and repeater technology, which paralleled research efforts in Britain.

Several companies currently have the capability to manufacture cable, including Norddeutsche Seekabelwerke in Germany, Cables de Lyon in France, Ocean Cable Co. in Japan, Standard Telephones & Cables in the U.K., and ITT Cable & Hydrospace Inc., and Simplex Wire and Cable Co. in the U.S. A variety of companies, including Western Electric, Compagnie Industrielle de Télécommunications, and Standard Telephones and Cables manufacture repeaters and terminal equipment. Thus several major industrial nations are able to manufacture and lay modern submarine telephone cables.

Satellite Telephony

The introduction of satellites was stimulated by a drive for national prestige and the space program rather than by a recognized need in the telecommunications field. The national Aeronautics and Space Administration (NASA), a government agency, developed the launch-vehicle technology, in contrast to the development of transoceanic telephone cable by private industry. AT&T's introduction of TAT-1 in 1956—although a major technical innovation in international telecommunications—stimulated few issues requiring an enlarged or redefined role for the government.

On the other hand U.S. satellite telecommunications generated a major debate about ownership—an issue in some way similar to the introduction of beam radio transmission in the British Empire in the late 1920s. Policymakers had to decide on an organizational framework that would provide the most efficient and least costly satellite system. U.S. common carriers sought ownership as a joint venture, arguing that economies of scale could be realized in making satellites an extension of submarine cables. The aerospace industry opposed this plan, demanding a role in management input; if that was not done, the development contracts would tend to go predominantly to firms that were affiliated with the common carriers. The common carriers saw no merit in aerospace participation; neither did the Federal Communications Commission.

With the passage of the Communications Satellite Act of 1962 a modified form of private enterprise was established without aerospace participation. A private corporation, Comsat, was formed, half owned by U.S. common carriers and half by the general public. Comsat was conceived as a common carrier, providing circuits to existing U.S. common carriers, such as AT&T, ITT, and Western Union.

The establishment of Comsat was followed by the launching of the first communications satellite (Earlybird) in 1965. Since then international cable links have continued to grow but satellites have accounted for an increasing share of the traffic volume. The relative growth of these technologies in terms of telephone circuits available is illustrated in Table 5–9. On the transatlantic route—the U.S.-European network—cables accounted for 69% of the circuits available in 1965. By 1973 satellites provided 85% of the circuits. The large circuit capacity of individual cable and satellite facilities causes considerable fluctuation in these percentage figures with each newly introduced facility.

D. Competition and Complementarity

Satellites compete with telephone cables in much the way radio competed with submarine telegraphy. As communications options these technologies are direct competitors; however, significant differences in the manner of providing their services produce complementary relationships as well. A discussion limited to the international communications industry itself suggests possible future implications based on purely heuristic similarities.

Increased national control of international communications may be a continuation of the need to resolve the instabilities of competition. As suggested in Chap. 2, national boundaries were the only barrier capable of arresting the consolidation of telegraphy into a monopolistic unity. The operation of present-day cable systems is even more complex. The government responses in Britain, France, and the USA to the competition by radio and the stresses of World War II led to more government control. State control of international communications facilities is now the rule, primarily because of political and strategic considerations. Also, technological advances have increasingly necessitated organized and cooperative international efforts, which made government participation imperative.

Decisions on the selection of the technology—cable vs. satellite—are influenced by the fact that these systems primarily operate as international telecommunications facilities. For example, Portugal, Spain, and Italy were in favor of laying TAT-5 because the cable would diversify their network system, they would derive revenues from the system as receivers of messages, and the system would provide direct access to the USA.

Security or the interception of sensitive information over either system is a negligible consideration because use of an encrypted mode is now standard practice. The critical element of security is the vulnerability of the system to disruption and availability of alternative modes.

System reliability is possibly the key distinguishing factor separating cables from satellites. No system is flawless, and the need for alternatives and back-ups continue to provide complementarity between the broadcast and cable options.

British policymakers were aware of that even back in the 1850s, when at least two routes to India were deemed necessary to insure dependable communications. Two Americans, Mackay and Bennett, dared not chance a loss of service on a single line if they wished to compete against the dominant transatlantic cable ''pool'' monopoly in the 1880s. Thus Commercial laid two cables in 1884.

Today the large circuit capacity of individual satellites and cables make reliability a crucial consideration. Satellite outages are more frequent and of shorter duration than cable outages. However, satellite repair is still impractical. Duplication and redundancy improve reliability and drive toward a complementary policy of using both technologies.

Economics of cables in comparison with satellites resemble that of the earlier telegraph period when radio was the competitor. Cables offer more economical service on shorter routes. Over longer distances, satellites (as radio before) provide service at lower cost. An exception to this rule may prevail for routes with very high traffic density. This exception seems to hold on landlines as well: microwave relays generally predominate, but waveguides are used for major trunklines with very high traffic density.

Satellites, like radio, have more rapidly increasing capacities and faster decline in costs per circuit. They also offer greater flexibility in routing and even the option of variable routing under command. Because satellite design life is less than cable's, there is a faster turnover of new technology—an especially attractive feature during a period of rapid technological advance such as the present.

Institutional factors weigh heavily in the selection of one technology over another. U.S. common carriers tend to support a balanced mix of both options. Common carriers tend to support cables because of cash-flow advantages from the rate-base increase resulting from cable ownership (in comparison to satellite leasing), the desire to have complete control of the technology (vertical integration of the communications system), reluctance to become dependent on Comsat, and concern over being relegated to a minor role in the industry by supporting the growth of another firm. In contrast, Comsat has not supported the authorization of further cable facilities, arguing that satellites offer superior flexibility, economy, and diversity.

Inherent limitations on either technology are different in kind. Cables face routing difficulties—landing rights, spatial congestion, and tectonic stability. The information capacity, measured in terms of ''bandwidth'' or a range of frequencies, is more constrained in cable technology than for satellites. Coaxial cables use polyethelyne or other substances to separate the central conductor from the outer shield and cylindrical conductor. Even the best of these insulating materials lead to some sluggishness at very high frequencies and become impractical at the microwave frequencies used for satellite relays. One implication of this limit is that television signals could be at first transmitted only with difficulty. Fiber-optical transmission with solid-state amplifers could essentially eliminate this barrier to information capacity.

Like all broadcast communications systems, satellites face the fundamental limit on the information-carrying capacity of the broadcast spectrum. Beamed transmissions provide more ''windows'' in the spectrum, but ultimately there are limits to using ever higher frequencies. The atmosphere itself absorbs microwaves, as it does visible light on rainy days.

At present, satellites have the information capacity or ''bandwidth'' advantage over cables. They can more easily serve broadband users such as television trunkline relays, video conferencing, high-quality facsimile, and trunkline computer or telephone links. The reliability of cable transmissions hold definite advantages for data transmission. If fiber optics are refined in the near future, cable capacities—bandwidths—may rapidly came to exceed those of satellites.

Therefore, one can see that the survival of cable options has always depended on the timing of innovations and hence the attractiveness to investors of the times. Loaded cables came before radio was perfected for international commercial service. Telephone cables using vacuum tubes, and later transistors, preceded the development of a satellite capability.

The future will certainly bring additional technological revelations if recent trends are at all a reliable indicator. National security and the maintenance of alternative service routes will preserve and protect the submarine cable industry. The relative rates of innovation in the two technologies are of concern. Satellites are developing much faster than cables. As long as bandwidth is available in the spectrum and fiber optics do not become a practical reality for submarine cable applications, satellites will retain their competitive economic edge. If the space shuttle and repairable satellite design prove practical, then satellite costs may fall even more. However, in the long run, the spectrum will become excessively crowded and cable will be the only reasonable option. Will the research, development, manufacturing, and maintenance infrastructure be present to meet that shifting demand?

Chapter 6

LESSONS LEARNED FROM A RETROSPECTIVE TECHNOLOGY ASSESSMENT

Introduction

THE RETROSPECTIVE TECHNOLOGY ASSESSMENT of submarine telegraphy has identified and analyzed the long-range consequences of this innovation in the hope that this analysis might provide some guidance in carrying out anticipatory assessments of technologies being developed today. The impacts of the cable are summarized below in Sec. A.

When it became clear in the course of the study that the impacts of submarine telegraphy were much broader and more important than had been thought at first, we tried to determine why, and whether the causes of this large cumulative impact had implications that could guide technology assessors and decisionmakers in selecting other emerging technologies that may also have serious impacts on a global scale. Section B addresses the reasons cable telegraphy stimulated broad social changes.

Section C considers major users of the cables. One approach to the difficult task of forecasting the possible impacts of a technological development is holistic imagination: to call to mind successively and reiteratively every human activity and every segment of the population and of societal structure and ask whether and how that area might be changed by the technology. A somewhat more conservative approach is to examine the steps through which the new technology is likely to make its way into the marketplace and reach wide utilization—the technology-diffusion approach.

Both approaches should be used; each has complementary advantages and disadvantages. The holistic-imagination approach is exhausting to assessors but seldom exhaustive of the subject, which must be the universe. The technology-diffusion approach assumes that social institutions will react to new opportunities as they have to similar options in the past. This rather mechanistic view tends to neglect the imaginative uses and, especially, the misuses and abuses to which people will put a new technology, which are often anticipated by neither the developers and entrepreneurs nor the assessors. If this approach can be made more realistic (which in this case also means more imaginative) it makes technology assessment's anticipation of impacts of technology more credible because it displays the paths by which we may move from "here" (today) to "there" (that point in the future at which the impacts will be realized as societal benefits or societal costs.)

In the first three sections, we deal with the impacts or consequences of submarine telegraphy as we discovered them by historical research, and with the reasons for those consequences; in other words the contemporary or 1976 (retrospective) technology assessment. In Sec. D we move to a second objective of our study, the historical assessment and the hypothetical assessment: What consequences did the Victorians foresee for submarine telegraphy, and how accurate were their expectations? If they had asked themselves the questions characteristic of technology assessment as it is developing today, might their expectations have been more realistic; and if so, would they have acted differently or made different decisions? We do not assume that technology assessment today is asking the "right" questions about social change; on the contrary, we are seeking better insight into the generic problems associated with social forecasting.

Section E pulls together the implications of our findings for anticipatory technology assessment. We are acutely aware that these conclusions—at least in so far as we have tried to generalize them beyond the case of submarine telegraphy—remain highly tentative until they are confirmed, modified, or rejected by other case studies of retrospective assessments. Nevertheless, the suggestions we make here can be useful to technology assessors now.

Section F summarizes the problems we experienced in applying technology assessment retrospectively. None proved unmanageable, and we hope that many more retrospective assessments will be undertaken.

A. Submarine Telegraphy Changed the World

The impacts of rapid interactive communication between North America and Europe, and later between all of the world's continents, significantly changed and expanded human activity on a global scale. Because submarine telegraphy was part of the rich tapestry of the Victorian era and the complex intertwining of technological development and social change in those years, its consequences may not have been fully appreciated by most of us in the 20th century.

Human Activity Changed in Scope and Organization

Submarine telegraphy created a world economy as its effects converged with those of steam shipping, railroads, and domestic telegraph lines. Cabled messages allowed the establishment of world markets in futures trading of commodities and the expansion of arbitrage transactions, led to significant reductions in warehousing and other middleman costs, reduced differences between merchant and consumer prices and between markets, and stabilized costs. In shipping, cables improved scheduling and ended the port-to-port search for cargos, giving the owner effective control over his fleet. This better control effectively expanded shipping capacity and made it possible to distribute consumer goods more efficiently. A transnational economy developed, marked by both large-scale (and eventually transnational) corporations and increased opportunities for competition by smaller commercial organizations and smaller nations.

Cables produced the beginnings of a world polity as well as a world economy. Political problems, trends, and events in all parts of the globe were tied together. Events in remote areas were sharply perceived as part of the pattern of the struggle for power between nations or blocs of nations. Each nation could more easily control its agents from the center of power where policy was formulated and diplomatic strategy coordinated. Responses to either threatening or conciliatory moves could be made and evaluated quickly. As communications became easier and cheaper, international organizations began to operate on a continuing basis and nations were able to cooperate in transnational efforts such as apprehension of criminals, control of epidemics, and disaster relief.

The public in many countries became newly alert to the consequences of their governments' actions in foreign affairs and in foreign wars. International news services were organized and even local newspapers began to carry foreign news. Beginning with the British blunders in the Crimean War, this publicity produced new pressures and new constraints on decisionmakers. Governments also soon realized that in times of hostility, useful information could be gathered from the citizens of an enemy country and quickly transmitted to their own military strategists. This unanticipated consequence of submarine telegraphy greatly increased the usefulness of military intelligence and created its offspring, censorship. Propaganda developed as governments realized that another nation's public, as well as their own, could now be manipulated by the conveying of false (or selected) information.

The ability to respond quickly to military threats, and the ability to mobilize and rapidly deploy armed forces over large distances, created both the opportunity and the necessity for a state of continual military readiness. Combined with the realization of the new influences of publics over decisionmakers, submarine telegraphy (together with railroads, land telegraphy, steamshipping, and powerful weaponry) must be considered a major contributor to the development of the concept of total warfare.

The dual effect of the new communications system—integration and improved control from the center, and increased participation from remote locations (centralization and dispersion)—was a characteristic effect of submarine telegraphy in many areas of human activity. It appears to spring from the nature of rapid two-way communication of information between people and activities who were previously connected only by correspondence over periods of weeks or months.

This double effect in turn created another double effect—more effective response to crises, and creation of new vulnerabilities. Central decisionmakers (whether commercial, political, or military) could quickly intervene in developing crises by transmitting orders. However, they were now subjected to pressures and constraints produced by public awareness of their actions. Participants at many locations (foreign service officers, generals in the field, business agents) contributed to central decisionmaking, but were themselves also subject to more continuous control. Systems of markets, of nations, and armed forces became more complex and interdependent when communications improved. They also became more vulnerable to disruption when breakdowns in these interdependencies occurred. The two themes—centralization-dispersion and control-vulnerability—crop up over and over as we consider the effects of submarine telegraphy.

By allowing a new level of integration and coordination, submarine telegraphy created new societal control systems. However, establishment of control systems, being a response to perceived risks, tends always to lag behind those risks and to require cumbersome institutional and power adjustments. It seems to be more difficult to establish societal control systems for a world polity than those for a world economy, perhaps because economic values are more universal than political values.

Submarine telegraphy, we have suggested, was a major factor in creating the institutions of the modern world—an interdependent world economy, a struggling world political network, the threat of total war. But in some areas where 19th century observers forecast enormous impacts, the effects were very modest. Victorians confidently expected that the Anglo-American value system, culture, education, and religion would be rapidly embraced by the "heathen" nations through the umbilical cord of the telegraphic cable. Although rapid communications did contribute to the homogenization of world culture, this transfer was far more interactive than the Victorian assumed it would be—and far less effective in producing common interests.

Educated citizens in the 1860s had also looked forward to a great stimulus for science. Submarine telegraphy contributed to advances in a few fields of applied sciences, such as oceanography, marine biology, and weather forecasting, but these contributions were relatively modest. The value of scientific exchange seldom depends on the rapidity of communication. But because of its business and military applications, submarine telegraphy forced the standardization of time zones and development of precision in measurement.

The more subtle social and psychological impacts of rapid interaction among continents may well have been fully as important as those summarized above. The perception of time and distance changed; for those with the responsibility for intervening authoritatively in remote events on behalf of the nation and for those who daily wagered their fortunes on a rise or fall of prices or the availability of a cargo in a distant market. But individual citizens were also affected. They became newly sensitive to the direct impact on their lives of battles, diplomatic transactions, crop failures, or business manipulations in countries of which they had been only vaguely aware. The rapid growth in the public's demand for news served as evidence of this new awareness and seemed to create both a sense of personal participation and, according to some observers, a sense of anxiety. Both may have contributed to our modern world view.

The Innovation Process Had Consequences Independent of the Technology

By the "innovation process" we mean here the many steps that must be accomplished between the time a new technology is first conceptualized or invented and the point at which it becomes available for use. These steps can include R & D, design, gathering of investment capital, pilot-plant operations, and many other intervening steps. Government actions such as subsidy, licensing, or regulatory requirements may be involved. As technical, financial, management, and political problems arise and are dealt with, these actions and decisions may set in motion consequences which are important—consequences aside from the effects of the technology itself.

Long before the Atlantic cable began to operate successfully, the struggle to plan and implement this technology produced consequences with a long-lasting impact on society. Electric communications was one of the first of the modern technologies to develop from the deliberate application of newly discovered scientific principles, rather than from inventive "tinkering." Such technologies effectively conveyed to the public the idea that solutions to practical problems and national needs could be sought and found through the application of science.

The cable experience showed that the risks and costs of the traditional inventor-entrepreneurial approach—full steam ahead, and solve problems as they arise—were too great in "high technology" enterprises. Cyrus Field's strategy of dealing with problems as they arose was ultimately successful. But along the way decisions made on the basis of financial and political imperatives alone proved to be costly blunders that delayed and endangered the success of the project.

The 1861 Commission of Inquiry, a cooperative investigation by industry and government, established the value of a more analytical approach to planning, of close involvement of technical experts, and of constant quality control. Each has since become a basic tenet of science-based industry.

The laying of the first long-distance cable was carried out by entrepreneurs with minimal government involvement. This approach was possible and practical at a time when the economy was rapidly expanding, profits were lightly taxed, and risk capital was relatively plentiful. However, both the need for and availability of investment capital were underestimated at the beginning of the enterprise, as in present-day contractor "overruns." As happens today, government money was sought to bail out the enterprise, but unsuccessfully. Field's determination and perserverance allowed him to raise new capital again and again, but the effort drove him irresistibly toward three consequences: diminishing his own share and control in the project, shifting of control from America to Britain, and the formation of monopolies. The way in which technology was implemented led to an industry structure which was vertically and horizontally integrated and which set a pattern for later point-to-point telecommunications industries. British dominance of the telegraph-cable industry had profound diplomatic and military consequences well into the 20th century. British control of the seas was strengthened for a time; Britain was able to control the flow of information to the rest of the world during much of World War I. Other European nations finally resorted to laying their own government-owned cables.

Few or no political innovations can be traced to the *initiation* and *implementation* phases of the technology. Public policymakers treated the unique technology in terms of existing categories and precedents (that is, as conventional mail-delivery contracts) even though these patterns were not fully appropriate; and as a consequence failed to make policy innovations that might have avoided or mitigated later social costs such as unchecked commodity speciulation, or unrestricted government censorhsip in wartime. Since public policymakers reconciled and reconfirmed existing national policies of economic-territorial expansionism, fiscal conservatism, and laissez faire, the impact of submarine telegraphy on public policy at the initiation stage was minimal.

In summary, the long-term impacts of submarine telegraphy brought about sweeping social change:

- it greatly stimulated the development of a world economy
- it began the development of a world polity
- it made a large contribution to the development of total warfare
- it involved mass publics for the first time in foreign and military affairs
- for society at large, it radically changed perceptions of time, distance, and human society itself

In addition to these impacts, the planning and the implementation of the transatlantic cable itself produced significant social consequences:

- making the public conscious of the relevance of basic scientific discovery to practical affairs
- pioneering a new approach to high technology enterprises, including phased planning, the

close involvement of technical experts, and the importance of quality control

• setting a pattern for the vertical and horizontal integration of modern communication industries

• establishing a long-lasting British dominance in world communications

B. Why Submarine Telegraphy Changed the World

This retrospective assessment sought to develop insights into the relationship between technological change and social change, so that we can better anticipate the consequences of emerging technologies. We therefore need to consider why submarine telegraphy produced such striking effects.

A New Capability for Society

In many anticipatory technology assessments, much of the time and effort is spent in rigorous analysis of the physical characteristics of the technology. The implicit assumption is that these characteristics (raw materials, energy use, land use, emissions, etc.) together with direct costs and economic pay-offs will produce the most important impacts. But the important consequences of submarine telegraphy did not result from the technical configuration involved, and the impacts would have been inadequately identified by an anticipatory assessment that concentrated on the technology itself.

Cables did not require large facilities or dedication of large tracts of land. The raw materials used were known, available, and in use. The volume of resources required, although large, did not perturb supply and demand, although meeting the demand for gutta percha did prove to have large effects on a local forest ecology, and probably for a generation on a local population. The ecological effects on the ocean floor were unimportant and shortlived.

When all the impacts are considered, most can be traced (1) to the ability to transmit information to remote locations in minutes rather than in weeks, and (2) to causing an action or receiving a response within the brief interval in which decisions, negotiations, transactions (and even military actions) may occur. The capability of interacting in this way on a transcontinental or global scale was a capability not previously available to human society.

In this context even the way in which the information was transmitted was by comparison of less importance. It mattered *relatively* little whether the message was delivered through cables or by wires strung from balloons or suspended from buoys, if these means could have been made practical and reliable. Even the facts that the communication was point-to-point rather than broadcast, or sent and received from fixed locations rather than mobile terminals, were less important than the fact that it could be transmitted rapidly. However, as we shall see, these factors did condition the usage and the social consequences in important ways.

A new capability for human society is a dramatic event. Sending the human voice over electric wires, the cracking of the atom, the first trip to the moon, a heart transplant, the recombining of genetic materials, the creation of "life" in a test tube are all technologies that may have significant societal impacts because they are unprecedented. But some societal capabilities are obviously more important than others.

Human society is made up of social systems—organized complexes of human interactions; the nation is (or was) the most overarching form of social system. Sociologists and political scientists sometimes describe the generalized conditions necessary to the stability, functioning, and maintenance of social systems as "functional requisites."[1] Integration and organization are functional requisites[2] and communication is necessary for integration. Thus an integrated international economy and a world polity (potentially a higher stage of social-system integration) could not develop so long as communications between national economies and national political systems was so slow and sporadic as to be, in many critical events and decisions, only a recording of past events. Submarine telegraphy, then, was important to the world because it provided not only a new capability, but one that served a functional requisite of society.

Unfortunately, though it may be relatively easy to recognize unprecedented new capabilities, it is more difficult to decide which are related to functional requisites of society, since social scientists do not fully agree on what all of them are. Generally, however, reproduction and life maintenance, food production and food and water distribution, and generation and distribution of raw materials are recognized as being among the functional requisites of human society; a new capability related to one of them (such as artificial photosynthesis for food or energy production) would be likely to have major consequences. Social technologies, such as a universal food allocation system, could also fit this criterion.

Criteria for wise selection among emerging technologies of those whose impacts are likely to be most significant (so that they most merit technology assessment) would be a useful guide because technology-assessment funds and expertise are limited. Our analysis implies some criteria for assigning priorities, but these implications must be carefully framed.

We do not think these findings imply, for example, that all communication or information technologies will have impacts of comparable significance. On the contrary, the societal benefits and costs of radio, overseas telephones, and satellite communications were less important than those of submarine telegraphy because they each provided a marginal improvement on the capability of rapid interactive transcontinental communication—although the improvements in message capacity were large.

We also do not imply that the physical characteristics of a technology are always of less importance in producing consequences than the service or product delivered or that they should necessarily be de-emphasized in technology assessments. In the case of the automobile, for example, emphasis on the benefits of the technology (personal mobility) obscured the potential consequences of the physical configuration which triumphed in the marketplace. Failure to assess the effects of physical characteristics (speed, fuel consumption, emissions) in a timely fashion represents inadequate social foresight.

We do draw two significant implications for anticipatory technology assessment:

- Emerging technologies that promise to provide an unprecedented capability for society should be considered strong candidates for comprehensive technology assessment. If the new capability is one that may affect a basic activity of society, the technology should be given very high priority.
- The suggestion has often been made that every comprehensive technology assessment should begin with a preliminary exploratory assessment in order to guide the allocation of time, funds, and staff effort within the study. The finding that the physical nature of the technology is sometimes of secondary importance supports this advice. An exploratory assessment would avoid the common mistake of overallocating time and effort to the technical description of the technology. It may suggest the need for additional behavioral and social scientists and fewer engineers on the research team, or other changes in staff assignment and resource allocation.

Networking Technologies

Communication systems and transportation system are examples of networking or systemic technologies—those which connect organizations and institutions into a system of related activities. The pattern of transactions which develops between nodes in the network (buyers and sellers, markets, cities, employers and labor force) tend to create interdependencies between them. The possibility of a system failure, or of external events that may disrupt the interdependence, then becomes a new vulnerability. For example, once a nation had developed a military posture depending on constant readiness to respond to information about threats developing in remote locations, its security came to depend on the ability to receive this information. The possibility of cable failure or (for other nations) the British power to shut off service was a great vulnerability; hence the drive to lay several cables to serve the same route, and later the imperative of maintaining cable service in parallel with satellite communications. Earlier, when the venture system was abandoned and control of tramp steamships shifted from captain to owner, the viability of the business depended on cable service to schedule cargo pickups at the right time and place; the seller also depended on the cable to schedule transport for his cargo. Networking technologies, then, usually have significant consequences for the areas they serve.

Once the capability of transoceanic communication had come into being, later modes of communication—radio, telephony, and satellite communications—had by comparison less profound consequences. However, for many uses these modes offered enough improvement or sufficiently different applications to replace the older technology of cables. Radio, for example, differed from cable telegraphy in two important ways: it could be used for either point-to-point or point-to-many communications, and thus provided a mass medium to compete with newspapers; and both transmitters and receivers could be mobile. Mobility allowed countless applications not possible with cables, especially military and marine applications.

The present study did not extend to close analysis of radio, telephony, or satellite communications.* In many areas it is likely that these later international communication systems had impacts that extended or magnified the impacts of submarine telegraphy. For example, multinational corporations and banks now balance their books daily, including their international transactions. Steamship coordination has a counterpart in air-travel coordination; the ICE-CAN telephone cable (to Iceland) was laid in response to the need for continuous secure communication with Iceland, which is a radio relay and radar contact point for transatlantic air traffic. International computer networks operate to coordinate air service. The involvement of the general public in the conduct of war and foreign affairs, which became significant when submarine telegraphy extended the influence of the mass media, reached a climax with public reaction to the televising of battles and bombing in Vietnam.

TABLE 6–1.—General application of the three functions of telecommunications: transmission, storage, manipulation of information (computation).

Transmission without storage or computation provides
- immediate access to distant location
- information exchange instead of physical transportation
- contact with mobile units

Storage alone provides
- compact and rapidly accessible information files
- economical retention of detailed records

Computation alone provides
- an alternative means of performing routinized activities, thus both reducing human mental drudgery and expanding capabilities in many new ways
- most capabilities of human work not requiring complex pattern recognition—if one can reduce an activity to a sequence of instructions, it can be computerized

Transmission and storage together provide
- flexibility to use information at a different time from when it is transmitted
- increased efficiency in the use of transmission capacity

Transmission and computation together provide
- coordination of activities over a wide area
- adaptability in the structure of the communications network based on electronically encoded instruction
- improved systems control through immediate detection of and response to activities

Computation and storage together provide
- individualized services requiring the manipulation of large amounts of detailed information at low cost
- detailed records of past transactions and computational support for planning and realistic appraisal of system behavior

All three (transmission, computation, and storage) provide
- monitoring and recording of activities over wide areas
- scheduling of activities over wide areas
- completely integrated computational and data manipulating systems

---

A retrospective technology assessment of the telephone has been done at MIT in parallel with this study under the supervision of Prof. Ithiel de Sola Pool.

Submarine telegraphy *transmitted* and *recorded* information; radio and telephone only *transmitted* it. Modern electronic telecommunications, especially computer networks, add two other functions: storage and manipulation (computation) of information. Table 6–1 suggests some of the applications made possible by these capabilities. A consideration of these capabilities suggests that thorough anticipatory assessment of computer networking may show that its impacts could also be significant on a societal scale.

Thus we conclude that the consequences of submarine telegraphy were world-changing in scope and significance because this service gave human society a capability it did not possess before, and this capability was closely related to a necessity of social systems, namely the integration of activities. Most communications systems will not have such dramatic impacts, since most will be incremental improvements over existing capabilities. However, any networking technology, including communications systems, is particularly likely to have important impacts.

C. The Mediation of Institutions: How Cables Changed Society

Most activities that have significant consequences for the larger community or society are carried out not by individuals acting at random but by individuals acting within institutions (that is, structures or organized stable patterns of human relationships such as families, interest groups, corporations, governments, etc.). The impacts of submarine telegraphy were most clearly manifested through the changes and adjustments of social institutions as they adopted the new technology and sought to use it fully and efficiently. Some of these institutions—commodity and financial markets, shipping companies and marine insurers, newspapers and newsgathering organizations, departments of government, and army general staffs—changed very rapidly as they grasped the new capability offered by cable technology. In some the changes in functions and internal structure were equally or more significant but occurred over a much longer period.

A very simple model helped in tracing these changes. According to this model an institution would first adopt the new technology to carry out its primary traditional functions more efficiently and at less cost. Next, in order to take full advantage of these efficiencies, the internal organization of the institution might change: the labor force reduced, retrained, or reassigned, resources reallocated, management restructured. If the new technology has capabilities not provided by the one it replaced (for example, computers as compared to mechanical calculators) the third step might be to graft new functions and activities onto the old institutions. Finally these changes might be in some cases sufficient to cause the radical transformation—or replacement—of the institution itself (Fig. 6-1).

These stages can be seen clearly in the response to the new communications capability of institutions concerned with diplomatic and military affairs. In diplomacy, cables were first used to supplement traditional means of communications—mail correspondence or couriers—and to provide a few specialized coordinating services. Against considerable apathy and resistance, several pressures gradually forced institutional adjustments. These pressures were the necessity of receiving and responding to urgent messages (emergency and crisis), the bureaucratic tendency to require regular reporting, and—as these brought about a growing volume of information—the perception of patterns and interrelationships between remote events that obviously required integrated or consistent national responses. Under these conditions policy control was gradually centralized at headquarters (although not before the institutions were threatened by breakdown from overload of information and responsibility in the headquarters staff). Headquarters staff was increased relative to representatives in the field and experts and specialists were added. Full utilization of the interactive capabilities of electric communication (capabilities not provided by correspondence) created new activities in the form of sustained and continuing international cooperative organizations and coordinated multinational diplomatic efforts. However, the diplomatic organization was not abolished or rendered obsolete as was the venture system of shipping.

In military institutions, somewhat the same sequence occurred, with cables first used only for transmission of critical information in isolated instances. The need to be apprised of and respond to unexpected threats, and the opportunity to deploy dispersed forces and coordinate logistics, helped bring about centralized control of military forces and stimulated the movement toward General

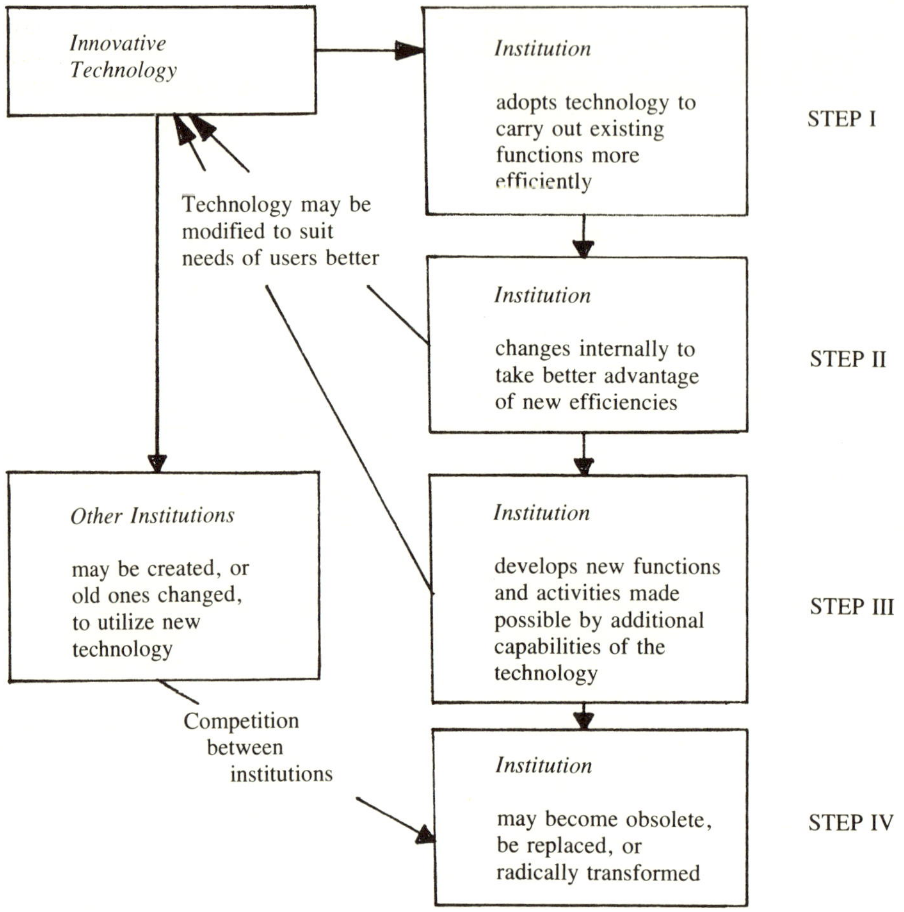

FIG. 6–1.—Impact of technology innovation on user institutions (after Joseph Coates, *Telecommunications Policy* 1:196–206, 1977).

Staffs. After armed forces had been reorganized and streamlined (a trend that was obviously also pushed by changes in transportation and weaponry, not only by improved communications), the capabilities of submarine telegraphy also supported or necessitated growth of many other functions ranging from military intelligence to military assistance for friendly countries.

In economic institutions (shipping, commerce, banking, etc.) adoption and adaption to the new technology moved so rapidly that all phases of this simple model seemed to occur nearly simultaneously. For example, cables were used immediately for buying and selling and for arbitrage; futures trading began quickly to replace spot sales of commodities, and within a few years an elaborate and organized system of futures markets had emerged. News distribution also changed very rapidly, although the long-range effects of this change on public attitudes took some time to develop. News services, which had specialized chiefly in commercial news, were rapidly transformed into international press organizations.

Why did the commercial institutions adapt themselves so quickly to a mode of operation that made efficient use of the technological innovations, whereas the diplomatic and military organizations changed relatively slowly? In all these institutions, information is of critical importance. All seek to achieve objectives through transactions. However, the objectives of economic institutions are immediate and aggrandizing (profit, growth, span of control); those of diplomatic and military institutions are long term and primarily conserving or protecting, and only secondarily—or intermittently—aggrandizing.

Both kinds of institutions seek to maximize their own benefits, but in economic transactions there is an assumption that both sides will benefit (sellers wish to sell, buyers to buy). Military/diplomatic transactions are different in that power is measured relative to power held by the other game player (i.e., both sides cannot become more powerful relative to each other at the same time).

Economic institutions seek to achieve objectives by risk taking, efficiency, and economy: speed is generally an asset, and the timeliness of information is nearly always more essential than its comprehensiveness or depth. Whereas economic institutions want to invest resources (funds) in order to reap returns, diplomatic/military institutions hope not to be forced to invest resources (power). They seek to achieve objectives by the threat of application of power. Comprehensiveness and depth of information is more important to them than immediacy except in (relatively rare) crisis situations.

Economic institutions must conduct transactions (meet challenge) frequently or continually to achieve their objectives. Military/diplomatic institutions seek to avoid the necessity for transactions (i.e., seek to preserve the status quo) so long as they are not actively challenged through aggrandizement by others.

The ways in which "success" is judged also differ. In commercial institutions success tends to be unidimensional—measured by profits (or by growth, which represents later profits). Failure is usually reversible. In diplomatic and military terms success is multidimensional, measured by complex trade-offs in position and power relative to others, whereas failure is sometimes irreversible, particularly in military transactions.

Innovations are in fact adopted or used by individuals within institutions. Economic institutions are managed by individuals whose affiliation with that institution is often temporary (tomorrow they may be affiliated with the competing organizations), but both the personal resources that individuals invest and the rewards which they seek are highly congruent with those of the institution. Diplomats and generals by comparison have in general a long-term and very secure affiliation with their institution, but the personal resources they invest and the rewards they can gain are different from those of the institution; investment in life-long service and experience is maximized by preventing rather than sponsoring change.

It appears (Fig. 6-2) that the institutions that adjusted to submarine telegraphy most swiftly and thus set in motion waves of social change were those whose objectives were aggrandizing and immediate; whose achievement of objectives could be measured simply, directly, and in very short time; which actively sought risks and opportunities to invest resources; and whose individual member-participants had motivations and rewards highly congruent with those of the organization.

Institutions slow to adapt fully to the new technology were those oriented toward risk-avoidance, whose measures of success were more subtle and less immediate and whose participating members had a long-term (safe) affiliation with the institution but motivations and rewards not necessarily congruent with those of the institution. These institutions were largely government bureaucracies. However, all bureaucracies are like that, not only government bureaucracies. As corporations become large and bureaucratized, the same characteristics may prevail. However, that is a speculation that goes beyond the boundaries of our study.

D. Nineteenth Century Technology Assessment: What Was Anticipated

There was no "anticipatory technology assessment" of submarine telegraphy, if by that we think of a conscious, systematic, integrated, and purposeful attempt to forecast its consequences. (Indeed, there was not even a systematic "technology feasibility" analysis that can be recovered from history, although that is clearly what Cyrus Field was doing when he questioned men such as Maury and Morse in the few days before he approached potential investors, and also what the 1861 Commission did four years later.) But contemporary decisionmakers, social commentators, and general observers were not uninterested or unaware of the potential effects on their society of this technological innovation. On the contrary, attempts to forecast its consequences were widespread;

FIG. 6–2.—Institutional readiness to accept innovation, in order of rate of acceptance from high (for scientific) to low (for diplomatic).

	Institutions					Individuals		
Institutions	Objectives	Mode of Operations	Requisites	Rate of Transactions	Character of Success	Resources	Affiliation with Institutions	Congruence of Motivations/reward
Scientific	Knowledge theory	Research	Select people, information, facilities	Long-term	Uncertain Failure expected and reversible	Creativity Training Knowledge Curiosity	Weak	High
Economic: Merchants Speculative Traders Bankers Shippers News Orgs.	Profit Growth Control of other orgs.	Transactions: Exchange of goods, funds, services, & information	Efficiency Economy Risktaking Speed	Constant High	Unidimensional Failure reversible Private	Money Information Risk acceptance	Short-term Low	High
Military	Increasing or protecting power	Threat Force	Readiness Firepower Size Mobility	Rare Low	Unidimensional Failure often irreversible	Training Rank Experience Personal attri-butes	Long-term High	Low to medium
Diplomatic	Conserving or increasing national power	Negotiation Promise/threat	National power: Economic, military, territorial	Intermittent low to high	Multidimensional Failure, sometimes irreversible	Experience In-depth understanding Personal attributes	Long-term High	Low to medium

we were easily able to gather recorded evidence from speeches, newspapers, journals, biographies and autobiographies, letters, official documents, sermons, placards carried in public celebrations, and popular songs and limericks.

As government decisionmakers debated support for the cable, they got their information and forecasts from promoters of the enterprise, and from the popular press or other sources of general information. We cannot be sure of the extent to which decisionmakers (private investors or government policymakers) either (a) consciously considered, weighed, and evaluated this informal "technology assessment," (b) were unconsciously influenced by it, or (c) relied entirely on their own internal assessment, which was no doubt conditioned or determined by the same attitudes, values, and information as that of their observers. (Decisionmakers in government today sometimes question the value of a comprehensive technology assessment on the grounds that they already integrate in their own minds the results of economic studies, technical studies, environmental studies, and social-impact analyses before making their decisions; the data sources, assumptions, weights, and values which go into their decisions are thereafter as inaccessible, and as nontransferable or communicable, as those of the Victorian decisionmakers are to us.)

The potential impacts that were most fully and accurately anticipated were direct and planned impacts—the costs and benefits of the cable itself (to investors, commodity and currency traders, and governments); and secondary economic impacts—very large increases in trade and commerce. Third- and higher-order economic impacts, such as reorganization and relocation of markets and erosion of British monopoly in the Far East trade, were much more poorly anticipated, if at all. In the middle of the 19th century, unchecked competition was viewed very favorably; had not Adam Smith concluded that competition would improve the supply of goods and services? But "competition" also meant that competitors were free to try to corner the market, secure monopolies, and otherwise wipe out competition if they were strong enough to do so. The Victorians failed to anticipate some of the social costs of cable telegraphy (at least to Britain) because they had not fully appreciated the risks of competition.

Questions of equity were repeatedly raised, but it is interesting to note that the individuals who raised these issues did so in the context of opposing government support for the cable, thus leaving the responsibility to the private sector; they were not advocating government intervention or the establishment of public control systems to achieve equity. Outside the economic sphere, anticipation tended to be less accurate, unrealistically optimistic, and parochial (e.g., the worldwide acceptance of Anglo-Saxon religion and values). Forecasts of social impacts especially were generalized and optimistic—which of course reflected the Victorian confidence that new technology, like science, was equivalent to progress, and that progress meant both increased material wellbeing and moral enlightenment. It is often difficult to categorize these social expectations as "accurate" or "inaccurate" because of their generality and ambiguity; yet there was clearly anticipation of impending social changes. For example, we found that an important impact of cable telegraphy was that political and military leaders quickly felt the public reaction to their blunders, and over the long term public awareness and involvement in national decisionmaking were significantly broadened. This was not clearly and unequivocably predicted by the Victorians, yet two quotations from 1863 must be recalled:

> And will it not mark the commencement of an era of the most serious revolutions—revolutions that through its means will, in all probability, be as bloodless as they will be important and startling by their rapidity and extent! (Introduction to the McClenachen Report, 1863.)

> The Queen, the Czar, the Emperor, must use it as they use the air that all breathe. Hence then comes a mighty conflict of antagonistic principles—an equalization of thought, the fraternity of mankind. Something is to topple over—but whose that *something* is, no Cassandra at the dinner-table can predict. (Journalist James Brooks, 1863.)

Surveying the voluminous 19th century expectations as a whole, we repeat—the emphasis is on benefits, both private and public, rather than on costs, risks, or uncertainties (for example, the accumulation of enormous monopolistic wealth, and the necessity for censorship, spying, and propaganda). Beyond the general faith in progress, there are other reasons that potential social costs were poorly anticipated. These reasons still obtain.

Social *benefits* were, and are, identified and put forward by the promoters of the technology to encourage support and investment. But the identification of social *costs* was, and remains, the province of busy decisionmakers with many more direct imperatives for favoring or opposing a policy. The policy issue was defined by the promoters.

Second, such proposals for positive government action were, and are, formulated in terms of achieving direct objectives and of minimizing direct costs (those costs which must be internalized). Other costs were by definition unintended and unplanned, and hence obscure. Externalities would also be broadly diffused over many publics and many overt and latent interest groups, and thus had less psychological and material impact on practical politicians and administrators. Policy intervention is generally on the basis of minimum necessity.

We have already pointed out that the forecasts of the 19th century observers (considered as technology assessors) were conditioned, limited, or even determined by the strong prevailing ambience of faith in technological progress. Their social forecasts were also abstract and hortatory in intent and in tone because they were embodied in sermons, speeches, and popular journals. But had the government commissioned an assessment, it would not have made much difference—the simplest of techniques used today in technology assessment were not yet developed. Trend extrapolation was, for example, in a rudimentary stage of development because collection and preservation of time series data was only a few decades old. Most information of an economical or social nature was private and nonaggregated, or at best was aggregated in a fashion suitable for the slow pace of information transmission. For example, prices of commodities were collected and preserved weekly or monthly rather than daily.

E. Implications for Anticipatory Technology Assessment

The conclusions we have summarized in this chapter have, we suggest, implications for anticipatory technology assessment.

Our analysis suggests, first, some useful criteria for selecting emerging technologies that merit comprehensive assessment. Funds and experienced analysts for technology assessment are limited; not all innovations can be thoroughly assessed.

''Networking'' or systemic technologies deserve serious consideration because they create new connections and new interdependencies among activities, organizations, communities, or regions. Such interdependencies can introduce both increased specialization and instabilities or vulnerabilities that are large in scale and general in scope. Communications systems, transportation systems, utilities, and other distribution mechanisms are networking technologies. Communications systems may be particularly important because they distribute information, and the possible applications and misuses of the information are subtle and complex.

Very high priority should be assigned to technologies that promise to provide an unprecedented capability for human society because we have little experience to guide us in coping with the impacts of such activities. Nuclear energy is an example from the recent past; genetic manipulation is an emerging technology of this kind.

Highest priority, we suggest, should go to technologies that promise to supply a new capability in areas that are basic to the organization and functioning of society; for example, food production or distribution, generation of energy and raw materials, and intelligence and control. Social technologies should not be overlooked.

There are also implications for planning and organizing technology assessments.

In some technologies, the physical characteristics of the technology itself or the way it is institutionalized, owned, or controlled, may indicate the most significant potential consequences; in others, such as submarine telegraphy, the configuration of the technology may have much less impact than the product or service it delivers and the way they are used. Both aspects may be very important for some technologies. The best way to avoid a misallocation of funds, time, and effort probably is to require a preliminary exploratory assessment at a low level of effort before a comprehensive assessment is undertaken.

We offer several suggestions to guide the initial search for possible impacts in anticipatory assessment.

The process of developing and implementing the new technology may have impacts that are significant and long lasting, independent of those which result from the technology in use. This process can be thought of as a series of technical, entrepreneurial, and political problems or challenges and alternative responses or solutions to those challenges.

The "technology diffusion" approach— identifying the institutions and activities most likely to adopt the technology first and their possible adaptations to its capabilities, can be very useful not only for identifying possible impacts but also for estimating a time span for their development. This approach can provide a credible scenario as to how significant impacts may come about.

A close look at converging developments in other technologies may help in estimating which indirect impacts are most likely, since probabilities of a social change driven (or made possible) by several technologies may well be cumulative. Steam shipping and submarine telegraphy reinforced each other in their impacts on the world markets, as railroads and landline telegraphy had done in national markets.

In evaluating impacts, our conclusions suggest that there are several biases to which assessors should be especially alert. Sensitivity to these biases allows a compensating attention to be paid to possible impacts that may be otherwise downgraded or undervalued.

The evenhandedness of the analysis may be unbalanced by the prevailing ambiance of the times—the popular or general attitude toward technology. In the Victorian era technological development was equated with progress; social benefits were emphasized, but societal costs (and uncertainties) were disregarded or considered unavoidable. In our era, on the other hand, preoccupation with potential social costs may lead to neglect of fruitful opportunities to develop physical technology to meet social needs, or to unrealistic demands for a "risk-free" society.

A converse bias arises because technology assessors get most of their information about the technology from its developers (whether entrepreneurs or government agencies), who are apt to stress potential benefits over possible costs or uncertainties, and possible uses over possible abuses of the technology. In laying the cable, the entrepreneurs consistently underestimated the risks and the direct costs of laying the cable while promising great benefits both immediate (e.g., economic development in Newfoundland) and long range (universal peace).

Technology assessors themselves provide a third bias; they prefer to emphasize what they are fairly sure of, particularly if they can quantify their estimates, rather than emphasize what is not or cannot be known. Yet decisionmakers especially need to be told the limits of the predictable—in other words, their real risks. We noted in the role-playing workshops and Delphi a strong and systematic *discounting* of indirect impacts (consequences separated from direct use of the technology by one or more intervening changes or adaptations). The more indirect the impacts, the lower their probability was assessed even though (a) the evaluators themselves had identified them as potential effects, (b) they had rated them as highly important, (c) they themselves pointed out that several trends or technological developments were convergently driving toward that same consequence, and (d) they sometimes knew that those impacts did in fact occur. If there is a general psychological tendency to underestimate the probabilities of indirect consequences, together with a general preference for emphasizing impacts that can be quantified (usually in dollars), assessments may be biased in several ways—especially toward direct economic impacts, particularly benefits, as compared with indirect benefits and costs, particularly those which must be expressed in behavioral or "social" terms.

There are also implications for the aspects of assessment that deal with identification and evaluation of policy issues.

Definition of the policy issues for submarine telegraphy was left to the promoters of the technology; and decisionmakers were content to deal with the new technology in terms of existing technological categories (without considering the uniqueness of the new technology), and in terms of reaffirming existing policies (without considering alternative policies). Thus, the U.S. government (and later other governments) came to depend for basic and necessary governmental functions, including national security, on services provided by a foreign-owned corporation. The government at first refused to participate in making international agreements relating to the cables because the system was privately owned. The British government, traditionally the bulwark of free speech, realized the implications of rapid transnational communications for national security only during the exigencies of approaching war, and had to respond in these desperate circumstances by creating authority for absolute censorship. Neither government had fully realized the differences between the cables, and other contract mail carriers for which they had a ready, fully controlled alternative in diplomatic or military couriers. Neither, therefore, reevaluated their existing policies of nonparticipation by the government in ownership or control of privately developed technology. The implication is that assessors should critically evaluate relevant policies as to their appropriateness and adequacy for dealing with the new technology.

We were in fact unable to recover and reconstruct, to an extent that was fully satisfactory to us, assumptions accepted by public decisionmakers, alternative decisions that may have been conceived, or trade-offs that may have been evaluated. We can only assume with a fair degree of certainty that they shared the expectations of diverse commentators as they can be gleaned from formal and informal documents. In contemporary decisionmaking, the necessary reconsideration and re-evaluation, on a continuing basis, of policy related to new technologies as they develop (for example, solar energy or recombinant DNA) is hampered by uncertainty as to which information and which analyses are available to the decisionmaking body at any one time, and how these databases (often not fully consistent or comparable) have been used in arriving at a policy.

The secondary research task in technology assessment is, therefore, of great importance—collecting, criticizing, comparing, and analyzing research reports to prepare a "standardized" database for decision making. The same summary base then provides a starting point for re-evaluation of decisions and policies as new information is obtained or uncertainties are resolved.

Finally, we assert on the basis of our experience in this study that *retrospective assessments of old technologies can provide productive insights and provocative hypotheses that can contribute to the further development of anticipatory technology assessment.*

F. Doing Retrospective Technology Assessment

Technology assessment (TA) is usually regarded as applied *futures* research because it is concerned with anticipating potential consequences of technological change.

The tasks involved in technology assessment, which provide a framework for this discussion, generally include (a) describing the technology in technical detail, including costs; describing competing or alternative technologies; describing necessary supporting technologies; and describing the corporate or other institutions that will manage the technology; (b) developing state-of-society assumptions as described above; (c) identifying the "stakeholders" and others who will be affected in one way or another by the technological development; (d) identifying impossible "impacts" or consequences; (e) laying out the policy issues involved, or the decisions that must be made; (f) explaining the possible effects of alternative decisions or policy positions; and (g) identifying the loci of responsibility and authority to make such decisions and policies.

Historians of technology, especially Bernard Finn of the Smithsonian Institution, had already thoroughly researched the development of the technology of submarine telegraphy and the laying of the transatlantic cable. Our task was to explain the consequences of these developments for people then and now. In considering the technology itself we were particularly concerned with understanding the scientific, technical, and management decisions that were made, alternative choices that were considered and rejected, and the basis for those choices. We were also especially concerned with the institutional context in which the technology developed. Since submarine telegraphy was an unprecedented technology (except for a few short-distance, shallow-water cables), alternative and competing technologies did not exist; the major supporting technology was gutta percha production. In other retrospective TAs there may be reasons for selecting a technology not previously studied by historians of technology, but the basic technical information on technologies developed within the last century is likely to be recoverable from museums and archives, augmented by diaries, letters, and autobiographies, personal recollections of inventors, and other historical sources. In this case study, we deliberately chose a technology for which the technical information needed was available in primary and secondary literature and had largely been gathered. In retrospective TA, therefore the *description of technology*, which often consumes much of the resources of anticipatory technology assessment, was straightforward and the major discipline and techniques required were those of history of technology and of science.

In anticipatory TA another task is making *state-of-society assumption*. This task provides (a) data about societal parameters that may be affected by the technology (as a baseline for evaluating the significance of potential impacts), and (b) analyses of long-range trends and consideration of changes that may occur in society and in other technologies by the time the emerging technology is utilized. Description of British and American society at mid-19th century was critical for our retrospective assessment. Historians as members of the core team carried a heavy burden not only in gathering the historical statistics needed but also in helping other members to understand the attitudes and values of the Victorians and (as far as humanly possible) to avoid embarassing anachronisms. The basic historical research itself was particularly difficult since as technology assessors we often asked questions that were not considered relevant in 1850–1900 and for which, therefore, there had been no reason to collect or preserve data. As one example, commodity prices and stock fluctuations were reported on a weekly or monthly basis before telegraphy, not on the daily basis that we wanted.

In addition to the historians on the team, we consulted a large number of other social, economic, diplomatic, military, and general historians and had several of them on our Oversight Committee. Often these resource people had not had occasion to consider the specific impacts of submarine telegraphy in their areas of knowledge, but invariably they were intrigued by our questions and helpful in coming up with the information and insights we needed.

The second or "futures" aspect of the state-of-society task required us to ignore our past and share the Victorians' view of their unknown future. It was easy to gather evidence from formal and informal historical records, of both general hopes and fears about the future and specific "expectations" or "forecasts" about the transatlantic cable. But we struggled constantly with epistemological problems as to how well viewpoints publicly expressed (by entrepreneurs, sermonizers, journalists, or Congressmen, for example) represent "real" values and—particularly—motivations. In historical research, moreover, it is fairly simple to identify decisions that were were made and options that were selected, but it is difficult to identify choices that were considered and rejected, or not explicitly considered at all. The inability to reconstruct (and so to evaluate) decisionmaking is, by the way, a problem even in contemporary policy formulation; one excellent reason for technology assessment is to make explicit the information, assumptions, and expectations on which decisions are based so that they can be continually reassessed.

In doing the state-of-society task for our retrospective assessment we also tried two innovative approaches that may merit further development. The first was to develop a theoretical framework for description of society to guide us in selecting baseline parameters. We called this the period profile approach. From the literature on social change provided by economists, sociologists, and political scientists, we extracted a set of processes or categories of social change (Table 6–2). From the extensive work that has recently been done on social indicators, we extracted a set of parameters (Table 6–3). Arranging these as a matrix, we had interstices from which we could select (and constantly modify) those which might be affected by transoceanic telegraphy and its direct consequences. Like most attempts to develop a rigorous "TA methodology" this one quickly proved to be too ambitious for the current study's time and resources, but our rudimentary attempt was useful enough so that we shall continue to work on it.

Our second innovative technique we called the "hypothetical assessment." An ad hoc technology assessment group consisting of the core members of the oversight committee, other experienced technology assessors, and relevant disciplinary specialists conducted two workshops and partici-

TABLE 6–2.—Categories derived from "societal change" literature.

Economic change

1. Accumulations of capital, including factories, machines, tools, equipment, inventories.
2. Increasing productivity per worker.
3. Growth of market interdependence.
4. Increasing specialization of labor, with decreasing share of labor devoted to agriculture.
5. Geographic and structural concentration of economic activity.
6. Application of scientific knowledge to economic problems.

Social Change

1. Functional differentiation of social units accompanied by changing or new value justifications for these units.
2. Growth in modes of social integration: associations, networks, clubs, etc.
3. Rationalization (formalization) of social procedures, decrease in informal intimate social relations.
4. Mobilization in terms of status and geographical position.
5. Growth of social disturbances as a result of lagging integrative mechanisms.

Political Change

1. Growth of rationalized national authority structure, increased identification with the nation state.
2. Growth of functionally specific government, national and local.
3. Broadening modes and manifestations of political participation.
4. Increased government involvement in societal functioning.
5. Growth of secular, impersonal regulation through legal systems.

Cultural Change

1. Increasing generalization and secularization of values.
2. Development of mass national culture.
3. Increase in individual creativity, especially as concerning scientific discoveries and inventions.
4. Growth of entrepreneurship and diffusion of significant inventions.

Individual Change

1. Present/future orientation, changing sense of time and space.
2. Desire to know, willingness to hold opinions on a variety of subjects.
3. Emphasis on low risk, calculability, efficiency.
4. Ingrained consumerism, technological optimism.
5. Achievement motivation, aspiration, commitment to work.

TABLE 6–3.—Categories derived from social-indicator literature.

Demography: density, rates, segments, mobility.
Employment: types, distributions, standards/quality, opportunity, associations.
Leisure: time available, organization, activities, customs, sanctions.
Production: rates, location, innovation, surpluses, organization, control.
Income: extremes/distribution, consumption rates, inequalities, prices, savings, fixed demands.
Commerce: means, agents, organization, constraints.

pated in a modified Delphi in which they played the roles of advisors to Her Majesty's Government in 1861 in considering the proposed laying of the transatlantic cable. These simulations provided many useful insights and helped us guide our search for indirect consequences of submarine telegraphy; they are described in more detail elsewhere in this report.

Other major components of TA—identification of affected parties, identification and evaluation of impacts, identification of critical policy issues and alternative strategies for resolving them, identification of appropriate decisionmakers and of their authority and responsibility—can be summarized more briefly. They called for matching the questions and concerns of technology assessment with the skills of historical researchers in finding, and hypothesizing, chains of cause and effect. While our *findings* as to "cause and effect" are open to criticism and alternative interpretations at every point, we have tried to make our reasoning about cause and effect as explicit and detailed as possible exactly for this reason. We invite and welcome this re-examination, which we hope will add to our own, and other people's understanding of the relationship between technological development and social change in the past and in the future. Our attempts to use models and theories borrowed from other developing areas of knowledge, such as ecosystems analysis, are particularly intended to provoke comments that will clarify and expand thinking along these interdisciplinary lines.

To summarize, retrospective TA appears to involve many analytical problems of both historical research and futures research in addition to those peculiar to technology assessment. Like other historians, we often found data inaccessible or nonexistent, and motivations and assumptions difficult to reconstruct. We found credible alternative scenarios as difficult to construct for the past as futurists find them to be for the future. Perceiving past events as ''inevitable'' is a trap as hard to avoid as expecting a ''surprise-free'' future. And extracting the consequences (whether past or future) of one technology from the background of ''exogenous events'' (such as the American Civil War) and of converging trends (such as the development of steamships, railroads, and landline telegraphy) is as hazardous for the past as it is for the future.

In addition, we had the usual problems of TA, which is still at an early stage of development: necessarily imprecise questions, borrowed and inadequately adapted techniques from older disciplines or those themselves still developing, and a universe of conflicting information to select from.

Despite all these problems we are ourselves convinced that retrospective TA is both necessary and productive; that it can and has increased our understanding of the relationship of technology change and social change; that it can and has provided some insights that will be of assistance to anticipatory TA, especially in providing criteria for selecting subjects and in allocating research resources; and that it has provided provocative hypotheses for future investigation. We hope that historians and technology assessors will combine their efforts in further retrospective assessments, and that research sponsors will make it possible for many retrospective assessments to be done. Each is only one case study, itself subject to much re-evaluation, but together many retrospective assessments can greatly improve our understanding of the world we live in and of its future. We, the retrospective assessment team, are prepared to assert that retrospective assessment is the most promising approach to improving anticipatory assessment; understanding of the past is essential for understanding the future.

Appendix I

UNDERSTANDING THE CONSEQUENCES OF SUBMARINE TELEGRAPHY and the Methodological Implications of That Understanding for Anticipatory Technology Assessment

By Robert Anthony

The many findings from the previous chapters are assembled here in order to clarify the relationship between the technological advance of submarine telegraphy and ensuing societal adaptation to that advance. Some insights can be gained only after a wide variety of bits and pieces have been organized so that a *Gestalt* emerges. The most important lessons learned have to do with the characteristics of the unanticipated consequences of technology, why these consequences were overlooked, and what can be done to reduce the vulnerabilities that will arise from future technologies.

Section A of this Appendix clarifies the relationships among the findings of the previous chapters. Section B examines these relationships from the perspectives of human geography, economics, and systems analysis. Section C considers the practicial and epistemological lessons for the conduct of both retrospective and anticipatory technology assessment.

Section A draws statistically significant conclusions from the summary tabulation of findings from the other chapters. (The tabulation is given in Appendix II.) The principal conclusions:

- The Victorians systematically perceived the benefits of submarine telegraphy for increasing the efficiency of activities, but they overlooked the vulnerabilities created by the widespread use of and dependence on the international cable network.
- The indirect consequences, good or bad, lagged the direct ones by an average of 30 years. The vulnerabilities lagged the benefits in efficacy by 20 years.
- Contemporary speculations reveal an awareness of many of the vulnerabilities overlooked by the Victorians, but both types of speculation still range far beyond what can be firmly established by detailed research.

Section B examines the patterns of impacts from submarine telegraphy in the light of other disciplines—economic and human geography, and systems analysis (specifically ecological systems models). The principal conclusions are:

- Improved communication scope provided by an international cable network promoted the integration of an international system of cities and places, and increased the scope, efficiency, and pace of economic, military, political, and cultural activities.
- The system of cities was characterized by the emergence of new high-level activities such as international news services and organizations, more diversified and specialized layers of activity along the vertical urban-rural continuum, and larger units for almost all activities.
- The vulnerabilities, especially those overlooked by the Victorians, can be understood on the basis of four systems stabilities principles adopted from systems ecology. Many of the responses to perceived vulnerabilities can be recognized as a generalized form of the adaptive strategies of systems reliability theory.

Section C presents the practical, epistemological, and methodological lessons for anticipatory technology assessment. The key points are:

• Ever-present practical limits to knowledge can obscure underlying inherent epistemological limits. These inherent limits must be recognized in the creation of a technology assessment method if the results are to be useful and warrantable in advising the policy process.

• Inherent limits to the ability to know about the future (or the past) arise from the way one comes to know something, from biases in evidence, and from the failure of "imagination" to perceive what *kind* of things there might be.

• Recognition of the distinction between what can and what cannot be known suggests a methodological innovation for technology assessment. Adaptive strategies can be formulated that bridge the gap between what is reasonably certain about the future and what is needed to assure acceptable results in the application of novel technology. The Victorians adopted several forms of adaptive strategies, but only remedially, after suffering many unintended consequences of submarine telegraphy.

A. Patterns from the Findings of the Report

What broad conclusions can be drawn from assembling the diverse findings of Chaps. 2–5 into a single analytical framework? More important, what conclusions can be drawn that can be generalized to assist anticipatory technology assessments?

Appendix II presents a systematic tabulation of the principal findings of the previous chapters; the statistical analysis of the significant patterns derived from that tabulation is shown in the present Appendix. The broad conclusions derived from that analysis are summarized here. The conclusions address three questions:

• Were there significant blind spots in the perceptions of either the Victorians or contemporary retrospective speculations about the consequences of submarine telegraphy?

• What consequences either emerged with time or by the indirect spread of the cables's influence?

• Are the answers to the above questions clarified by a consideration of the impacts of the cable according to the characteristics of extended scope, stepped-up pace, improved efficiency, or increased vulnerability for a wide range of activities?

Foresight and Hindsight

Since the central purpose of technology assessment is to support decision processes that guide the introduction of technology into societal context, retrospective technology assessments, performed to improve the TA methods, must deal explicitly with the attitudes, values, and legitimizing norms shaping past decisions as well as with tracing the historical impacts. The present retrospective TA incorporated two perspectives of those attitudes, values, and norms—one based on foresight* and another based on hindsight.**

---

*Foresight was derived from scholarship, by reviewing the writings and recorded deliberations of contemporaries of the events leading to submarine telegraphy or participants guiding its development. The advantage of hindsight is that the retrospective view provides a documentable historical record of both delayed and indirect consequences of a major technological achievement. A correlative disadvantage is the difficulty of reconstructing the attitudes, perceptions, and unused options of the decisionmakers in the past. Only written records survive; many motives and many options were not recorded. Also, many explicit identifications of possible impacts by perceptive Victorians may not have had significant influence on the decisions being made. But one can suppose that the assumptions, the style of thinking, and the range of topics considered worthy of recording reflected the mind set of the decisionmakers. Thus, any broad identification of the attitudes, values, and norms that shaped the development of submarine telegraphy requires a liberal exercise of judgment before specific impacts can be connected with prevailing sentiments and commentary. For this reason, only very distinct patterns were taken seriously.

**Hindsight or a present (1976) view was provided by an ad hoc technology assessment team—composed of technology assessors, historians, and experts from a variety of other disciplines. They were asked to play roles of Victorian assessors in two workshops and through participation in a written Delphi questionnaire. As expected, the analysis of their responses (Chap. 4) shows that their judgments were quite similar both as role players in 1861 and in retrospect from 1976.

Throughout this section the hypothetical consequences of submarine telegraphy from the contemporary (1976) workshop are designated as the "Delphi impacts." The documented views of those living in the period of the cables' creation, development, and spread are designated as the "Victorian expectations."

Comparisons of contemporary views with Victorian ones can reveal shifts in the expectations about the effects of technology on society resulting from the experience gained between Victorian times and the present.

A rather liberal interpretation of the Victorian expectations and the Delphi impacts was adopted so that any failure to match an historical event with either the hindsight or foresight views would be significant. Moreover, the breadth of the exploratory research given in Chapters 2–5, combined with the broad statement of the impacts and expectations, reinforces the confidence in these failures to form matches. Comparisons made on this basis reveal several important patterns:

(a) Both the Victorian expectations and the Delphi speculative-impact list captured the majority of the cable's consequences found by historical research. The Delphi did somewhat better but not significantly so (Table I–1). The few consequences missed by both the Victorians and the Delphi tended to be the more specific, those difficult to match even with the liberal interpretation used. More detailed breakdowns of the types of impacts are needed to reveal patterns and yield insights into the causes of the oversights; these breakdowns are given later in this section.

(b) The wide-ranging speculations from the Delphi workshops extended well beyond both the historically verified impacts and the Victorian expectations. Within this range of contemporary speculation, the Victorian expectations matched rather well with the actual historical impacts (Table I–2). Does the large number of speculative impacts neither expected nor found imply that the Delphi workshop approach is a very inefficient tool for identifying impacts? Or does it imply that the historical research techniques are quite limited in verifying reasonable suppositions?

(c) The workshop technique was shown to be inefficient but also valuable for suggesting reasonable consequences that extended beyond "researchable" questions based upon today's techniques. Restriction of the range of Delphi speculation to impacts deemed most *likely* by the participants eliminated a great majority of the impacts neither found nor expected (Table I–3). The remainder appear to be complex suppositions typical of modern (1976) impressions of the relationship between technology and society, which are often beyond the capability of research techniques to verify. Such contemporary speculations would have been anachronistic in 1861. Among these remaining impacts were: the substitution of communication for transportation; changes in the substance of communication; more modernity in education, a science and technical emphasis; and an altered image of America in Europe (beyond Britain). Another Delphi impact in this class—stimulation of the growth of large cities—will be shown to have almost certainly resulted from international telegraphy.

TABLE I–1.—Identification of historical impacts by Victorians and retrospective speculation.

		Victorian Expectations			
		Expected	Not Expected	Totals	Percentage
Delphi Impacts from Workshops	Found	47	16	63	82
	Not Found	9	5	14	18
	Totals	55	22	77	
	Percentage	71	29		100

There is no evidence for a systematic blind spot shared by both the Victorians and the contemporary assessors in trying to anticipate the full range of historical impacts since no correlation appears. A chi square of 0.43 would occur in 49% of all random trials.

TABLE I–2.—Matches with the Delphi survey impacts from historical research and Victorian expectations.

		Victorian Expectations		
		Expected	Not Expected	Totals
Delphi Impacts from Workshops	Found	33	8	41
	Not Found	8	25	33
	Totals	41	33	74

The Victorian expectations were very consistent with the historical findings when matched within the full range of Delphi speculations. (The correlation of Victorian expectations with historical findings has a chi square of 23.4 for one degree of freedom—a one in 1000 chance.) Nearly as many Delphi speculative impacts were not found or expected (25) as were both found and expected (33).

TABLE I–3.—Matches with the Delphi survey restricted to only *likely* impacts from the Delphi survey.

		Victorian Expectations		
		Expected	Not Expected	Totals
Delphi Impacts from Workshops	Found	22	5	27
	Not Found	5	8	13
	Totals	27	13	40

The correlation between being found and expected from Table I–2 remains even for the "likely" Delphi impacts alone (40 out of a total of 74 speculative impacts). However, the correlation is not as strong; a chi square of 5.57 for one degree of freedom occurs at random once in 50 trials.

(d) Impacts considered likely in the Delphi and expected by the Victorians but not found in the historical research were largely sweeping and optimistic generalities that would be difficult either to verify or to reject in most practical research settings. Examples are increased international understanding, leading to smaller likelihood of wars; growth of knowledge about other countries; and increased prestige for the English language. Even if there is good reason to believe that the cables did (or did not) contribute to these important impacts, the *extent* of the cable contribution is still more difficult to determine.

A direct implication of these results is that contemporary speculative assessments can raise complex questions about impacts well beyond the ability of present research capability; yet these impacts are viewed as sufficiently likely (and important) to warrant consideration in decision processes. In fact, many of the above Delphi questions could only have been asked today because they were based on the relatively advanced theoretical understanding that resulted from improvements in research techniques over the last century. *So improved understanding appears to advance powers of speculation beyond the range of secure knowledge.*

Direct and Indirect Consequences of Submarine Telegraphy

One of the central concepts of technology assessment is that of "indirect" or "higher-order" impacts. The most immediate aspects of a new technology—technical feasibility, economic practicality, and market demand—are nearly always examined. Consequences that result from the adaptation of the societal or environmental context of the technology after it has been implemented are the "second-order" impacts. These and higher-order impacts may not be considered so carefully. Retrospective technology assessment provides a unique opportunity to pursue the sequences of probable cause and effect emanating from a long past technological event.

To give concrete meaning to the concept of indirect or higher-order impacts, a scale of "remoteness" was defined. (See Chap. 4 for the definition of this scale as applied to the analysis of the Delphi survey responses.)

1 The new technology opens up novel applications for its promoters and users.

2 Promoter and user institutions are reorganized so as to make better use of the new technology; in some cases new institutions arise or existing ones fail.

3–4 Shifts in British and American relationships and shifts in worldwide patterns of activity resulting from international submarine telegraphy spread well beyond user and promoter groups or their control.

5 Changes in values, attitudes, cultural patterns, and the norms that render future actions legitimate or unaçceptable are seen to result from the widespread use of the cables.

One would expect that the more remote impacts would take longer to be felt than the less remote, but that need not be so. A technology could augment an established cultural trend or value shift, or could influence an activity already undergoing rapid change so that the sequence of cause and effect progresses much faster in these areas than in other impact areas. In other words, "remoteness" refers to the directness of cause and effect and not necessarily to a chronological scale.

When each of the impacts, from the summary tabulations of Appendix II, was assigned to one of the categories on the remoteness scale, the average year in which the impacts in each category occurred did form a uniform progression forward in time—the more remote were last (Table I–4 and Fig. I–1). It took an average of 30 years after the direct impacts for the most indirect impacts to become evident.

Different types of activities were affected by submarine telegraphy at different rates. Shipping

TABLE I–4.—Time sequence of impact maturity vs remoteness from direct causation.

Remoteness category	Year of the impact: Pre 1850	1850–1875	1875–1900	1900–1925	Post 1925	Totals	Avg. year	RMS years
1	1	7	2			10	1865	13
2	1	11	5	4		21	1877	21
3–4	2	12	4	5	2	25	1881	28
5		7	4	8	3	22	1895	27
Totals	4	37	16	17	5	78	1882	26

Likelihood that the trend is spurious is less than 0.1%, corresponding to a chi square of 22.8 for 3 degrees of freedom.

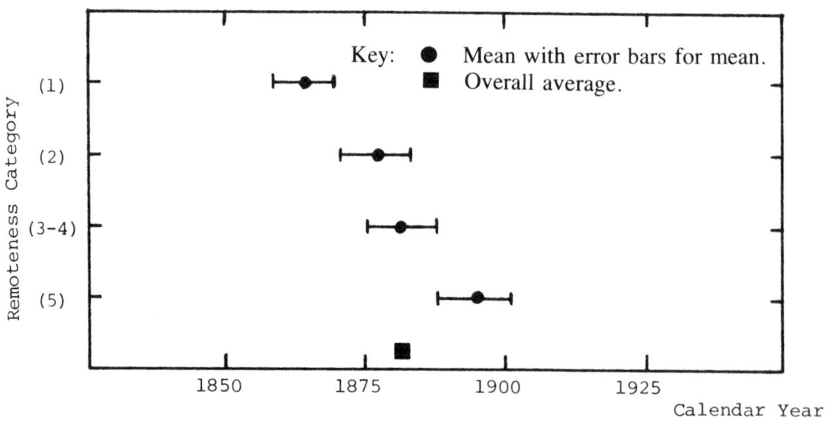

FIG. I–1.—Mean of each cluster of impacts for successive degrees of remoteness.

and commerce were among the earliest to change. Scientific discoveries and the "social impacts" on the perception of time, international law, etc., also came early; these were predominantly "cooperative" or "supportive" activities in the worldwide scheme of things. Diplomacy, war, and national security were next to exhibit the consequences of improved international communications. The news services were *created* by the telegraph; their early development predates submarine telegraphy as the land telegraph systems were spreading across Europe. Thus the most significant impacts of the news services were indirect and among the last to be felt. (See Appendix II for details.)

News had the most significant influence on attitudes, values, international culture, and the norms by which decisions were made. These indirect impacts of submarine telegraphy, mediated by changes in news distribution, were very much overlooked by the Victorians.* The overlooked impacts fell into three broad classes: the drive of the American press for independence from both the British press and the U.S. government; the increasing sensitivity of national decisionmakers to the attitudes of their citizens and to their images abroad; and the intelligence and propaganda uses of the cable in the time of war.

The Delphi captured these impacts—modern observers are more sensitive to the role and importance of news services.**

If different types of activities are ranked according to how directly the cables influenced them, again on the remoteness scale of 1 through 5, a somewhat different pattern emerges. In spite of being less immediate in time, the effect of the cables on diplomacy were more direct than those on commerce. In other words, most of the diplomatic impacts were manifested within the institutional structure of that activity but took longer to occur—diplomacy responded to, but did not initiate or cause change. Impacts related to war and national security greatly affected world patterns outside the sphere of their own activity. Science and social cooperative ventures affected man's view of himself and the world; their ultimate influence would have been felt much less directly than those from diplomacy.

Several more or less expected, yet provocative, patterns emerge from a comparison of the remoteness of an impact with the likelihood of it being anticipated by the Victorians or identified in the Delphi. Common sense would say that even expert judgment becomes imprecise and less reliable as increasingly remote consequences are contemplated. This supposition is borne out by the findings here.*** *Many impacts were remote and long delayed; these impacts were less well perceived by the Victorians and the contemporary Delphi workshop participants*. The analysis of the Delphi in Chap. 4 showed that the respondents' assessment of the *likelihood* of impacts occurring owing to the cable fell off with remoteness. However, their assessment of the *importance* of these remote impacts was as high as the more direct ones.

If these results are generalized to apply to anticipatory technology assessment, they cut two ways. Indirect impacts would appear to be both overlooked and discounted during the deliberations leading to the implementation of a new technology. But many of the most *important* impacts were among those overlooked. This is a double bind for the technology assessor and decisionmaker alike—how can one be expected to act upon imprecise, uncertain, and incomplete perceptions even if there is a sense of great importance attached to them? This problem is discussed further in the methodological section.

---

*Eight of 11 historical impacts were overlooked by the Victorians; this was unlikely to be a research bias.

**Eight of 11 historical impacts were approximately identified in the Delphi. Finding these impacts in the Delphi illustrates the effectiveness of attempts at role playing in a retrospective assessment.

***The Victorians did progressively worse in identifying impacts as they became more remote on the scale; they ranged from around 80% found for the immediate impacts in category (1) to only 48% in the most remote category (5). Similarly, the Delphi impacts in the direct category (1) were all verified by historical research but those in more remote categories (2, 3–4, 5) were not. Only 57% were found in categories 2, 3–4, and 5.

All the Delphi impacts expected by the Victorians but not found by historical research were in the most remote categories. These impacts were predominantly social and cultural in nature and generally reflected excessive optimism. The greatest disagreement between the Delphi list of possible impacts and the historical findings came at the remote end of the scale and for cultural impacts.

The Optimistic Bias in Forecasting: Efficiencies vs Vulnerabilities

The benefits of improved communications are almost always "obvious"—except to an occasional Thoreau. Improved communications increase the efficiency of most activities. They can generally be used to reduce perceived risks.

The detrimental impacts of improved communications are less intuitively obvious. Disbenefits may include not only the possibility of increased control by a dominant central party, but also the destabilizing effect of increased and diversified participation. The former liability may necessitate regulation, mutual agreements, or appeal to a higher authority. The latter disbenefits are more difficult to perceive as new interdependencies and relationships arise from the use of communication systems. The principal example from this study is the spread of news services and the greatly increased participation of the "public" in what had been exclusively diplomatic and military decisions. The detrimental aspect results not from greater participation, but from the gross instability that can arise, bringing greater loss to all but a few. In general, these detrimental consequences represented significant and seemingly unnecessary penalties of technological advance.

*This section shows that the Victorians repeatedly overlooked the detrimental consequences of international cable telegraphy while they perceived most of the benefits of increased efficiency.** Detrimental consequences were often *systemic* in nature—affecting many parties without their perceiving why until too late. These detrimental consequences are called "vulnerabilities" here; a few examples will clarify the meaning of this term.

When shipping and marine insurance rapidly adopted cable communications to coordinate cargo deliveries, the ensuing intensely competitive bidding for cargoes by brokers created a false impression of excess demand, which stimulated excess shipbuilding. Prices were driven down and threatened to bankrupt the shippers. This unexpected marketplace instability was halted by the formation of shipping conferences—private arrangements among shippers which acted like government regulation to stabilize trade and serve the majority of customers.

When futures markets, arbitrage, increased credit, reduction of middleman costs, redistribution of warehousing, and the assumption of risk by speculators in commodity markets greatly improved the efficiency of international trade and commerce, Britain's trade monopoly in the Far East eroded. The restructured markets soon required standards for commodities and regulations for the conduct of trade. Banks began requiring hedged purchases before giving loans. These impacts were adaptive responses to novel vulnerabilities.

As to war and national security, the constant military readiness allowed (and necessitated) by cables may have helped precipitate World War I.[1] The influence of the citizenry, its morale, domestic censorship, and control of news going to neutral nations were other vulnerabilities created by the dependence on cables.

Weather forecasting and oceanography, on the other hand, were areas in which cables *reduced the vulnerability to natural events*. Hurricanes were readily perceived natural risks unaffected by the capability of rapid human communication.

Comparison of the Victorian expectations with the historical consequences from the summary tabulations reveals a strong optimistic bias toward searching out the potential improvements in efficiency but overlooking the vulnerabilities (Table I–5). The Victorians missed only one increase in efficiency out of twenty-one identified in the impact list. They missed eighteen vulnerabilities, or

---

*One could argue that if a detrimental impact were not overlooked, then it could have been avoided. Thus, it would not have become a detrimental impact. If this line of reasoning were correct, it would render the argument that the Victorians repeatedly overlooked the pending detrimental impacts of the cable either circular or tautological. That is not the case. Detrimental impacts, or vulnerabilities, were most often different in *kind*; they were systemic. Systemic benefits were also overlooked. In other cases detrimental impacts were penalties paid to live in a world restructured by the use of cables. Such impacts were often anticipated, but were nevertheless unavoidable.

57% of a total of thirty-one.*

A perusal of the summary tabulations found in Appendix II shows that efficiencies always predate vulnerabilities in each activity sector—commerce, diplomacy, news, etc. Overall, an average of twenty years elapses between efficiency improvements and the emergence of vulnerabilities (Table I–6). This table also shows that efficiencies predate vulnerabilities within each

TABLE I–5.—The optimistic expectations of the Victorians.

		Type of Historical Impact		
		Increased Efficiency	Increased Vulnerability	Totals
Victorian Expectations	Expected	20	13	33
	Not Expected	1	18	19
	Totals	21	31	52

The chance that the above pattern would occur at random is less than one in 1000; it corresponds to a chi square of 13.1 for one degree of freedom.

TABLE I–6.—Efficiency and vulnerability due to submarine telegraphy according to time and impact remoteness.

Remoteness scale	Efficiency or vulnerability	Pre 1850	1850 1875	1875 1900	1900 1925	Post 1925	Totals	Mean year
(1) Users and promoters	Eff.	1	6				7	1859
	Vul.			1	1		2	1900
(2) Institutional adaptation	Eff.		4	1	1		6	1875
	Vul.		3	5–1	1		9–1	1882
(3–4) International relationships & activities	Eff.		5	2		1	8	1878
	Vul.		2	3	3		8	1891
(5) Attitudes, values, and decision norms	Eff.							
	Vul.		1	3	5–3	3	12–3	1908
Totals	Eff.	1	15	3	1	1	21	1871
	Vul.		6	12–1	8–3	3	31–4	1892

Note: Negative numbers correspond to impacts which represent a *reduction* of vulnerability. Vulnerabilities lag efficiencies with the confidence of 5 parts in 1000 that the order would be reversed in our data, which corresponds to a chi square of 14 for 3 degrees of freedom.

*The Delphi also identified the increases in efficiency, but only four of the vulnerabilities were overlooked. Thus the likelihoods of finding efficiencies or vulnerabilities were not significantly different. Delphi participants, of course, benefitted from hindsight, despite their efforts to place themselves back in time. Also, the Delphi participants may have been more sensitive to the importance for exploring the possible negative repercussions than were the Victorians, with their prevailing optimism. Since the historical research itself was guided by the same modern thought process reflected in the Delphi, it may have been blind to additional areas of impacts that fall outside of 20th century concerns and experience. This may be an unavoidable hazard of both the anticipatory and retrospective technology assessment. It is addressed in the last section of this Appendix.

category of the remoteness scale. The most direct impacts were predominantly increases in efficiency and the most remote were exclusively vulnerabilities; see the totals column of Table I–6.

Vulnerabilities may be somewhat more spectacular and therefore easier to recognize in the historical research. Nevertheless, the numerous detrimental impacts produced by submarine telegraphy through delayed, circuitous, and poorly perceived avenues suggests an important lesson for anticipatory technology assessment.

From an analysis of the effect of international cable telegraphy on efficiency and vulnerability, three broad characteristics emerge.

(1) Increases in efficiency nearly always come first as existing practices are improved; vulnerabilities develop later as competitors, institutions, nations, and attitudes change.

(2) Increased efficiency seems to come more directly; vulnerabilities tend to be indirect consequences.

(3) Vulnerabilities are much more difficult to identify or anticipate than improvements in efficiency.

The message is clear: users and promoters adopting submarine telegraphy achieved significant efficiency gains for their activities. Much later the pervasive indirect influence of the international cable network produced many important systemic vulnerabilities. The next section offers some explanations for these findings.

B. Insights from Other Disciplines

The findings of the last section can be understood from the perspectives of several established disciplines. Attempting to explain the findings with established theories suggests generalizations of the retrospective study that apply to anticipatory technology assessments. Each of these established theories was developed long after the international submarine cable system was essentially complete, so that this understanding could not have been shared by the Victorians and could not have affected their decisions.

The increased scope and improved efficiency of many activities and the quickened pace of life brought on by the cables can be understood from the perspective of economic and human geography—specifically from the accelerated development of an international system of cities and places.

Understanding the unexpected vulnerabilities requires a different perspective—examining the adaptive nature of interactive processes, especially competitive ones, as pictured in the disciplines of systems analysis and systems ecology.

The emphasis here is on understanding *how* the historical impacts came about, especially the unexpected ones. This approach would contribute to, but is different from, determining whether foresight and general theoretical understanding of the implications of technology for society can be greatly improved, a topic discussed in the final section on methodology.

An International System of Cities: The Economic and Human Geography Perspective

Economic and human geography is based on a recognition of the importance of spatial accessibility on the structure of the economic marketplace and human activities in general. Transportation of passengers or commodities clearly adds to the cost of marketplace transactions among spatially separated locations. Similarly, maintaining contact among dispersed peoples or the widespread participation in decision processes benefits from improved communications. By the mid-1930s Christaller and Losch had formulated the Central Place theory to explain the patterns of towns, cities, and metropolitan centers based on the optimum locations for viable economic activities.[2] Since then their work has been greatly expanded and generalized.

Central Place theory rests on three concepts: (1) all activities must have a minimum size to be viable; (2) each activity has a minimum range that encompasses enough clients who can gain access at reasonable cost to support it; and (3) these two concepts imply that a hierarchial structure of towns, cities, and metropolitan areas will emerge.[3]

Different activities characterize each level of the urban hierarchy as shown (Fig. I–2). Retail trade, physicians' services, and churches are ubiquitous and characterize local places, and (along with agriculture or recreation) characterize rural towns. Wholesaling, warehousing, and manufacturing are more centrally located at or near urban places and cities. Metropolitan centers have the broadest reach and the most unique activities—news centers, communications hubs, corporate headquarters and government offices, banking and insurance, and specialized professional services.

Submarine telegraphy's contribution to the articulation of an international system of cities largely explains its impacts when characterized by increased scope, efficiency, and pace of activities. Cables were directly instrumental in collecting maritime insurance information, tying commodity and money markets together, and controlling the spread of epidemics. Transportation efficiency was improved by operators arranging cargoes for ships, directing troop movements, and scheduling diplomatic travel. These impacts illustrate that the cables acted directly to improve spatial accessibility, and, according to the Central Place theory, would be expected to draw the cities and places more tightly into an international system.

Greatly increased spatial accessibility made possible new higher-level urban activities centered in the leading cities of the world. News services and international communications cartels are leading examples of highest-level activities. Novel international organizations, cooperative scientific ventures, and universal standards of time centered nearest the world's leading city, London, are additional examples.

New highest-level central functions tended rapidly to become monopolistic. Examples are news services, ownership of communications networks from the hub, newly formed international organizations, and the extension of British naval dominance. This tendency toward monopoly might be understood if we recognize that these new functions had only recently become viable by rising above the minimum thresholds of size and range; there was not yet enough "market" to support more participants. As development continued some of these monopolies fragmented: NYAP broke its Reuter's alliance, competing cables developed, and British naval dominance of the seas eroded.

Established activities of trade and commerce, diplomacy, and military preparedness articulated many new levels up and down the hierarchy (Fig. I–2). This is a form of vertical specialization resulting in increasd *diversity* of participants. In each case, the transformed activity was more efficient than the previous form.

Lower levels of the hierarchy were drawn into the integrated network and the increased scope of competition induced specialization in agriculture (and probably in mining and manufacturing, although we did not investigate these areas). This increased specialization leads to increased size for individual participants and reduced diversity on a local scale, but to increased diversity on an international scale.

Along with increased opportunity for commerce and discourse came a general increase in the pace of life, especially characteristic of urbanizing regions.

Telegraphy reinforced the effects of related "integrative" technological developments of the period—steam shipping, steel hulls for ships, railroads. The mutual effects of these technologies on urban growth were probably much greater than the sum of their individual contributions.*

According to the theories of Allen Pred and others,[5] central places grow primarily because of their advantageous location in the communications patterns. Emerging activities and promising technological innovations are either created in the urban centers, or their inventors and promoters gravitate to the urban centers for support in the form of investment capital, specialized expertise,

*Quantitative studies of the urban hierarchy show that city sizes increase geometrically with increasing numbers of distinct types of activities or "central functions."[4] That is equivalent to saying that each new function added to a central place increases its population by a multiplicative factor rather than simply an additive increment. Generalizing this relationship to an international scale, one would expect that submarine telegraphy, steam shipping, steel hulls for ships, and related land-based technological innovations of the period agumented one another multiplicatively to stimulate the growth of the emerging international system of cities. That would be a worthwile topic for further research, since it permits one to separate the independent effects of several mutually supportive developments. It might provide a framework within which to study such questions as the potential impact of multinational corporations and international computer networks on the present international system of cities.

URBAN HIERARCHY/ Activity*	COMMERCE AND FINANCE		DIPLOMACY & MILITARY ORGANIZATION		NEWS AND COMMUNICATIONS		NOVEL CONSEQUENCES
	Before	After	Before	After	Before	After	After
INTERNATIONAL INSTITUTIONS Multinational organizations METROPOLITAN CENTERS, REGIONS News & communications centers Central offices or headquarters Banking and insurances Professional specialists URBAN PLACES AND CITIES Manufacturing Wholesaling and distribution Warehousing LOCAL PLACES OR RURAL TOWNS Retail and trade Physicians Churches Agriculture	British trade monopoly in England Merchant houses dominate trade & credit	Lloyd's of London is a heavy user of cables & distributor of maritime information Arbitrage dominates security markets Futures markets develop in producer nations' cities Warehousing disperses to producing nations Transportation brokers arrange for shipping Credit extends directly to small business Nations tend to specialize in various crops or commodities	*Both diplomatic corps & military:* Small central staff Distributed manpower near potential crises with heavy responsibilities on local authority	Britain extends world naval dominance Central general staff develops Central staffs of Foreign Office & State Department grow & specialize Diplomatic decisions must be reviewed at central office Troops & fleet concentrated into powerful strategic centers Logistic coordination permits effectic access to dispersed locations More frequent reporting from field staff Military intelligence grows many-fold	Principal city's newspaper dominates Telegraphy patchy with many competing firms	National news services develop & form international alliances John Pender builds an international cable empire centered in London National telegraph systems become monopolies	International organizations develop Oceanographic study becomes international cooperative effort
URBAN CHARACTERISTICS OF ACTIVITIES (E) Efficiency (S) Scope (P) Pace		(E) Middleman costs drop (E) Warehouse stocks decline (P) Worldwide market price differences fluctuations decrease (owing to increased pace of transactions		(S) Apprehension of criminals by cabling ahead, control of epidemics, & coordination of diplomatic travel		(S) International conferences establish legal principles for operation & assignment of liabilities for cables	

*This figure is an approximate mapping of activities in descending order of level in the urban hierarchy (first column), not an exact item-by-item representation.

FIG. I–2.—Scope, pace, and efficiency of submarine telegraphy form a pattern consistent with the further development of an international system of cities and places.

managerial skills, or concentrated markets. Thus urban centers are first to implement and benefit from the increased producitivity of invention and innovation. It is therefore not surprising that Cyrus Field was soon drawn to London, the leading world center of the time. There he obtained support when Americans had contributed only token participation. Eventually the development of a worldwide empire of submarine telegraphs was accomplished by John Pender and his London-based firms. In this century the USA has come to dominate international communications technology and industry as U.S. economy and influence expanded to world preeminence.

All the historical impacts in the categories of scope, efficiency, and pace can be understood by this one integrating theme—the continued expansion of an international system of cities stimulated by the development and spread of submarine telegraphy. This complex picture justifies the affirmative recognition of the Delphi impact "Stimulation of the growth of cities." It appears that research does plod after speculation.

Understanding the Unexpected Vulnerabilities: The Systems Perspective

A generally beneficial revolution in communications capability, brought on by submarine telegraphy, also produced a number of unwanted consequences or "vulnerabilities." Many of these vulnerabilities were entirely overlooked by the parties which created and promoted cables. How did these vulnerabilities arise? Why were they overlooked? Can the lessons learned be generalized to future decisions affecting the introduction of technology into societal context?

The value of retrospective technology assessment hinges on the ability to generalize the lessons of the past so that they can reasonably apply to the future. Since the past is unique and the future is unprecedented in its specifics, only the broad systemic features of past lessons can be generalized. But these systemic features must be interpretable in sufficient detail to enlighten the policy process, both public and private, that governs the introduction and management of novel technologies in a societal context.

We show that many of the consequences that were both undesirable and unexpected were systemic in nature and that the world view of the Victorian promoters precluded their recognizing these impacts. While many of the elements of the world view persist today to cloud our perceptions it is possible to identify specific systems principles that might have suggested these vulnerabilities before they occurred. These same systems principles counsel us that the future is inherently uncertain, that knowing the source of these uncertainties is worth while, and that there are identifiable classes of precautionary actions that reduce the risks inherent in uncertainty.

What Is Meant by "Systemic Effects"?

Systemic effects are the combined results of the interactions of many interdependent participants. Three distinct "systems" are considered here: economic transactions forming an international marketplace; diplomatic and military relationships among both developed and colonial nations; and the diverse, often competing, clusters of political and cultural attitudes, values, ideas, and ideologies. These last are reflected in the public pressure on military decisions and the Victorian expectations that Western culture is naturally superior and would come to be accepted.

In the present context, "systemic" also implies that no single party is in charge. Marketplaces are made up of a network of transactions without a central controlling body; even diplomatic and military bureaucracies produce decisions from diffuse and convoluted interactions among a large number of participants. The formality of "someone being in charge" becomes less meaningful as the bureaucracy grows more complex. Usually no single party can have a detailed or holistic view of the events as they happen. It is typical for participants to play their roles unwittingly, innocent of the systemic consequences of their actions, rather than with malice or intent.

Table I–7 lists most of the important systemic consequences of submarine telegraphy taken from the summary tabulation of Appendix II.

Several systemic consequences were essentially beneficial, for example increased international trade and cooperative international organizations. Two of these beneficial systemic effects were overlooked by the Victorians. The reduced risks and costs resulting from futures market trading were not perceived until after 1900. The emergence of an international system of cities of increased productivity was not understood until the 1930s.

Several systemic consequences were viewed with indifference by the Victorians. Today one would be more concerned and most likely view such consequences as detrimental.

Consequences clearly detrimental to the sponsoring society or organization were appropriately called vulnerabilities; see the lower portion of Table I–7. They represent threats to the persistence, viability, or stability of an entire system and its participants. Impacts such as these are felt by all parties, whether or not they played an active role in initiating the difficulty.*

TABLE I–7.—Systemic effects of submarine telegraphy.
Summary from Appendix II, except items marked by asterisk, which are taken from other chapters.

Anticipated by sponsors	Unanticipated by sponsors
(S) International volume of shipping continues to grow; (E) scheduling of shipping greatly improved—effective capacity increased.	(V̄) Futures markets reduce price fluctuations, risks of trade and commerce, and reduce overall middleman costs—beneficial role of speculators not perceived until after 1900.
(E) British naval dominance greatly enhanced; British army and navy reorganized for greater strategic flexibility and reduced tactical vulnerability.	(E*) Improved market efficiency, scope, and pace produce a worldwide system of cities and places with greater interdependence than anticipated by the cable promoters.
(E) Coordination of diplomatic activities—travel, control of epidemics, apprehension of criminals, etc.—greatly increased; (S) General Staff organizational format spreads to worldwide military planning, strategy, and tactics; elaborate military intelligence develops to support central strategy planning.	
(S) International organizations proliferate; international cooperative ventures develop in oceanography and meteorology; (V̄) hurricane warnings organized effectively throughout the Gulf of Mexico region.	
(S) Diplomatic headquarters staff grows faster than field staff; (E) specialists added; and (P) pace of decisionmaking quickens.	(V*) Telegraphy rapidly becomes a national monopoly in each country; in spite of the Adam Smith ideal of competition and government disengagement, British nationalize their domestic telegraphy to quell disputes among newspapers and complaints from the public.
(P) International news services bring information to public while events are still happening; now public can follow and even affect the ongoing decision process; (E) news services become national monopolies with interlocking international agreements (and, in Europe, close ties to national governments).	(V) In spite of their support for the cables, the British Far East trade monopoly and dominance of the large merchant houses is eroded by competition from foreign and smaller firms soon after cable linkages are secure.

*Systemic instabilities and unpredicted change pose ample cause for concern and much anxiety. However, a significant degree of uncertainty, variety, and unpredicted fluction about the status quo has also been shown to be necessary to systemic wellbeing in a number of settings. Large variations were thought to weed out less productive participants by Clément Juglar in his analysis of market panics and cycles. Similar relationships have been found in ecological studies—environmental variations weed out species that are detrimental to the ecology as a whole.[6] Von Neumann and Morgenstern's theory of games shows that often a "mixed strategy" with unpredictable responses by opposing parties is required to produce mutually acceptable results.[7] This procedure underscores the subtlety and difficulty of achieving an adequate understanding of systemic effects and the potential benefits or disbenefits of novel and promising technologies.

Table I–7 (Cont'd)

(P) Pace of life quickens worldwide (observed whether or not understood); (V) perceived pressure to respond to distant events with attention, adaptation, or intervention.	(V) Brokers of shipping capacity emerge, overbid, and inflate perception of real demand causing shipowners to overbuild capacity and face bankruptcy—shipping conferences stabilize transactions.
(V*) Soon after radio becomes capable of international telegraphy it threatens to undermine revenues of cable telegraphy and divert all investment in cable growth—mergers between radio and cable interests are induced by governments to preserve cable's uniquely secure qualities.	(V) Cables are supported as mail contracts acceptable to promoters but that neglects the quickly developing national dependence along with the contingencies, dependencies, or policy actions in response.
(V) Censorship or interruption of cable service during war is recognized as a contingency otherwise unenforceable—hence a right of the belligerents.	(V) Increasing time pressure and scope of responsibility creates complex and diffuse headquarters staff operation in diplomatic corps, reduces confidence in the decision process.
	(V) Public news services exert powerful influence over British government during Crimean War; strategic information leaks to Russians; and U.S. locate Spanish Fleet from Madrid press reports—new channels of military influence not previously considered.
	(V) Constant readiness encourages suspicion and helps precipitate World War I.
	(V*) Brotherhood and international peace among men *does not follow* from increased contacts, communication, or international "understanding" as expected.
	(V) British plan and execute elaborate military, diplomatic, commercial, and domestic censorship during World War I; U.S. gets 90% of news of Germany from Britain and France.
	(V) British crack German code during World War I; reveals systemic vulnerability of excessive trust in a countermeasure and not being able to verify security.
	(V) The concept of total war on a global scale arises from economic interdependence and the importance of civilian attitudes in national military mobilization; war is now directed against the popular will to fight and economic viability as well as the military forces.
	(V) By World War II nations recognize the importance of having independent international news media due to increased public participation and influence on decisions.

KEY
E: Effectiveness increase
S: Scope increase
P: Pace increase
V: Vulnerability increase
V̄: Vulnerability decrease

An important characteristic of the systems affected by submarine telegraphy is their susceptibility to drastic and irreversible changes brought on by seemingly innocuous and apparently incremental actions. Systems exhibiting abrupt transitions of behavior are said to show "threshold effects." In economic, political, and cultural systems just a few participants or a gradual change in normal relationships can lead to conditions in which all participants are caught up in a radical structural alteration—suddenly and without an obvious precedent or warning. These changes are unplanned, often detrimental to the majority of the participants and aptly called vulnerabilities induced by the more gradual process of technological and institutional change.

Economic, political, and cultural systems also share several properties that act to obscure the presence of thresholds until the sudden changes take place. The interdependences among participants are so convoluted and complex that tracing causal sequences—A caused B which caused C, etc.—becomes unproductive, frustrating, and purposeless. The participants cannot perceive thresholds that lie well beyond their normal range of experience. Responsibility can become incongruent with authority, and thus produce misunderstanding and recriminations which further cloud issues and block more effective action. The resulting vacuum of understanding is usually filled by the ideas of leading thinkers of previous eras. Too often these ideas are accepted unquestioningly, misunderstood, unqualified, and transfigured into the "common wisdom."

Why Did the Victorians Overlook Many of the Vulnerabilities?

It would be improper to imply that any one reason "explains" why the Victorians overlooked many of the vulnerabilities indirectly produced by submarine telegraphy. There are at least three plausible reasons, each of which would have been sufficient in itself to have obscured the impacts of the cables.

The first was simply *overconfidence* in the countermeasures taken to protect against recognized or dimly perceived vulnerabilities. The Germans' confidence in the impenetrability of their codes during World War I is an illustration of this attitude. It repeatedly led them into catastrophe at crucial junctures in the War.

The second is *that systemic effects* are hard to identify with confidence. It took forty years to recognize the beneficial effects of futures markets and the speculators in those markets. Is it surprising that detrimental systemic effects also went unnoticed?

The third reason was probably the most significant. *Prevailing attitudes* toward open competition, and unfettered interactions, were favorable, instabilities notwithstanding.

Adam Smith's influential conception of free market competition providing the highest quality at lowest prices pervaded the times.[8] Smith warned against all forms of intervention into the free market: cartels, monopolies, corporations run by managers as well as manipulation or control by the state.[9]

Thomas Malthus, while employed by the East India Company in the early 19th century, espoused free trade, à la Adam Smith, for the subcontinent of India.[10] He borrowed from the naturalists to construct his famous hypothesis that population increases geometrically* but improvements in food production will increase only linearly.[11] That was the first recognition of the fierceness and potential instability of competition among participants who are individually capable of multiplicative reproduction or exponential growth. Malthus's colleague David Ricardo went on to show how this competition would crush laborers' wages as landowners prospered from the rents derived from increased food production.[12]

Thus the prevailing views of competition were that it brought perfectibility—ever-improved performance from the participants—and that it was an inevitable and inescapable part of the balance of nature. Victorians both expected and accepted that some parties will succeed while others fail; but the risks and responsibilities were on the shoulders of those parties. Under these views, technology assessment or an exploration for "unintended" impacts would be truly anachronistic. Whatever resulted from free and open competition was best, inevitable, and "natural."

Despite the efforts of the British government to induce competition among telegraph companies, both the transatlantic communications route and the domestic systems became monopolistic. The government had to buy out and nationalize the domestic system to quell controversy, abandoning in this situation the principles of free trade and perfectibility through competition.

Some Key Difficulties of Describing Systemic Vulnerabilities Within Economic Theory

During the 19th century such men as Antoine Augustin Cournot (1838), William Stanley Jevons (1871), and Vilfredo Pareto (1897) reduced the bewildering complexity of the ideal marketplace to a set of mathematically tractable propositions.[13] This accomplishment enabled economists to assess the position of their clients with respect to the economic milieu on the assumption that free and open

---

**Geometric growth* results from biological reproduction, from compounded interest, or from the spread of disease or even rumors. The underlying mathematical principle is that in each successive time interval Δt the growing population will have increased in proportion to both its size in the preceding interval and the rate of growth r:

$$N_{t+\Delta t} = N_t + r\Delta t N_t = N_t\,(1 + r\Delta t) \qquad \text{(for one interval } \Delta t)$$

$$N_{t+n\Delta t} = N_t\,(1 + r\Delta t)^n \qquad \text{(for } n \text{ intervals } n\Delta t)$$

or in a continuous-growth approximation, we have the differential equation:

$$dN_t/dt = rN_t$$

which represents exponential growth:

$$N_{t2} = N_{t1} \exp\,[r\,(t_2 - t_1)]$$

transactions among participants had produced an optimal distribution of prices, commodity production, etc. Unfortunately, the conditions for this optimum presume a stable market. *Therefore, the very definition of the marketplace precludes this mathematical formalism from being applicable to the study of extreme instability.*

Mainstream economic thought also has difficulty in representing the value of information and hence the value of the communication of information. Nobel laureate Kenneth Arrow, studying the economic value of producing information (in this case inventions), recapitulates the three conditions for Pareto optimality as they relate to information-based activities.[14]

Appropriability refers to the necessity of assigning (and transferring) ownership to commodities offered in a market. Information, as a commodity itself, is valueless until disseminated. But once it has been disseminated it quickly becomes common property and ownership is lost since its further use is not easily traced.

Divisibility refers to the only means of establishing a price based on the desire of purchaser to pay for various quantities of the commodity. Thus, a price paid for 10 bushels of wheat establishes the market value of 100 bushels. The *information* contained in a price quotation for a wheat futures contract is not divisible. A prospective purchaser of the quoted commodity cannot assess its usefulness before purchasing it. Only large quantities of similar information approximate the condition of divisibility (such as that provided by a stock market, a news service, or an intelligence system).

Perfect information refers to the absence of uncertainties. Uncertainty—such as the lack of awareness of lower prices offered by other sellers, the inability to assess differences in quality among similar commodities, or imperfect credit due to differences in personal familiarity—reduces the degree of optimality (or "efficiency") of a market. Arrow points out that *information is defined as the reduction of uncertainty*. To assign a value to information or communication of information, economists must relax the condition of perfect information that permitted them to define the marketplace in the first place.

It would be unfair to economists to say that they have not or cannot treat uncertainty, information, or potential vulnerabilities.[15] Inventory theory, game theory, and time-series analysis of market fluctuations are examples of techniques for dealing with such questions. However, it is fair to say that the client-oriented perspective—the necessity of obtaining quantitative solutions to problems posed by clients—has seriously restricted the types of mathematical approaches used by economists.

A Broader Systems View

In order to study systemic instabilities, one must adopt theoretical models and mathematical representations capable of exhibiting unstable threshold behavior. The systemic vulnerabilities produced by submarine telegraphy can be understood with the help of a few very widely interpretable systems models. The models' simple construction from nearly universal properties is responsible for their generality and potential importance to the methodological development of technology assessment.

The systems of greatest interest in technology assessment share the Malthusian property: they are capable of exhibiting exponential growth. The economic enterprises of firms can grow by compound-interest returns on investment. Diplomatic and military exchanges can produce a geometric increase in territorial acquisition or colonial empires. Rumors, political sentiments, and scientific theories can spread like epidemics; the group of informed or infected parties increases by recruitment from the general population.

The rapid exponential growth of any individual participant is limited either by the readily available resource base or by competitive pressures from other participants sharing the same resource base. Whenever one technology replaces another (such as steam power substituting for sails in shipping) the limited resource base is the extent of the market (for shipping). For a colonial empire the "available resources" are the organized markets for which commercial gains cover the costs of colonial administration. For political or cultural norms the ultimate resource base is the intercommunicating polity of common interest.

Competition frequently limits each participant to a small fraction of the total resource base. Several firms producing the same commodity limit each other in a market place. Adjacent national empires most often inhibit further colonial expansion. Ideologies, political coalitions, and scientific theories are seldom without competitors for followers.

a. Competitive exclusion refers to a process in which the encroachment of one competitor on others is unconstrained and leads to the exclusion of some participants from the system. The general principle states that competitive exclusion always occurs unless there are some differentiating characteristics within the competitive context that act as a basis for distinguishing the niche of each participant from all others.[16]

Numerous examples illustrate the importance of achieving stability through differentiation. Geographical distances between competitors in the marketplace stabilize competition among similar types of firms; that is the basis for economic geography. The insensitivity of telegraphy to distance explains the instability of competitive interactions among telegraph firms. A dominance hierarchy of military power is often sufficient to stabilize diplomatic interactions. Lesser parties need only demonstrate it is not worth the effort of the dominant parties to exercise excessive influence over them. Social class, differential access to information and education, or organizational membership stabilize the competition among ideologies and values.

Two of the major vulnerabilities arising from international submarine telegraphy have all the characteristics of competitive exclusions. News services had such overwhelming advantages of size and scope that only national borders provided sufficient differentiation to stabilize competition among them (as with telegraphic services). News services became monolithic, with Reuters having the clear advantage of being centrally located within the British-controlled international cable network; the U.S. and European news services formed agreements with Reuters to share news reports. Eventually this dependence on a single source for news of diplomatic significance proved unacceptable.

Whenever two military powers have comparable forces—within the uncertainties of military intelligence—and there appears to be a potential of either dominating the other, competitive positions become unstable and produce the preconditions for the breakdown of diplomatic relations leading to war. Lewis Fry Richardson showed that a tit-for-tat arms race culminating in World War I began as early as 1908.[17]

Before World War I, the forces allied with Germany had resource comparable with those allied with France. Once the combined military budgets of all parties exceeded £194 million, a Malthusian increase of further expenditure continued for five years, until the outbreak of the war.* Richardson interprets this quantitative finding as the result of unstable competition in which both sides become increasingly alarmed at the other's armaments and are induced to augment their own. Thus the

*The following Table and Figure adopted from Richardson's work shows the arms budgets and the striking regularity of the exponential divergent growth for the prewar years.

Defense budgets in millions pounds sterling

	1909		1910		1911		1912		1913
France	48.6		50.9		57.1		63.2		74.7
Russia	66.7		68.5		70.7		81.8		92.0
Germany	63.1		62.0		62.5		68.2		95.4
Austria-Hungary	20.8		23.4		24.6		25.5		26.9
Total $= U + V$	199.2		204.8		214.9		238.7		289.0
Time rate $= \delta(U + V)/\delta t$		5.6		10.1		23.8		50.3	
$(U + V)$ at same date		202.0		209.8		226.8		263.8	

Here δ signifies 'take the annual increase of' whatever symbol follows next.
From Monog. Supplt. No. 23 of *Brit. Journ. Psychol.*

potential threat of aggressive offensive actions is increased and it becomes increasingly difficult to interpret the opposition's arms as defensive.

Once hostilities had begun, the ability to support a war effort depended on the economic strength of an alliance and the morale of its citizenry. Because these factors rose to an importance comparable with that of battlefield strategy, economic vitality and citizen morale became targets of war. The world wars of the 20th century became total wars, embracing all factors contributing to the stability of a belligerent's position. Many of the interlocking relationships that stabilized the transactions among the international systems of cities in peacetime became battlegrounds—shipping, communications, trade, and commerce assured sources of raw materials, etc.

The systemic nature of total war was even more evident during World War II. However, by that time radio had become a co-equal with cable telegraphy and the impacts of communications had spread well beyond the focus of this report.

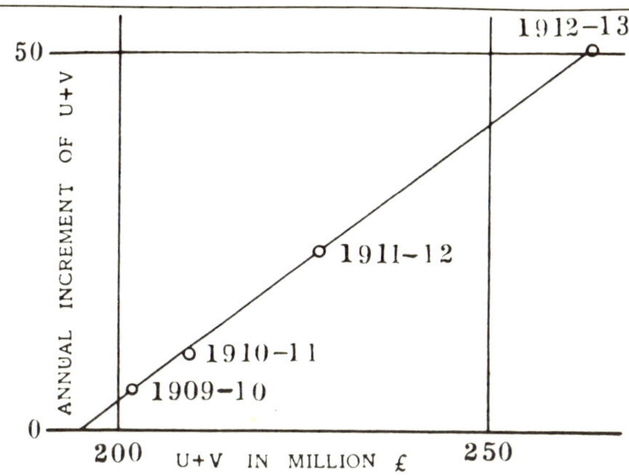

From L. F. Richardson *Mathematical Psychology of War* (1919). In British copyright libraries. (See also *Nature* 135: 830–831, 1935.) The diagram first appeared in *Nature* 142: 792–793, 1938.

Richardson's mathematical model shows the armaments of the opposing sides grew at a rate k once the other side had exceeded an unthreatening limit, say, U_o for one side and V_o for the other:

$$d(U-U_0)/dt = k(V-V_0) \qquad d(V-V_0)/dt = k(U-U_0)$$

If we add and subtract these two equations, we gain insight into the systemic response of the symmetrical relationship of intimidation in an arms race:

$$\frac{d[(U + V) - (U_0 + V_0)]}{dt} = k[(U + V) - (U_0 + V_0)]$$

$$\frac{d[(U - V) - (U_0 - V_0)]}{dt} = -k[(U - V) - (U_0 - V_0)]$$

The second equation says that the gap between the armaments of the two sides tends to close until the arms difference equals the small difference in the respective tolerances of the two sides to viewing the opposition's arms as defensive. The first equation predicts exponential growth of the excess armaments beyond the tolerance levels:

$$\frac{d(U + V)}{dt} = 0.7(U + V - 194)$$

for the arms race leading to World War I, where the amounts are expressed in millions sterling. Because of this tendency to grow above the threshold of £194, these equations are unstable. (Note that armaments of less than £194 would diminish; the equations are totally unstable.)

b. Delayed responses or the inability of participants to keep pace eventually leads to an instability. Sometimes these delayed responses are broad enough in scope to classify as systemic vulnerabilities.

Even though competition for finite resources limits growth, the encounter with those limits, or the opportunity to make gains by encroachment on competitors, may not be immediately evident. Very general theorems of systems analysis state that the stabilizing effects of mutual action and reaction will almost certainly be lost if the reactions are delayed too long or the actions are sped up without a comparable increase in the pace of reactions.[18] This principle would explain the widely felt pressure to respond and the accompanying increase in the pace of activities following the laying of cables.

British shipping companies overbuilt capacity because former mechanisms for estimating shipping demand were inadequate. The lag in perception after saturation of the market for shipping led to financial loss to all the shippers.

The arms race preceding World War I increased the potential threat of rapid military conquest. Constant preparedness was an attempt to provide security through an assured ability to react. One explanation for the actual outbreak of hostilities was that the perceived ability to react could no longer keep pace with the threat of an aggressor's capability to attack. The fear of being second caused premature belligerence—before the true facts could be ascertained.

Futures and arbitrage markets most clearly reveal the effect of quickened pace as a prerequisite for competitive viability. Faster recognition and greater certainty of price differences among markets increased transactional pace and volume. This speed-up greatly reduced price fluctuations among distant markets and led to greater systemic efficiency. (It is unclear whether tying international markets together provides reduced vulnerability to *worldwide* panics and depressions or not.)

c. Too many relationships almost always produce a destabilizing effect. In complex systems with many participants, each participant can effectively relate only to a very small fraction of the other participants. The technology of international submarine telegraphy produced an unprecedented growth in the number of types of relationships among participants in systems of worldwide scope. This growth produced unforeseen vulnerabilities. Very general analytical findings suggest that complex systems made up of participants capable of Malthusian growth are unstable if either too few or too many relationships exist among their participants.[19]

At one time it was common wisdom that more complex systems were less vulnerable to disruption. The argument was that they had greater resources to cope with the unexpected and more internal relationships by which responses could be formulated. One could also imagine that a disrupted complex system might simply revert to a more stable formative configuration. These views are rapidly dissolving with increased understanding and experience to the contrary.

Another possible explanation of the common-wisdom view that more relationships produced greater stability is that this is how familiar mechanical analogs behave. Everyone knows that a barn built with insufficient crossbracing is likely to collapse suddenly during a storm. Certainly this structural "system" is made less vulnerable by the addition of more braces—the creation of more working relationships among its component "participating members."

As one would expect, marketplaces and political systems are quite different from barns. A more appropriate analog (in which the creation of new relationships can result in divergent growth, not structural stability) is the familiar example of the audio amplifier whose microphone is inadvertently turned toward the loudspeaker and picks up its own output; the resulting increase in sound level is again picked up and amplified—and so on, until the capacity of the amplifier is saturated. A hand placed over the microphone cuts the destabilizing relationship and quiets the noise. By analogy, arbitrary or unprecedented creation of new relationships within systems capable of participant growth is overwhelmingly likely to produce an instability.[20]

Systems with components or participants that can grow are called *active systems*; systems in which the components absorb outside influences are called *passive systems* (such as the barn). This distinction takes on fundamental importance because physical systems that retain their properties in isolation obey the second law of thermodynamics, which requires that all "active" behavior be only transitory. However, open systems with a constant source of free energy or "fuel" from outside can maintain "active" properties indefinitely. Free energy from the sun has sustained the active system of biological processes for over four billion years.

Cable communication created new and stronger connections among nations, colonies, and military powers. The response of governments was to increase the size and complexity of central bureaucracies charged with diplomatic and military decisionmaking. But these bureaucracies and their component departments were also capable of divergent growth or instability, as was the international system they were to manage or control. General theorems like the above are often perceived intuitively, but imprecisely, long before they are formally recognized. That may explain the general concern with the growth of complex diplomatic and military bureaucracy even during the early 20th century. Exacerbating this concern were the infrequency and limited opportunities of the military to test or demonstrate their capability (or incapacity).

The emergence of international news services created powerful and unprecedented relationships—direct channels connecting the public to world events and to the decisions in progress. This development proved to be an unexpected liability for controlling the leakage of strategic information and an unprecedented constraint on military decisions. Only after repeated painful experiences did the British act to diminish the impact of this new relationship. They may have over-reacted by censoring all cables and all traffic during World War I. This study was too short to look for a possible postwar loss of confidence in the British management and control over so much of the world's commercial, diplomatic, and domestic news traffic.

A secondary implication of the relative sparsity of relationships within stable complex systems is that an unprecedented requirement for rapid adaptation will most likely necessitate the close cooperation and mutual response of several participants that have no established direct working relationships. In other words, the sparse relationships which do exist within complex systems are there because *past* experiences have proved them to be necessary. Political (and legal) actions in the British and U.S. traditions are especially constrained by the past through the creation of precedents to legitimize actions. That may partially explain why the U.S. Congress treated submarine telegraphic communication as a mail service in contracting support for a successful transatlantic cable. Controversy surrounding this action focused on the growing differences between the North and South rather than the novel aspects of rapid communications or the vulnerabilities that might result from its use.

d. Excessive temporal variability in the systems environment causes an erosion of the complexity and also the diversity of both participation and types of activities. Submarine telegraphy reduced the temporal variability and uncertainty of international transactions. That permitted specialists to replace generalists, but it also created a new dependence—and hence a potential vulnerability.

A companion misconception to the notion that more interrelationships promote stability is that greater complexity produces more predictable environmental conditions. This misconception may be a misguided belief that complexity produces averaging effects, hence greater predictability. Accumulating evidence suggest the opposite—greater certainty in the environment permits and encourages greater complexity (for self-organizing systems whose participants can exhibit Malthusian growth). The converse does not appear to hold—greater complexity does not necessarily lead to reduced variability in the system's environment.[21]

An illustration of this principle is the erosion of the British Far East trade monopoly and the decline of the large British merchant houses. International telegraphy permitted much improved certainty in world trade. Prices, demand for goods, credit worthiness, and crop quality at distant locations were known rather than being a gamble. Therefore specialists, serving specific market functions, could compete for credit, buyers, warehousing needs, and sources of supply that had been the exclusive territory of the large British merchants.

More timely and accurate knowledge of a wide range of events also permitted the organization of coordinating and planning bodies on an expanded scale—the growth of diplomatic and military central bureaucracies depends on the constant flow of large volumes of relatively certain information inputs. Similarly, the growth of large cities or systems of cities depends on reduced uncertainties, fluctuations, or instabilities—rather than being simply the agent of their reduction. The continuance of stability, limited variability, and the flow of dependable information is a vital necessity to the many participants in these complex systems.

Before World War I, the world depended on cable communication for continued economic well-being through international trade and investment. During the War, the British enacted economic sanctions and communications barriers for firms in neutral nations trading with, or otherwise supporting, the Germans. Cables also carried news of the battle and shaped attitudes and sentiments toward the opposing forces. The British censored news and controlled most of the reports reaching the U.S. These were largely unanticipated consequences of the degree of dependence on the cable that had developed over the long period of economic freedom preceding the war years.

e. *Defensive actions taken to assure reliability* are penalties assumed by cable users to avoid perceived vulnerabilities. Unfortunately, many of these defensive actions reflect the remedial measures taken after difficulties have proved them to be necessary. One is never certain that unprecedented events may not exceed the ability of these defensive actions to cope.

Defensive actions are also systemic effects identified in the summary tabulations of Appendix II and shown here in Table I–8. These actions were taken to assure the reliability and verity of information provided by submarine telegraphy.

Reliable and accurate exchanges of information soon became a necessity to the maintenance of the systemic efficiency, complexity, and specialization created by the cables. Two generic strategies are shown in Table I–8—actions that increase the security of established or critical connection, and those which provide redundant or alternative connections in case some of the critical connections prove to be faulty. For both strategies the analog of shoring up a barn *is* appropriate. It can be done either with stronger timbers or by additional crossbracing. These basic strategies are the same whether applied to the purposes of message transmission, transactional or coordinating activities, or improving the verity of background information contributing to decision processes. These defensive actions are the basic strategies of reliability theory.[22]

The first column of Table I–8 gives examples of increased security and redundancy based on the "technological fixes" of encoding messages or providing more than one cable route. Less familiar

TABLE I–8. Increased reliability as a strategy for reducing vulnerability.

Generic strategy	Message transmission	Coordination & transaction	Background or support for decision processes
Secure communi-cations linkages	Codes used to inhibit false or tampered market quotations, offer diplomatic secrecy, and military security.	Chicago Board of Trade establishes regulations for futures trading. Commodity standards assure quality in transacted goods.	National news services reduce local news biases. Procedures established to prevent diplomats in the field from manipulating the choice of mode of communications to gain greater control over head office.
Redundant or alternative linkages	Multiple cable routes used to assure a communications link. Submarine cables are maintained even after radio offers more economical service.	Hedging required by banks for loans on commodity futures. Naval vessels and troops are distributed among "strategic" centers.	New York Associated Press breaks agreements with Reuter's of London for use of Reuter's coverage of South American news and U.S. reports to S.A.; NYAP also declines offer by U.S. State Dept. to share diplomatic cable. Before WWII all nations become sensitive to the need for independent news coverage.

examples of these principles are found among the formal procedures adopted by institutional structures—regulations, standards, hedging several purchases, or dispersing military liabilities. Greater confidence in the news reports and commentary was achieved by institutional mechanisms based on these same principles.

Ecological Models

The five systems principles used here to improve the understanding of the impacts of submarine telegraphy were adopted from the rapidly developing discipline of systems ecology. Only the most general and broadly interpretable principles were selected. The conditions necessary for these principles to apply are nearly always satisfied by most of the systems of interest to technology assessors. The principles:

- Numerous participants interact in a complex web of relationships in which no single participant can or does "control" the behavior of the others, yet the participants depend significantly on one another's actions.
- Individual participants or members of the system can exhibit Malthusian growth if permitted.
- Growth of each participant is limited primarily by resource availability, which is further constrained by competition with other participants.
- Environmental variation and sheer complexity create inherent uncertainties for each of the participants; these uncertainties require the participants to adopt a range of adaptive strategies in order to remain viable.

The systems principles can be classified according to the fundamental characteristics of space, time, and uncertainty (Table I–9).* The classes of systemic instability cover the first four of the five principles. The fifth is any large-scale failure in the application of systems reliability strategies. For each of these principles the failure mode is characterized by an abrupt threshold transition. The way in which instabilities come about are described in the last row of Table I–9.

These principles are the basis for several quantitative models and theories in systems ecology. To each of these principles and models corresponds a number of parameters and variables that can be interpreted in the technology assessment context: for example, the diversity of participation, the degree of environmental fluctuation, and the characteristic response times for actions or reactions. It is practical to obtain these parameters and variables in most technology assessments; however, we find in retrospect that this type of information was often considered less important by the Victorians and data are often unavailable. Still, the particular consequences of submarine telegraphy most often overlooked and most often considered detrimental could be understood in the light of these principles; we suggest that they are a useful set of heuristic tools for examining the potential impacts of technology in anticipatory studies.

Systems ecologists have probably pursued the development of these principles and analytical tools because natural ecologies are so complex as to defy detailed representation and because many ecological systems exhibit extreme variability—factors of ten to one hundred in population over a year are not uncommon. This devotion to holism has led some to examine the parallel between economics, social systems, and ecologies.[23] There is a controversy concerning the nature and extent of this parallel.[24] This controversy was side-stepped in this Appendix by the parsimonious selection of a few very appropriate principles and their application in abstract form without reference to the detailed interpretations given in a natural ecological setting.

C. Methodological Lessons

The generalized patterns of findings from the experience of the Victorians with submarine telegraphy, and the insights gained from other disciplines, provide a number of methodological

*The fundamental characteristics selected here parallel Kant's three types of "a priori" knowledge: time, space, and causality. Uncertainty is Kant's "causality," since causality could not be perceived by the cognitive agent as a distinct type of relationship if there were not some circumstances in which "noncausal" relationships occurred; hence there must be uncertainty a priori.

TABLE I–9.—Classes of systems instability and reliability strategies.
CLASSES OF SYSTEMIC INSTABILITY

	Competitive Exclusion	Delayed Responses or Inability to Keep Pace	Too Many Relationships	Excessive Temporal Variability Destroys System Complexity and Diversity
	Competition for a common resource base or niche among two or more participants of similar functional strategies is unstable.	Stabilizing feedback works only when it comes in a timely fashion. When delayed too much after it is needed, it can destabilize.	Too many new relationships will almost certainly produce a destabilizing result.	Complex specialized systemic relationships depend on a high degree of environmental predictability.
Space Co-existence and inter-relatedness	Competition is among co-existing participants within the spatial dimensions of the system. Competitive exclusion results in one participant losing his space or "territory."		Relationships can either promote stability of a "dynamic equilibrium" or produce divergence from that equilibrium; rapidly and arbitrarily creating many new relationships is overwhelmingly likely to produce instability.	Complexity, diversity, and specialization of participant functional strategies (roles) are inherently spatial concepts—they refer to co-existence and mutuality as well as competition.
Time Sequential events and feedback responses		Each participant acts or reacts according to the surrounding conditions. If responses are too slow, the opportunity will have passed or the threat will have overtaken them.	Newly created relationships are new paths of feedback. Beneficial growth is a form of unstable positive feedback, but unexpected/ unplanned feedback is likely to be detrimental.	Temporal variability leads to uncertainty whereas spatial variability often provides distinctions among niches. Thus temporal variability tends to reduce diversity, whereas spatial inhomogeneity encourages diversity.
Uncertainty Signal-to-noise concept basic: "Uncertainty" is the noise which precludes participants from organizing their worlds.	A "common" resource base and the competitors with "functionally similar strategies" are defined in terms of the competitors' ability (or inability) to *discriminate* among the small differences which are always present. Competitive exclusion results when competitors can no longer discriminate among distinguishing features due to environmental variability or shifting postures in the competitive game.	Uncertainties in the relationships among resources and other participants tend to complicate or obscure the choice of when and how much to respond.		Uncertainties and unpredictable variability requires a wider adaptive range and hence less specialization among individual participants.
Failure Mode Characteristics How does the instability come about?	Shares of the resources commanded by either competitor swing up and down randomly until some participant(s) is (are) eliminated.	The transition from a timely response to a late one is often very abrupt; delays often worsen rapidly once the threshhold is passed.	Transition to instability can be very abrupt, but may or may not affect the entire system, rather than just a few participants.	Sudden onset of unanticipated circumstantial events can rapidly destroy much of the system's specialized diversity.

Table I–9 (Cont'd)
COMPONENTS OF SYSTEMS RELIABILITY STRATEGIES

Secure relationships.	Redundant relationships of similar function.	Respond faster or reduce need to respond faster.	Information support to assess where one stands.
A chain is only as strong as its weakest link; strengthening the weakest link is a reliability strategy.	A steel cable of many wire strands continues to function even if some strands fray.	Living systems are kept in balance by constant corrective action: a standing person corrects for small perturbations of his balance.	The many attributes of complex-system reliability (and threats to it) are only partially perceived and indirectly assessed.
Individual relationships are made more secure and reliable.	Alternative relationships are established to accomplish the same purposes in case other relationships fail.		What are the alternatives? What are the characteristics of established relationships?
		Respond faster before the situation gets out of control. Provide emergency resources to use while a response is being prepared to an unexpected situation.	What is the nature of the situation? How fast a response is needed?
How secure is secure? What means are there to assess the security of given links? Where should trust be placed?	Are redundant relationships really independent? How many such relationships are enough?	Does a faster response create a problem for others who must *react* to the the original "response"?	What are the indirect implications of responses? Uncertainty is the limit to information—they *define* one another.
Failure of any systems reliability strategy leads to abrupt systems collapse.			Poor information can lead to continual misinterpretation of the root causes of difficulty.

lessons for the conduct of both retrospective technology assessments and anticipatory technology assessments. This section focuses on the most fundamental of these lessons.

Since whatever exists today developed from, and depends on, what was yesterday, a retrospective view could be expected to provide insight into how the future will develop from the present. Understanding how that happens has long been a dream of historians and scientists alike.

The Victorians only dimly perceived the implications of international submarine telegraphy. Today much of what they overlooked is more immediately on men's minds. Yet new, seemingly unprecedented technology and international relationships pose more complex questions. Can these complexities be penetrated by the abstraction of the lessons of the past into patterns interpretable in the future?

This section describes a few of the *practical* and *inherent* limitations to gaining insight into the future. Simply recognizing that "analysis," by whatever method, is severely limited in providing unequivocal predictions or forecasts suggests that an alternative approach should be sought and refined.

The rudiments of an alternative approach are found within the reactions of both the Victorians and early 20th century stakeholders in the development of submarine telegraphy. These alternatives are found in their *adaptive responses to cope with uncertainty*. They used the strategies of stability theory in the broad context of institutional, political, and societal uncertainties as well as technical ones for coping with unforeseen contingencies.

The use of even more general adaptive strategies can be refined into a systematic methodology. Traditional forms of analysis would still be used to reduce the range of logical possibilities, to define the possible actions of stakeholders, and to establish what can be known. Both the practical and the epistemological limits to knowing the future must be explicitly stated in the analysis.

The contingencies within these limits to forecasting are then to be covered by carefully planned adaptive strategies. The ability of such strategies to cope with contingencies can be assessed in the present.

This procedure avoids the paralyzing effect and endless controversy that would result from our founding a technology assessment study on answering questions about the future that are impossible to resolve. Such an approach is closer to the mechanisms found within nature's own "adaptive systems"—the natural ecosystems.

These ideas are next developed in greater detail.

Practical Limitations to Both Knowledge and Analysis

Retrospective technology assessment and anticipatory technology assessment are forms of applied research and policy analysis. Like all forms of research or analysis these studies are quite finite—they are limited in manpower, duration, and in the availability of data. These are familiar constraints to all researchers; they need no further comment here.

Practical limits often fall heavily on both retrospective and anticipatory technology assessment because their range of concern is so broad. Reducing the problem scope and discriminating impacts of the focal subject, in this case submarine telegraphy, from its broad context is a complicated task when the focal subject is intimately woven into the fabric of history.

Events and impacts resulted from the mutual influence of several technologies—steamships, land telegraphy, railroads, improved weapons, and even social technologies such as improved management techniques. There are analytical tools for distinguishing the individual contributions of several influences acting simultaneously—factor analysis, comparative studies, etc. It is possible that a complete quantitative description of the Victorian period would permit a factor analysis of each technology's contribution. The simultaneous influences of submarine telegraphy and other developing technologies of the period probably multiplied one another's effects rather than merely adding an independent increment. The success of a factor analysis would hinge on finding situation in which one or the other technology acted to a greater degree than the others. But such an approach was well beyond the scope or intent of this study and of most other exploratory assessments.

Inherent Limits to Hindsight and Foresight

Taken to extremes, practical limitations can reveal inherent limitations. For example, if time and effort were not practical constraints then the simple fact that the analysis is only part of the whole problem context—and in some cases the analysis effects the subject of its attention—would limit the precision of either hindsight or foresight. But these are ultimate limits, seldom approached in real analyses outside the physical sciences.

Knowledge is useful and sought after because the great majority of experience forms regular and identifiable patterns.* Much of the apparent "uniqueness" of unfolding sequences of events is precedented in some way and can be anticipated to a degree. This reality is the basis of organized existence and life itself.

If additional analytical effort could provide assurance of an ever-diminishing cloud of uncertainty surrounding future events, then there could be a "science" of futurology. This science would identify and forecast ever more precisely the regular patterns formed by experience. Uniqueness and unexpected events would appear more and more ordinary within this expanded framework of analysis.

Far-reaching consequences can hinge on seemingly insignificant twists of fate. Long-standing regular patterns that have come to be expected may be radically altered or shattered altogether as a result. This potential sensitivity of the course of events to a myriad of details creates an inherent source of uncertainty and limits the penetration of analytical understanding into the future.

An illustration from the early days of submarine telegraphy shows how simply reversing the time sequence of two events could have radically altered the course of social and political history thereafter. Whether or not that would have been reflected in a redirection of technological and economic trends is even less certain.

Anticipation of future events based on previous knowledge subsumes the concept of causality. Event A causes event B, which then causes event C. The events and their relationships may be precisely understood or only vaguely perceived. In either case the more or less orderly patterns of experience and historical trends lead us to believe that there are causal relationships whether understood or not. By the reversal of the order of two essentially independent events, their mutual consequences are very likely to be altered.

During the American Civil War, a U.S. naval vessel intercepted the British merchantman *Trent* to remove two escaping Confederate officers. This action so enraged the British Government that U.S. Minister Adams claimed the British would have declared war on the Union if they had not been delayed by weeks in their diplomatic communications. In other words, had the 1858 transatlantic cable continued to function, the course of American history and the development of the North American continent could have been radically altered. Cyrus Field argued differently. Had there been a cable, rapid diplomatic negotiation would have precluded the need to board the *Trent* in the first place. Who is to say which man is right?

*It still seems remarkable that the progression of technological improvements to submarine cables and the increasing capacity of the transatlantic telegraphic routes would follow a smooth progression over 50 years. In a similar fashion, national economies grew steadily over the past 200 years, if one is willing to average over the momentary set-backs. Human population, books published, energy consumption, and other significant features of human activity are known to follow remarkably regular patterns. Perhaps the great variety of human activity leads to consistent, general, statistically predictable trends; then we might possibly know enough to supply our needs for wise decisionmaking. This assumption of steady progression is the basis for technological, economic, and social forecasting (enterprises of notable but limited success).

Technology forecasting should be the easiest of all the prognostication tasks since the properties of physical objects are by far the most independent of their history and the most easily analyzed. But technology forecasting is not primarily about *things*; rather, it is concerned with *ideas* about things. The invention of magnetic telegraphy brought something new into being—it was a collection of ideas about how to transform copper into wire, iron into magnets, chemical energy into electrical, and interruptions of current into messages. Yet from a broader perspective the idea of the magnetic telegraph might be viewed as the inevitable consequence of increasing knowledge about these component technologies.

This one illustration, and many more like it, shows that the most delicate and untraceable alteration in causal sequences can have profound effects. Since we shall never know everything, the future will always be inherently uncertain in important regards.

Technology assessment therefore intrinsically depends on anticipating the *unexpected possibility* as well as extrapolating the expected continuation of current trends. The more unusual the event the less likely it is that it will be found by intuition based on experience. More systematic and abstract approaches may hope to embrace novelty by heuristically connecting it with more general patterns of experience. But this approach has no ultimate assurance of success.

All that we know, or think that we know, comes from the past. It comes in the form of hard, durable evidence; it also comes as relatively untraceable prescriptive norms, values, and ways of thinking passed on and refined through the generations.

Retrospective technology assessments are systematic attempts to learn from the more durable evidence—what did happen as the result of a technology that has had time to run its course? Many of the limitations of the retrospective approach arise from irretrievable loss of the less durable inputs to the decision process underlying the prescriptive norms.

Examples of the difficulties posed by the loss of transient information abound. A particular alternative course of action may or may not have been considered in a given situation, but only a few were recorded. Did the prevailing attitudes and the writings of the participants characterize the real thought process that guided their decisions leading to international submarine telegraphy? What influence did personalities,* transient memories, and chance encounter have? If it is important to distinguish group decision processes from those dominated by individuals, can one determine retrospectively which mode dominated a particular decision?

Answers to these questions can only be guessed in the historical context. A web of inference is constructed in the hope of providing enough evidence to make it convincing.

Even durable evidence representing attitudes and norms of the past is biased by the selective recording of popular views over less favorable ones, the emphasis on the actions of elites, the selective destruction of seemingly trivial information, and the differences in education or circumstance of the various parties. These distortions are inherent since only a small fraction of what takes place can ever be recorded and the criteria used to preserve information depend on what was considered important at the time it was recorded. Thus many additional questions that would be researchable today with our broader information bases are not researchable retrospectively.

Questions whose answers depend on transient attitudes and prevailing norms *can* be addressed with comparative ease and accuracy in *anticipatory* assessments. The parties are available for comment. Conversely, the *substantive consequences* are inherently uncertain for the reasons given before.

Therefore what is knowable about the past is different in kind from what is knowable about the present or the future. Some things can be known about the present that cannot be known as precisely about the past. Both types of knowledge contribute to understanding the relationship between technology and society.

Analysis can assure logical completeness but cannot guarantee comprehensiveness. The analytical categories of "Britain will go to war" and "Britain will not go to war" are logically complete, but by no means comprehensive. Comprehensiveness hinges on perceiving the full range of what *kinds* of things there are. Only after all the possible categorical distinctions have been enumerated can logic be applied to achieve completeness in the enumeration of impacts.**

---

*Certainly Cyrus Field was a persuasive businessman, yet his personality traits nearly scuttled the lobbying attempt to gain support from Congress for a transatlantic line. Why was his style of persuasion not effective in the political forum? This example illustrates the degree of complexity of understanding the *decision process* controlling the introduction and guidance of a technology in a social context.

**In a practical sense several independent perspectives are used to determine whether "convergence" has been achieved. That is, each perspective is contributing little not found by the others.

The speculations of the contemporary workshop participants provided a check for this study. Speculation

The Victorians exhibited repeated failures in imagining distinctions dependent on systemic features. There is little reason to believe we do much better today. This must be considered the most pressing frontier of basic research today—what *kinds* of systemic behavior can complex adaptive systems exhibit?

Imagination is essential to the enumeration of the possible distinctions and kinds of things there can be. Unfortunately, there is no formal method to replace imagination. There are crutches to assist imaginative explanation, but there is no assurance that they will ever produce a formal algorithm.

If both hindsight and foresight are inherently uncertain owing to essential limits on our ability to know—what then are the alternatives for technology assessment? What kind of analytical approach makes sense to use?

Adaptive Strategies to Cope with the Inherent Limits to Foresight*

The Victorians responded to the initial failures at laying long submarine telegraph cables by establishing a study commission; that is unremarkable in itself. The commission studied the technology in detail and confidently pronounced submarine telegraphy was mature enough to be as successful in the future as it had been disastrous in the past. Analysis can often establish technical feasibility. But the remarkable recommendations were those of management approach and guidelines for assigning the risks and responsibilities to be shared by contractors and investors. This was a rudimentary recognition of the adaptive pattern of responses expected from various parties under alternative institutional arrangements.

As shown in the last section, the two basic systems reliability principles—improve component reliability and provide network redundance—were use in a variety of contexts to cope with a wide range of potential vulnerabilities and uncertainties.

Nevertheless, according to the historical research done in this study, nearly all the major adaptations aimed at reducing vulnerabilities were performed remedially. There may be some research bias influencing this finding since crises, failures, and instabilities draw attention as more spectacular consequences than the disasters that were avoided. But the crises and failures befalling the Victorians could have been anticipated in kind if not in detail. Rather, remedial actions to correct for these failures of foresight were most often performed hurriedly, under stress, after the major losses were suffered, and with less than the best results.

Adaptive stragegies to cope with potential or experienced vulnerabilities can be implemented in a preventive fashion. The general principles used to interpret and understand systemic vulnerabilities offer a means of heuristically searching for and anticipating future pitfalls. To be useful in technology assessment, anticipation need not imply either "prediction" or even the computation of "likelihoods." It need only create an awareness among stakeholders and permit kinds of bounds to be placed on the range of contingencies. Thus, anticipation provides a means of estimating whether the proposed combination of adaptive strategies and variety of options will be sufficient to blanket the contingencies should they arise.

The effectiveness of adaptive strategies to improve reliability is limited by two factors. There is

can (and did) extend well beyond researchable topics and reflects the sketchy limits to "reasonable" subjects for study. There is no guarantee that speculation is in any way complete—there may be culturally determined perceptual biases. But speculation does provide a sampling of the universe of possibilities according to a wide range of current perspectives as a check on the biases of the research team.

*Two distinct interpretations of the term "adaptivity" must be distinguished. The first interpretation refers to the "adaptivity" shown by the promoters, users, and societal context in deriving the benefits of increased certainty, faster response, and greater scope. This "adaptation" is probably driven by the necessity to keep one step ahead of the competition as much as it is by foresight. But once this adaptation has altered established practice, vulnerabilities arise both from the dependence on maintaining the new patterns of activity and from the creation of unprecedented relationships. Recognition of, and response to, these vulnerabilities produced the second form of "adaptivity"—actions intended to secure vital relationships, assure the verity of information, and reestablish the upset balance of interactive forces. It is this second meaning of adaptivity that is implied by the term "adaptive strategies" used in this section.

a practical limit on how much "adaptivity" can be afforded; all forms come at a price. A more fundamental limit is inherent uncertainty encountered in assessing how much adaptivity is enough. Usually, experience accumulated over many years determines the trade-off between cost and adaptivity. Yet significant technological change, social change, or shifts in the internal structure of participation quickly render the lessons of experience obsolete.

Unprecedented change requires more general selection of adaptive strategies than simply adding redundancy and securing critical relationships. Other examples include a system of priorities to coordinate the allocation of resources if there is an unanticipated scarcity, or a methodical warning of stakeholders of their vulnerabilities under plausible contingencies so they can organize their own responses.

The role of analysis would be two-fold in supporting a technology assessment methodology based on identifying and covering residual uncertainties with adaptive strategies. The first role of analysis would be the traditional one—use established theories and technology assessment techniques to constrain the range of plausible consequences and identify possible impacts as sources of uncertainty. Take this analysis only as far as the uncertainties permit.* The second role of analysis is novel. Enumerate the full range of long-term and short-term interpretations of the basic strategies of reliability theory and other hedging techniques.** Determine whether these techniques can cope with the plausible contingencies.*** Select the most promising and perform more detailed policy analysis on them.†

This approach avoids many of the epistemological problems surrounding precise forecasting of future events.‡ It more closely resembles the adaptive stratagems exhibited in natural ecologies—explore widely and keep as many options open as possible. Since the unexpected future events may foil the common strategies, the atypical maverick may often prove to be the salvation of the species.

---

*Avoid assuming away a wide range of plausible contingencies in order to obtain a "researchable task" according to some pet analytical tool. This is a major pitfall of interdisciplinary research. Each disciplinarian selects assumptions to fit the needs of his research methods, leaving it up to the others to justify or adopt those assumptions. The results or components of analysis are seldom compatible on large interdisciplinary projects.

**The principle of redundancy from reliability theory is implicit within the U.S. Constitutional principle of checks and balances. It is also implicit in the free market, to the degree that there is real choice of selection among buyers and sellers. In other words, there is ample precedent in the U.S. for the adaptive strategy approach.

***Excessive use of adaptive strategies can reduce "undesirable," "unexpected," or "avoidable" variation and impacts to the *detriment* of the entire approach. Normal diversity of conditions and events provides the *test cases and experience* that offer confidence in both the identification of vulnerabilities and the effectiveness ofnthe strategies.

†Attitudes and values enter into the selection and application of adaptive strategies. For example, test cases raise the issue of equity. If adaptive strategies are to work, they must be used as such and not as political or competitive weapons—the rules are yet to be worked out.

‡A caveat: the very success of an adaptive strategy makes it *appear* superfluous—the potential crises or impacts it was intended to divert are not experienced. It would be natural to improve "efficiency" by reducing the costs absorbed in maintaining the adaptive option. However, many of the detrimental consequences arising from submarine telegraphy arose abruptly, too abruptly to reinstitute a neglected strategic option. This situation creates a paradox: if we know we need an adaptive strategy we have not got it, and if we have it we forget we need it! This trap must be avoided. Unfortunately, most performance measures used by industry and government stress "efficiency" and neglect adaptivity. That must be changed if the adaptive strategies are to assume their proper place in coping with what can be only dimly perceived.

Appendix II

SUMMARY TABULATIONS OF THE FINDINGS OF CHAPS. 2–5

The diverse impacts of submarine telegraphy and the expectations for this technology were described in Chaps. 2–5. Before statistically significant patterns among these findings could be discerned, they had to first be tabulated in a systematic fashion. This summary tabulation is given in Table II–1.

Seven major subdivisions of Table II–1 parallel the impact sectors of Chap. 3: shipping, commerce, diplomacy, war and national security, science (which includes weather forecasting and oceanography), social [which includes international law and time (standards and perception)], and the news.

Summary statements of impacts are listed in the left-hand column. Bulletted headings designate impact areas; dashes are impacts from submarine cables. Asterisks designate related events, which were included to reinforce and complete the picture presented by the impacts of telegraphy but were not included in the statistical analysis. The capital letter symbols to the left of the dashes for each impact (E, S, P, V and $\tilde{V}$) are codes representing the following impacts of the cable on the affected activity:

E: increased *efficiency*

S: increased *scope*

P: increased *pace*

V: creation of *vulnerability* or response to perceived vulnerability

$\tilde{V}$: *reduction in vulnerability*

The wide middle column of Table II–1 presents both the time frame of the impact and the type of impact on a five-point scale. Some impacts occurred in a single year, others developed over a period of time as shown by a band in this column. The five-point scale represents the directness with which the cable caused the impact. This "remoteness" scale is described in more detail in Chap. 4 and briefly in Appendix I.

In brief, the steps along this scale are:

1. Options made available to user and promoter groups.
2. Structural or institutional changes in the user groups to utilize the cable better or adapt to its presence.
3. Changes in British-American relationships, especially owing to the transatlantic cables.
4. Changes in worldwide patterns of activities induced by the international cable networks.
5. Subjective changes in values and attitudes, especially those which affect broad decision processes.

The third column, titled retrospective Delphi, indicates the closest matching exploratory impact statement from the contemporary Delphi workshop. The prefix numerical code is the remoteness assignment given to the Delphi hypothetical impact statement in Chap. 4. The capital letter and number following the colon is the code for the specific Delphi impact from among those listed in Chap. 4. (The Delphi topic categories are economics, military, political, social, and cultural.) The remoteness category assignments for the impacts found in the historical research were quite

TABLE II–1.—Summary tabulations.

IMPACTS AND CONSEQUENCES OF SUBMARINE TELEGRAPHY (S.T.)	1850 . 1900 . 1950	RETROSPECTIVE DELPHI	VICTORIAN EXPECTATIONS
■ SHIPPING (AND MARINE INSURANCE)			
• Improvement in shipping intelligence			Advantage in scheduling ships and insurance information perceived immediately 1:E13; 1:E14
E – Lloyds of London uses cables immediately and heavily	—//—— (1)	1:E13	
E – Lloyds builds signal stations along coastlines of major routes	· (1869) (1)	1:E13	
* Lloyds establishes radio communication with ships for additional intelligence	—— (1) *		
• Effective capacity of shipping fleet increased			
* Ships with steel hulls and steam engines replace wooden hulled ships, providing greater speed and larger capacity	1850 —— (1) —— 1910		
E – Owners exercise long-distance control of selection of cargoes, routing, and scheduling; venture system declines rapidly	1860 —— (2) —— 1884	1:E13	End of Venture System predicted by Peter Cooper, 1863 1:E13
E – Scheduled (liner) shipping benefits from coordination of passengers, schedules	1830 1866 —//—— (1) ——	1:E13	

• Competition increased, leading ultimately to cooperative industry action			
S -Shippers build more capacity in anticipation of increased availability of credit to merchants world wide	1856 ——— (2)	2:E14	
S -International shipping continues to increase worldwide	——— (4) ———	4:E11	Certainly expected by merchants 4:E11
S -Greatly increased traffic to India following completion of India link	• (4)	4:E11	
V -Brokers of shipping capacity emerge; overbid; inflated perception of capacity	1870 ——— (4) ——— 1890	4:E9 4: E8	
(V)-Shipping conferences emerge to control competition by establishing areas and rate agreements	1820 ——— (4) ——— 1890	4:E3	Shipping conferences expected because they began so soon. 4:E3

Table II–1 (Cont'd)

IMPACTS AND CONSEQUENCES OF SUBMARINE TELEGRAPHY	1850 . 1900 . 1950	RETROSPECTIVE DELPHI	VICTORIAN EXPECTATIONS
■ SCIENCE			
• Weather Forecasting			
* Ship's logs used to understand the patterns of weather development (the circulation patterns around hurricanes); this did not provide warnings	. (1820's)		Great impetus for science and education widely forecast. 5:C7
S – Arriving ships in Boston or New York submit weather reports as warning to Britain; not successful enough to continue	. (1885) (1)	3:P10	
∿V – U.S. gets hurricane warnings from Cuba; tracing of hurricanes by 1878, fully developed system by 1901, from 1890 on no hurricane struck U.S. coast without warning; saved $38 million in property damage in one case alone.	1878 — 1901 (2)		Precursors tried, shows interest in cooperation for weather forecasting.
S – Stimulation of international cooperation; advances in instruments and techniques.	———— (5)	4:S11	
S – First International Meteorological Conference in 1872	. 1872 (5)		
* Radio was needed to provide enough observation points in the mid-Atlantic to make weather warning and forecasting practical	1910 — 1930		

- Oceanography

Event	Date			Ref.
* Matthew Maury surveyed ship captains to get what had been proprietary navigative information--presented at first International Maritime Conference, Maury discovered "Telegraph Plateau" from Atlantic soundings before cable proposed.	1838 1853			
* U.S. government continues oceanographic studies to support navigation under the Commerce Clause of the Constitution	.(1807) (2)			
S -British and Americans cooperate in soundings for transatlantic cable	.(1858) (3)			3:P10
S -Mid-Atlantic ridge discovered because of cable laying work; the Azoic theory of marine biology falls due to life found on deep sea cables; underwater earthquakes produce turbidity currents discovered by cable interruptions	.(1856) Ridge .(1860) Azoic Falls (5)		.(1933)	4:S11

Table II–1 (Cont'd)

IMPACTS AND CONSEQUENCES OF SUBMARINE TELEGRAPHY (S.T.)	1850 1875 1900 1925 1950	RETROSPECTIVE DELPHI	VICTORIAN EXPECTATIONS
■ TRADE AND COMMERCE			
* Large merchant houses invest in development of international cables	.(1858-66) (1)	NOT 2:E1 NOT 3:E2 3:E17 2:E19 2;E23	More competition for trade and commerce among heavy shippers E9--4
• Organization, reorganization, and relocation of markets			
E - Futures markets expand, crowd out spot trading; continuous agricultural market created	——— (2)	2:E23	Edwin Morgan emphasized the value of cables to U.S. position in world grain markets (1863)
E - Commodity markets tend to relocate in producer nations	1861 ——— 1883 (2)	3:E17	
E - Warehousing disperses to producer nations and independent warehousing businesses emerge	——— (4)	3:E17	
E - Arbitrage markets grow up (by 1870, 87% of London Exchange transactions are for arbitrage	——— (4)	3:E18	
V - Chicago Board of Trade establishes regulations on commodity trading	.(1869) (2)	(2:E23)	

Impact	Timeline	Source	Commentary
•Reduction of risks, fluctuations, and costs			
E -Warehouse inventories reduced as scheduling is improved	——— (1)	1:E15	Farmers expected to have more control over sales - Senator Mallory, 1857
P -Fluctuations in price in world-wide currency and commodity markets reduced as information exchange and adjustment is speeded up	1855 ——— 1875 (4)	3:E6 (4DK)	A journalist wrote that an equalization of stock prices and commodities the world over would rightfully restore commerce to that of placing goods where they were needed rather than a game of chance
V -Commodity standardization and grading developed to support futures markets	.(1870)	(2:E23)	
E -Middleman costs sharply reduced	——— (4)	1:E15	
(V) -Service of futures traders in underwriting risks recognized as socially useful	1900 ——— (5)	5:C12	Senator Tooms of Georgia argued against cable as benefiting speculators while filching public
V -Hedging required by British bankers for commodity transactions	1900 ——— (2)	2:E23	
V -False quotations stopped by use of codes	——— (1)	1:S3 1:S4	
•Increased Competition			
E -New credit mechanisms replace Bills of Exchange, providing increased opportunity for small merchants, increased competition and diversity in lending institutions	——— (4)	4:E9	Merchants will be caught by market equalization and heavy shippers will find more competition
V -British trade monopoly in Far East erodes	(1870's) (4)	4:E9 (2:E19)	
E -Agricultural production specializes as world markets coordinate exchange among nations/regions			U.S. advantage perceived

Table II–1 (Cont'd)

IMPACTS AND CONSEQUENCES OF SUBMARINE TELEGRAPHY (S.T.)	1850 . 1900 . 1950	RETROSPECTIVE DELPHI	VICTORIAN EXPECTATIONS
■ SOCIAL CHANGE			
● International Law			Spread of Anglo-American culture, values, religion, etc.
S -Conferences to establish international agreements on the use and control of submarine telegraphs; International Telecommunications Union formed in Paris to agree on international telegraphy.	.(1871) Rome .(1875) St. Petersburg .(1884) Paris (4)		5:C14
V -Censorship and interruption of relay transmission of messages detrimental to the nation--during war was recognized as a right--otherwise unenforceable. British and French did censor during WWI.	(5)	3:P1 (5)	5:C13 (5)
V -Restoration of cables to neutrals from beligerants and indemnities for loss after war agreed upon at Hague.	.(1907) (4)		
V -Agreements tailored after mail-ship precedents; novelty of technology did not produce new principles.	1858 (2)		
* Radio requires agreements to allocate the finite capacity of the broadcast spectrum--	1910 (2)		

Item	Date / duration	Source	Comment
• Perception of Time			
* Britain had adopted a uniform time standard by 1848.	.(1848) (2)		Awareness of time growing throughout Victorian era, even before telegraph
P-Increased sensitivity to global time differences	1858 ——— (5)		
V-Perceived pressure to respond to distant events with attention, adaptation or intervention	1866 ——— (5)	(1:S15) 5:M6	
P- International Meridian Conference in Washington, D.C. to select an international time standard; 27 nations sent delegates. Submarine telegraphy centrality partially assisted selection of Greenwich Meridian as prime.	.(1884) (4)	(5:C13)	
P-Pace of life quickened in civilized world.	——— (5)		"Quickening of pace of life" predicted with nervous diseases increasing -- observation by Dr. George M. Beard and Thoreau.
• Generalized increase in public attention to and influence over national and international actions.			
V -See all categories above.	——— (5)	1:S15 (5:M6)	

Table II-1 (Cont'd)

IMPACTS AND CONSEQUENCES OF SUBMARINE TELEGRAPHY (S.T.)	1850 . 1900 . 1950	RETROSPECTIVE DELPHI	VICTORIAN EXPECTATIONS
■ DIPLOMACY			
• Use of cables to supplement and extend routine correspondence			Routine uses expected and early.
E - Coordination of diplomatic travel, passport clearance, etc.	(1866)———— (1)	4:P5	
E - Ceremonial and courtesy functions	(1866)———— (1)	4:P5	
E - International cooperation to apprehend criminals, control epidemics, aid in natural disasters, etc.	———— (4)	4:P5 4:P6 4:P7 4:P8	North American Review anticipates scientific apprehension of criminals. 4:S5
• Development of quick response and continual readiness			
P - Secretary of State has round-the-clock telegraph clerk, is "on call" for emergencies	.(1874) (2)		Mixed review of Trent Affair during American Civil War--C.Field felt the whole affair would have been avoided;Lord Lyon felt it would have brought Britain into the war against the Union.
S - Dept. of State telegraph staff grows: - five full-time clerks - teletype - crisis center - sophisticated equipment	.(1910) (4) .(1939) (4) .(1950) (4) .(1960) (4)	2:S17	
V - Security of cable messages becomes concern, advanced codes developed	1876———— (2)	1:S4	
V - Multiple routes for reliable communications become essential	———— (2-4)		

Item	Timeline	Source	Comment
•Development of Specialization			
V - Overload of information at headquarters threatens viability of foreign office procedures in Britain, U.S.	——— (4)	2:S17	Career diplomats testified in Britain that cable's impact wouldn't be great; it would be useful but a mixed blessing. (This characterizes the ensuing experience rather well.) 4:P5
S - Headquarters staff grows more rapidly than field agents, functional specialists increase (U.S. & Britain)	1870 ——— 1890 (4)	4:P5	Career diplomats foresaw a greatly increased burden on central staff. 2:S17
V - Decision-making becomes complex, diffused through headquarters staff	——— (2)	(2:S17)	
•Gradual shift in decision-making from field to central headquarters		4:P5	
V - Diplomats in field manipulate choice of communication modes to encourage or discourage intervention of headquarters authority.	——— (2)		Obsolescence of foreign service predicted--(did not happen quite that way). 4:P5
E - Central repository of information, coordination, shifts responsiblity	1930 ——— (2-4)	4:P5	
• Growth in international interactions and participation			
P - World Wars I and II produce 10-20 fold increase in cable traffic; after wars traffic remains several times greater than before	.WWI (4) .WWII		
S - Development of continuing international organizations	1910 ——— (4)		North American Review anticipates growth of international organizations.
S - U.S. financial contribution and personnel representatives to international activities becomes significant.	1945 ——— (4)		

Table II–1 (Cont'd)

IMPACTS AND CONSEQUENCES OF SUBMARINE TELEGRAPHY (S.T.)	1850 . 1900 . 1950	RETROSPECTIVE DELPHI	VICTORIAN EXPECTATIONS
■ WAR AND NATIONAL SECURITY			
E -British naval dominance enhanced by British control of worldwide cables.	1870 — 1910 (4)	3: M1 2:M7	British dominance expected by many. 3:P3
* Radio provides alternative communication links, British dominance challenged	1910 — (4)	*	
• Increased mobility and flexibility: changes in military strategy		2:M3	Government support and priority given military for cable use demonstrate perception of their importance; same went for radio when available.
E -First military/logistical use of cable to cancel troop movement	.(1858) (1)	2:M3	
E -Troop redistribution (concentration at strategic points with regional mobility	1861 — 1887 (2)	3:P3	
S -British naval fleet redistribution	1904 — (2-4)	(2:M3)	
•Constant readiness - centralized control - new vulnerability			
E -Increasing centralized control of logistics	1870 — 1918 (2)	2:M3	
S -Development and acceptance of General Staff concept	1870 — (2)	2:M3	
E -British army reorganization	1909 — (2)	2:M3	Concentration of power an outgrowth land strategy based on telegraph and railroad.
V -Constant readiness helps precipitate World War I	.WWI (5)	5:M6	

Item	Timeline	Source	Comments
● Growth of importance of military intelligence			
V -Development of cryptography and cryptoanalysis.	1880 ——— (2)	1:S4	
V -British crack German codes in WWI but German's never find out.	1914 —	1:M2	
V -Development of elaborate military intelligence operations in U.S. and Britain.	1882 ——— (1)	5: M4	
● Censorship			
V -Public influence over military affairs and enemy utilization of public information beginning with Crimean War.	1854 ——— (5)		
V -British SIS formulates plan for extensive government control of information	.(1904) (5)	3:P1 (5)	Military quick to learn impact of telegraph messages during war.
V - British domestic and imperial cable censorship	.(1914) (5)	3:P1	
V - British intercepts all commercial cables to and from neutral countries	.(1917) (4)	3:P1	
V * Concept of <u>Total War</u> reflecting international economic inter-dependence, importance of civilian attitudes and capability for national/transnational mobilization in short time	——— .WWI ——— (5)	5: M5	Many forecasts increased peace and brotherhood due to communication. 5:P9; 5:C16 Decrease in British/U.S. hostility also forecast. 3:P10

Table II–1 (Cont'd)

IMPACTS AND CONSEQUENCES OF SUBMARINE TELEGRAPHY (S.T.)	1850 . 1900 . 1950	RETROSPECTIVE DELPHI	VICTORIAN EXPECTATIONS
■ NEWS			
• Development of large news organizations, changes in news industry			
* Land telegraph undercuts delivery of news by train; London Times loses their monopoly on international news distribution revenues throughout England - stimulus for nationalization of land lines in Britain	(1849) (1); (1869) (1)	———	
* Reuters, Havas, and Wolff develop news services. N.Y. Associated Press forms agreement with American Telegraph Company	* (2)	2:E1 4:E3	It was obvious to everyone that the news would be an early user of the growing submarine telegraph system.
P - Rapid growth of major news services as public increasingly demands current international news	(2)	2:E22 (1:S15)	Telgraph expected to replace postal service; more thoughtful realized telegraph was too restrictive.
E - Reuters and NYAP form an alliance which makes London the undisputed center of world news	(1893) (3-4)	(3:P10) 4:E3	
V - NYAP breaks agreement with Reuters to cover South America and supply news of U.S. to S.A. to balance image of U.S. abroad--radio makes independent foreign news coverage economical	(1934) (3-5)	(3:P1) (5:C9)	

• Increased public influence over decisionmakers			
V - British attack on Sebastapool during Crimean pressured by public opinion; British suffered loses due to public press leaks of strategic information picked up by Russians during the Crimean War.	. (1355) (5)	1:M2 5:M4 5:M6	
V - News service reports reduce bias in reporting to appeal to broader audiences.	——— (5)		News vendors or "inventors" will suffer loss. (1: 53) Gossip and trivia predicted as major content.
V - U.S. becomes increasingly sensitive about its world image from 1930's onward	——— (5)	5:P6	
V - By World War II all nations recognized the importance of having independent news media due to increased public participation in (influence on) decisions	.WWII (5)	5:P6	
• Vulnerability to government control			
V - U.S. locates Spanish Fleet in Santigo harbor through private news services operating from Spain	. (1898) (4)	1:M2	
V - Nearly all news of Germany to U.S. comes through Britain during World War I	.WWI (3)	3:P10	
V - British censor news during World War I	.WWI (5)	3:P1	
-NYAP declines State Department offer to use diplomatic cable to South America at reduced rates thus resisting trend of news services toward becoming an official voice of their government	. (1917) (5)	(3:P1)	

consistent with the assignments to the Delphi impacts. However, categories (3) and (4) were combined for the statistical analysis presented in Appendix I because the Delphi concentrated on the transatlantic cable, whereas the impact research of Chap. 3 ranged far beyond that.

The final column, entitled Victorian Expectations, presents the anticipatory statements found in the historical research revealing the Victorians' expectations for the cable's consequences. Fairly liberal interpretations were used to provide matches between the actual historical impacts and Victorian expectations because these two sets came from such different sources that the identification of a topical overlap must involve judgment. By and large the impacts were more specific and focused than the expectations, as is to be expected. Nevertheless, even a hazy perception of what is likely to happen is valuable in technology assessment and for avoiding difficulty, so that a liberal interpretation is warranted.

Some Victorian expectations mentioned in Chap. 4 are not listed in Table II–1 because the corresponding impact was not found in the historical research. These additional expectations were included in some of the statistical analyses; their results are insensitive to small variations in their number and do not require painstaking tabulation here.

Table II–2 supports the statements made in Appendix I about the time order ranking of the seven major subdivisions of Table II–1. The final composite ranking is the average of two alternative component rankings, obtained by comparison of the portion of the impacts that occurred before/after 1875 and before/after 1900, respectively. The slight differences between these two rankings going into the composite are insignificant; there is only a 5% chance of the two rankings being independently generated.

TABLE II–2.—Time sequence of impact maturity by activity sector.

	Pre-1850	1850-1875	1875-1900	1900-1925	Post-1925	Fraction of Impacts After: 1875 %	1875 Rank	1900 %	1900 Rank	Combined Rank*
Shipping	3	5	1			11	1	0	1.5	1
Science		4	2			33	3.5	0	1.5	2
Commerce		12		2		14	2	14	4	3
Social		6	2	1		33	3.5	11	3	4
Diplomacy		5	4	4	2	66	6	40	5	5.5
War & national security		5	3	6		64	5	43	6	5.5
News	1		3	4	3	91	7	63	7	7
Total	4	37	15	17	5					

*Ranks at 1875 and 1900 are consistent at the 5% level of confidence; therefore, the combined rank is meaningful.

Appendix III

ANALYSIS OF THE DELPHI

Tests Applied to the Responses on the Delphi Survey to Verify Their Utility in Support of the Conclusions in Chap. 4

It was recognized that the analysis of a retrospective Delphi impact questionnaire survey would require a different approach from that for the typical prospective Delphi. Because of the relatively small sample size (25 respondents) and variety of topics and questions, the analysis was performed by hand. Promising leads were pursued by means of short-cut statistical approaches. All results presented here are statistically significant to the degree of significance given in each case.

The presentation of analytical results proceeds in a natural fashion from questions of interpretation and consistency of the survey instrument to the more interesting ones concerning the principal results for retrospective technology assessment of submarine telegraphy in particular, and the methodology of technology assessment in general. In this progression the topics covered are:

- interpretation of impact statements
- consistency of individual responses
- consistency of responses among three participant groups
- comparison of the 1861 perspective with that of 1976
- comparison of the importance of impacts with their likelihood
- enumeration of the most likely and most important impacts

Interpretation of the Impact Statements

The broad range of topics and possible consequences covered in the Delphi and the need for brevity led to quite generalized statements of the impacts. These generalized statements did produce some interpretation ambiguities for respondents. Three illustrations are given below. Yet, on the whole, the quantified responses were consistent as will be shown in the next section.

First "value of British naval supremacy reduced," was an hypothetical military impact. The two historians on the research team, Drs. Bernard Finn and Thomas Jaras, concluded that the value of British naval supremacy would actually *increase* rather than decrease. Their reasoning was that the cable and the Navy were symbiotic—the cable helped coordinate the Navy to make it more efficient and naval supremacy was a prerequisite to preventing the disruption of cable communications during conflicts. They assessed this impact as very *unlikely* even though its converse would have received a high score if it had been asked. When a *converse* is not available as an impact to be selected, one might erroneously interpret a low score on an impact as indicating little perceived *relevance*.

Second, the development of codes and cryptography could be viewed as a countermeasure for new types of criminal behavior, such as theft of information and purveying false information. Yet almost all respondents assessed the probability of the development of codes and cryptography as nearly certain, whereas novel criminal behavior was considered unlikely. The simple hypothesis evidently did not hold, since either codes were considered a completely effective deterrent to crime or codes were considered simply as a means of increasing the cable's transmission efficiency. As one can see, the interpretation remains open.

Finally, three seemingly diverging impacts were rated very likely and quite important if they occurred: more competition in world trade, increased concentration or growth of monopolies, and increased British dominance in trade. Delphi participants apparently did not regard these two developments (increased competition and increased concentration) as contradictory. In this perception, they were of course correct: the expanding scale of world trade allowed the growth of large corporations and oligarchies, with continued British dominance for a significant period; and at the same time the growing marketplace articulated positions for a great many small merchant firms to enter the market. The growth of large newsgathering organizations was accompanied by an increased number and diversity of newspapers and journals. On the whole, markets became more efficient and the numbers of participants also increased.

Consistency of Individual Responses

The utility of a Delphi survey depends on the aggregation and comparison of individual, quantitative assessments of topics otherwise difficult to condense and evaluate. In order for aggregation and condensation to be warranted, each participant's responses must be internally consistent and the numerical values assigned must have a common scale of interpretation.

Visual inspection of the responses revealed that some respondents tended, on the average, to give lower (or higher) numerical values for the impacts—what we might call a pessimistic (or optimistic) bias relative to the group as a whole. One respondent tended to assess the impacts from his own discipline above the group's estimate; this same individual tended to be pessimistic about the remaining topic areas. These differences can be expected and could reflect legitimate differences of opinion. If one intended to increase the precision of the *relative* assessments of impact importance, one would have to calibrate such differences to a group mean. This procedure was not warranted in the small survey performed here.

The degree of convergence or divergence of opinion for individual impacts can be tested by a comparison of the distribution of responses to the distribution expected if only the group average were significant. Greater clustering than that expected represents convergence toward an implicit consensus; greater disparity of responses represents a polarity or inconsistency of opinion. This test was performed on the four impacts most likely, the four least likely, and the four nearest the overall group mean. The dots in Fig. III–1 represent the maximum-likelihood distribution under the assumption that the average is the only significant information in the distribution. The most likely four and the four nearest the mean show evidence for clustering to a convergent opinion; the clustering about the center of the overall mean case is statistically significant at the 3% level of confidence. One sees no evidence for divergence or inconsistent responses in any of the three cases.

The best check of consistency for individual respondents came from a comparison of two nearly identical impact statements under two different topic areas: the fifth military impact, "less likelihood of warfare, because of increased international understanding," and the ninth political impact, "increased international understanding, less likelihood of wars." Ten out of 25 participants gave identical responses to both impacts even though they covered nearly the full range of estimates of likelihood—from onc to nine. The average difference from perfect agreement was only 0.75 units on the ten-point scale.

Consistency of Responses Among Three Groups

Six of the 25 participants were historians; two of these six (Finn and Jaras) were members of the research team. Comparison of the responses of each of these three groups, even though it shows them to be not entirely independent, gives useful assurances that the assessments were consistent and that particular knowledge of the historical period would not produce greatly different results. The analysis demonstrated consistent trends in evaluations among the three groups. However, the six historians estimated the impacts as more likely than did the total group, whereas Finn and Jaras considered them, on the average, less likely; that was termed an optimistic or pessimistic bias toward cable causation. Finally, the smaller the respondent group the wider the range of opinion variation. This marked feature is referred to as "polarization."

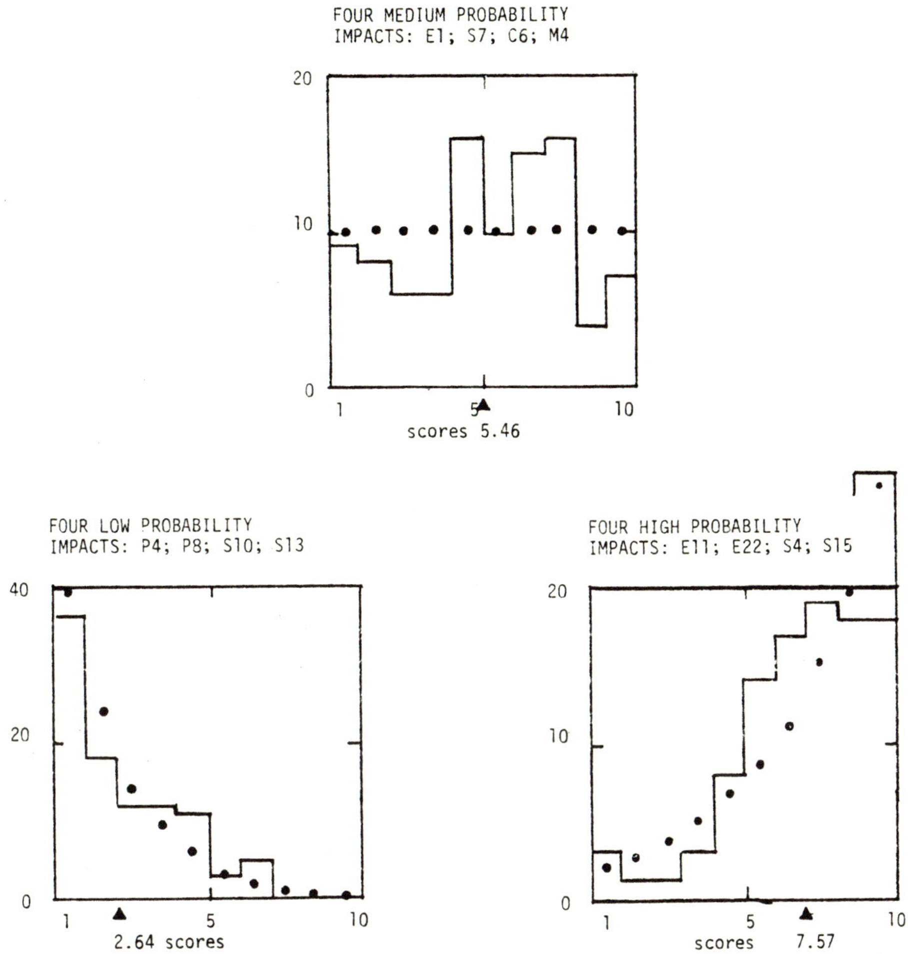

FIG. III-1.—Comparison of distribution of responses, with maximum-likelihood distribution, given only the mean.

Derivation, based on the work of E. T. Jaynes (*cf.* Katz, *Principles of Statistical Mechanics,* Freeman, 1967).

Let P_i = probability of response i. Assume that only $\langle m \rangle = \Sigma_i\,(i \bullet P_i)$ is given. Then the maximum likelihood distribution is $P_i = \exp[(-\lambda \bullet i)/Z(\lambda)]$ for $\frac{1}{2}(-N+1) \leqslant i \leqslant \frac{1}{2}(N-1)$, where Z is a normalizing constant as is λ.

$$\text{Let } \beta = \exp(-\lambda/2) \text{ and } A = \sum_{i=\frac{1}{2}(-N+1)}^{-\frac{1}{2}} P_i,\ B = \sum_{i=\frac{1}{2}}^{\frac{1}{2}(N-1)} P_i;$$

$$\text{then } \beta = (B/A)^{1/N},\ \lambda = (-2/N)\ln(B/A),\ Z = \frac{\beta^N - \beta^{-N}}{\beta - \beta^{-1}},\ \text{and } \langle m \rangle = \frac{1}{2}\left(N\frac{\beta^N + \beta^{-N}}{\beta^N - \beta^{-N}} - \frac{\beta + \beta^{-1}}{\beta - \beta^{-1}}\right)$$

As $\beta \to 1$, $Z \to N$ (the total number of bins) and $\langle m \rangle \to 0$.

Note. Maximum likelihood definition: P_i such that $\Sigma_i\, P_i \log P_i$ is maximum, under the constraint that $\langle m \rangle$ is given and fixed.

Figure III–2 shows the comparison of the total group to the historians' average assessments for each of the 74 impacts in reply to the principal question, "In 1861, what do you think is the probability of this impact occurring as a result of the cable, or with that as a contributing causal factor?" Three significant features are evident from this Figure. The general tendency of the points to cluster along the diagonal leading from the lower left to the upper right demonstrates the degree of consistency between the two groups. In fact, the variation *about* the trend line shown is only one-third of the variation *along* it. The shift of the average values for the two groups to the left of the exact center of the Figure indicates the relative optimism of the historians—0.6 units in this comparison. The upward slope of the trend line is greater than the 45° angle of the principal diagonal; this result indicates a polarization of attitude for the historians with respect to the total group. Over the entire scale of ten points, the historians evaluate impacts by two units less-to-more-likely than the full group.

Similar comparisons between the full group and the average of Finn and Jaras's opinions as well as between this average and their consensus exhibit the same three features (Table III–1). The smaller groups have consistently greater polarization. Polarization is expected from group discussion* as

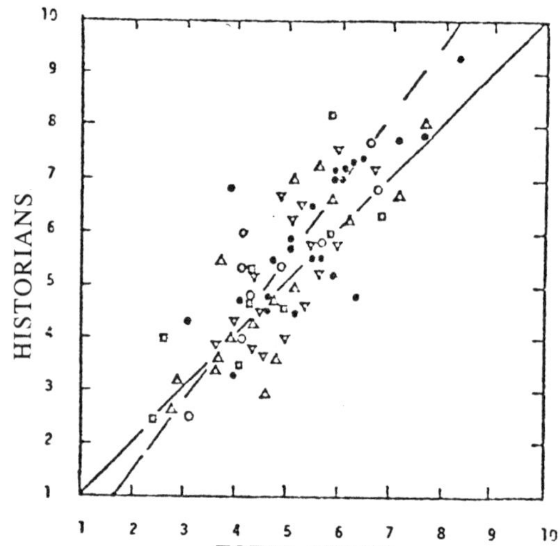

FIG. III–2.—Comparison of mean responses for responses to Question A, historians vs total group: ● economic, ○ military, □ political, Δ social, ∇ cultural.

TABLE III–1.—Comparison of responses by various groups.

Groups compared	Variation about and along trend line (units)		Optimism/ Pessimism	Polarization
	About	Along		
Historians to Total	0.6	1.8	Optimistic by 0.5 unit	Positive by ± 1.0 unit full scale
Finn/Jaras (Average) to Total	0.8	3.0	Pessimistic by 0.5 unit	Positive by ± 3.5 units full scale
Finn/Jaras (Consensus) to (Average)	1.0	0.0	Remained the same	Positive by ± 1.0 unit full scale

Pessimism/optimism significant at the 5% level of confidence. All other differences are significant at the 0.1% level in comparison to the null hypothesis.

*D. G. Meyers and Helmut Lamm, "The polarizing effect of group discussion," *American Scientist* 63: 297–303, 1975.

with the consensus between Finn and Jaras. However, the simple averaging of individual differences would reduce the variation in the average response for larger groups. Interpretation of the survey average responses in absolute terms would be meaningless, yet they are quite consistent and significant in relative terms.

Comparison of the 1861 Perspective With That of 1976

These gaming and Delphi exercises were part of an hypothetical technology assessment performed as if it were set in the context of the original decision to lay the transatlantic submarine telegraph. To what extent could the Delphi participants distinguish the retrospective 1861 perspective from their contemporary 1976 vantage point? This question can be examined quantitatively for the total group, the historians, and the consensus between Finn and Jaras.

For each of the 74 impacts, Question A was asked retrospectively: "In 1861, what do you think is the probability of this impact occurring as a result of the cable, or with that as a contributing factor?" Question D is quite similar but asked from a contemporary perspective: "If you believe that this impact (event or trend) did occur, to what extent was the cable a causative factor?" Even though Question D was contingent on answering question A with a score of five or more, the Delphi gave useful estimates for both questions.*

Again the comparison was performed by plotting each group's average reply to the two questions on a scatter diagram (Fig. III–3). This plot permits visual inspection for anomalous clusters or deviations and provides the basis for a statistical assessment of the likelihood of the patterns that emerge. Deviations from a homogeneous random distribution can be seen by a bordering of the principal diagnoal band containing the majority of the points. The degree to which two questions are related is revealed by the excess number of impacts falling in the end regions of this band—15 above and 19 below. This degree of spread along the band would occur randomly in less than 0.1% of all random trials. Seventy per cent of the total variation would be removed by elimination of the elongation. In other words, the views of trained historians toward the likelihood that the submarine cable caused the impacts is on the average very similar from either a retrospective or a contemporary perspective.

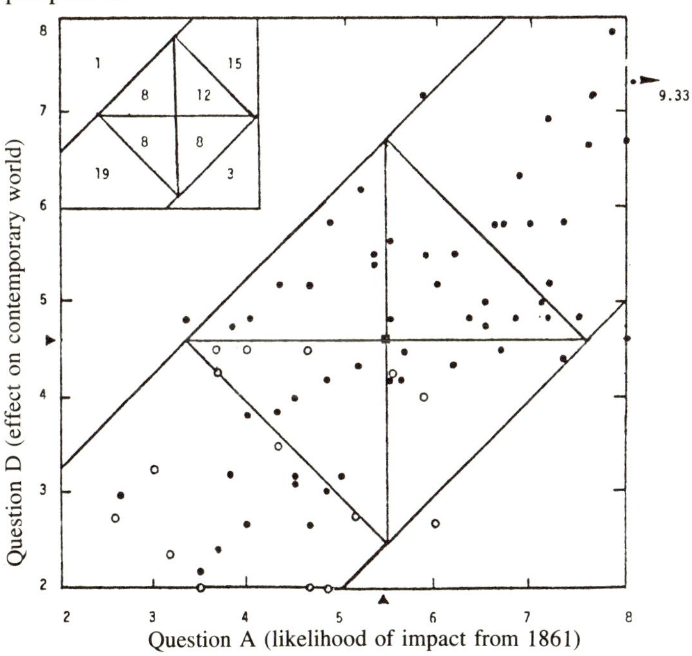

FIG. III–3.—Near-total correlation of Questions A and D for historians group.

*However, the analysis of the less likely impacts was abbreviated for the total sample—only 43 out of 74 impacts were analyzed; 39 had average scores of five or more.

TOTAL GROUP:

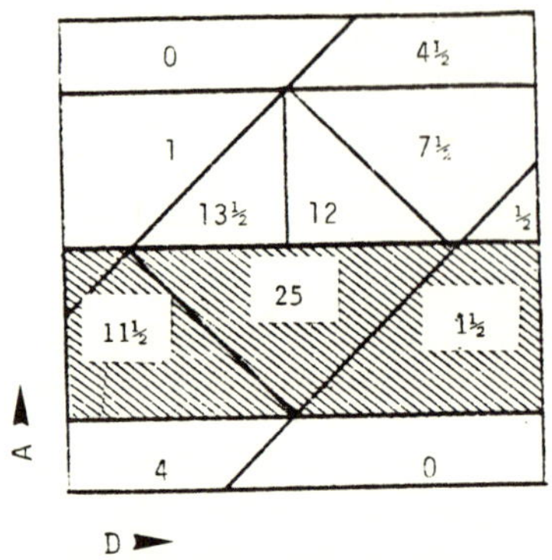

Correlated at the 0.1% level of confidence. (Shaded band not investigated; inside numbers are the expected events not included in the analysis.)

HISTORIANS:

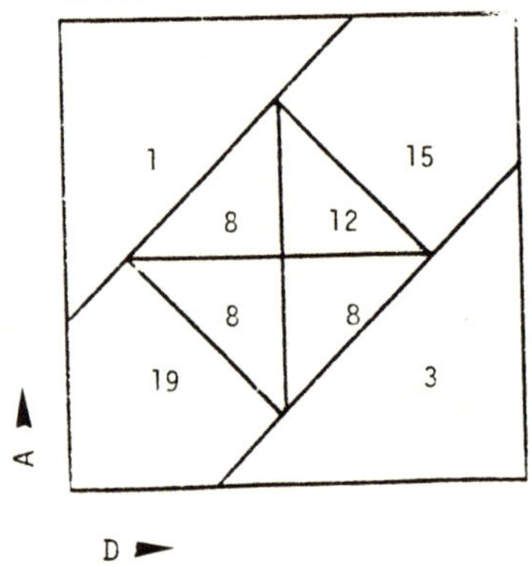

Correlated at well beyond the 0.1% level of confidence with 70% of the variance explained by the "identical question" hypothesis.

FINN/JARAS:

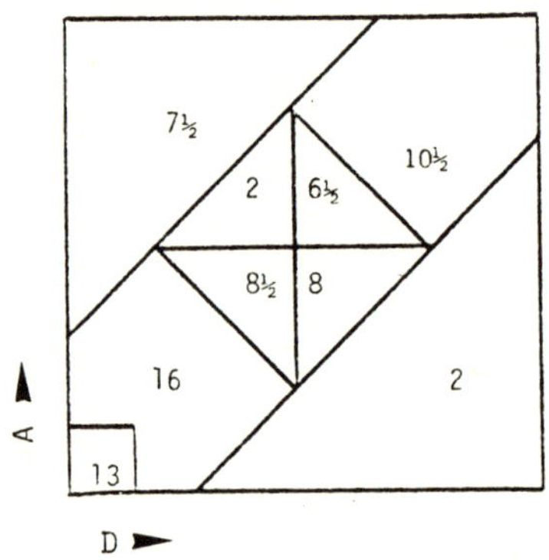

Correlated at well beyond the 0.1% level of confidence with the 13 deleted from the lower left corner. The residual asymmetry is inherent in the data pattern, which has a much larger scatter than the previous cases.

FIG. III–4.—Correlation between 1861 and 1976 perspectives toward impact likelihood.

The same relationship between retrospective and contemporary views emerges from the total group averages and from Finn and Jaras's consensus. Figure III–4 shows the diagram summaries similar to those given in Figure III–3. The boxed numbers in the cross-hatched portion of the diagram for the total group represent the expected locations for the impacts not analyzed. But the statistical measure of significance is based on only the analyzed impacts. Finn and Jaras considered thirteen impacts irrelevant; these impacts were eliminated from the overall estimate of correlation. Even with these impacts out of the sample, there would be only a 0.1% chance of observing the asymmetric count pattern. Therefore, according to the views of the participants, these two questions are very similar.*

*For this reason, the remaining two questions under the 1861 perspective were not considered important enough to pursue further.

One must recognize that there are at least three possible explanations for this relatedness:

1. Even highly trained historians asked to project themselves into the 1861 setting cannot free themselves of their contemporary understanding of the cable's impact.

2. Estimates of impact likelihood are so self-evident that group judgments are an adequate tool in their evaluation retrospectively or prospectively; one must simply be willing to explore the impacts in a systematic fashion.

3. The impacts selected for evaluation reflect the contemporary bias toward effects that are considered likely according to historical perspective and contemporary experience. Thus questions that *would not have been* asked are included; questions that *would have been* asked are exluded. Agreement in impact evaluation is thereby guaranteed by the impact selection process.

Comparison of the Importance of the Impacts with Their Likelihood

One might expect a bias to operate between estimates of likelihood and those of importance. The contemporary view of the degree to which the cable was viewed as a *causative factor* (Question D above) was compared with the perceived *importance* of the impact given in response to Question E—"How important was the impact (the event or trend) in making the contemporary world as we know it (regardless of whether the cable was a causative factor in its occurrence)?"

This comparison was conducted in the same fashion as that between Question A and D above. Figure III–5 shows the block diagrams corresponding to the three groups: the total of 25 participants,

TOTAL GROUP

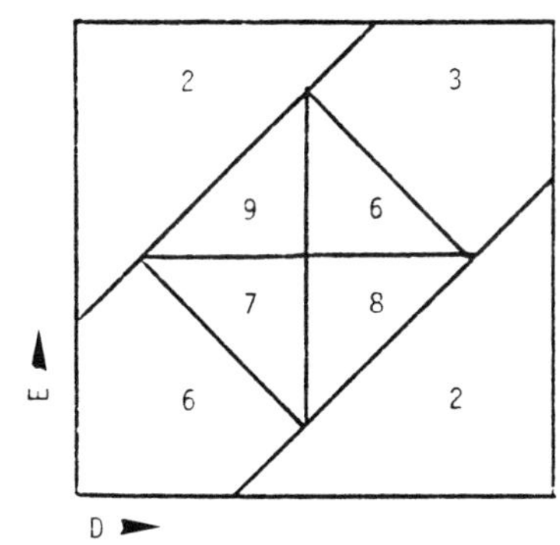

The 43 impacts are shown. There is no evidence of correlation.

HISTORIANS

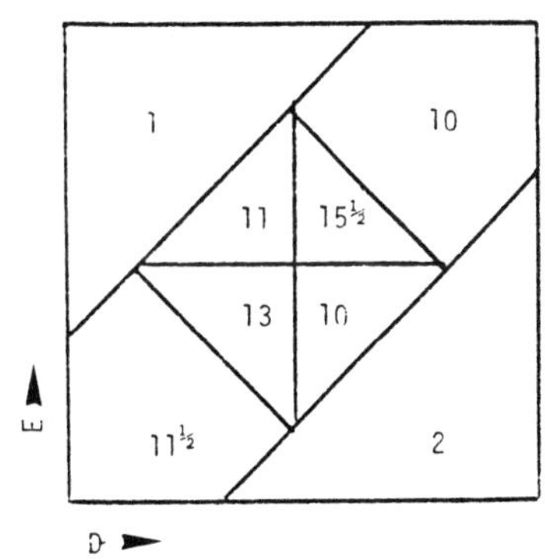

Correlated at the 0.1% level of confidence. However, 10 impacts with limited responses are more strongly correlated along the lower left diagnoal.

FINN/JARAS

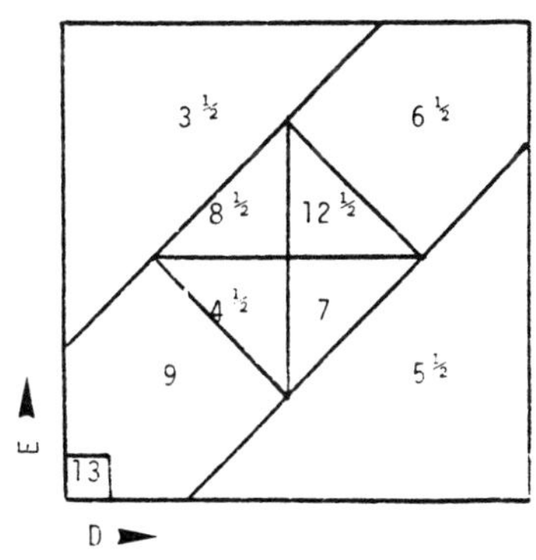

Without the 13 "irrelevant" impacts in the lower left corner, this pattern is likely 30% of the time on a random basis.

FIG. III–5.—Comparison of responses on Questions D (likelihood of a cable causation from a 1976 perspective) and E (the assessed importance of the impact to the modern world).

the six historians, and the Finn/Jaras consensus. Only the historians exhibited a significant correlation. However, if the 13 low-relevance impacts had been included in the Finn/Jaras consensus, they would have produced a strong, yet spurious, correlation. If there were two clusters of impacts—relevant and irrelevant—for the historians as well, the averaging has rendered them inseparable.

In conclusion, there is less correlation between the estimated degree of cable causality of an impact and the impact's importance than between the 1861 and 1976 views. Only the historians show a relationship, and that could be spurious.

Enumeration of the Most Likely and Most Important Impacts

The specific impacts evaluated as most likely and/or important were summarized in Sec. 4–D. Impacts were grouped in five categories: economic, military, political, social, and cultural.

On the average, the economic impacts were significantly more likely according to participant assessment (Fig. III–6). This result is interesting because economics was the first social science to emerge and was the most quantitative from the beginning.

Applying a cut to the distribution of average impact likelihoods for each of the three groups produced from ten to twenty impacts in each topic category according to the list of Delphi questions. The six impacts selected by all three groups were considered the most likely. At random an agreement on six out of 74 impacts with 10–17 choices for each group would have a probability of much less than 0.1%. All three groups agree very closely (Fig. III–7).

The three groups agreed on 7 out of 43 impacts as most important (Fig. III–8). Of the remaining 31 less likely impacts not analyzed for importance in the total sample, only one was considered important by both the historians and Finn/Jaras; that was, ''more prestige for English commerce and trade (Britain and America).'' Once again, the chance of such general agreement on seven impacts out of 43 is less than 0.1%.

Only three impacts were considered both likely and important by two or more groups. The random expectation of concurrent selections in this situation is for 2.6 events to be held in common; three or less would be expected in over 50% of random trials. If evaluations of importance were correlated with those of likelihood to the extent that the historians exhibited, one would expect six to be in common between the two lists; finding only three would occur in only 15% of random tirals. If the correlation were only somewhat stronger, eight would be expected and finding three would be inconsistent at the 5% level of significance. Therefore, the concurrence between those considered most likely and those considered important argues for the independence of these two evaluations. However, the small sample size does not provide a conclusive test.

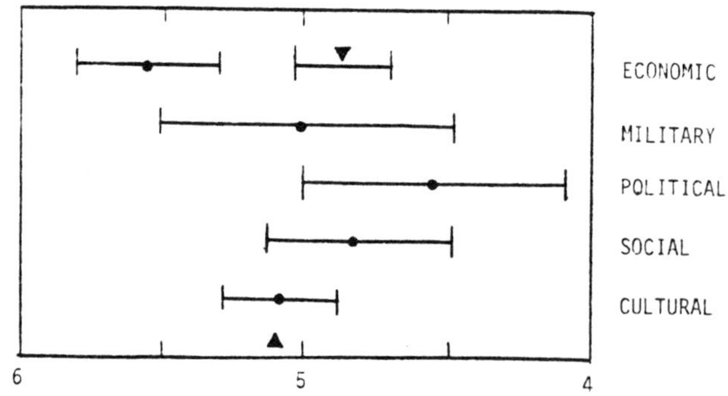

FIG. III–6.—Comparison of estimated likelihood according to subject area: ● mean of subject area, ▼ mean of all subjects *excluding* economics, ▲ mean of all subjects.

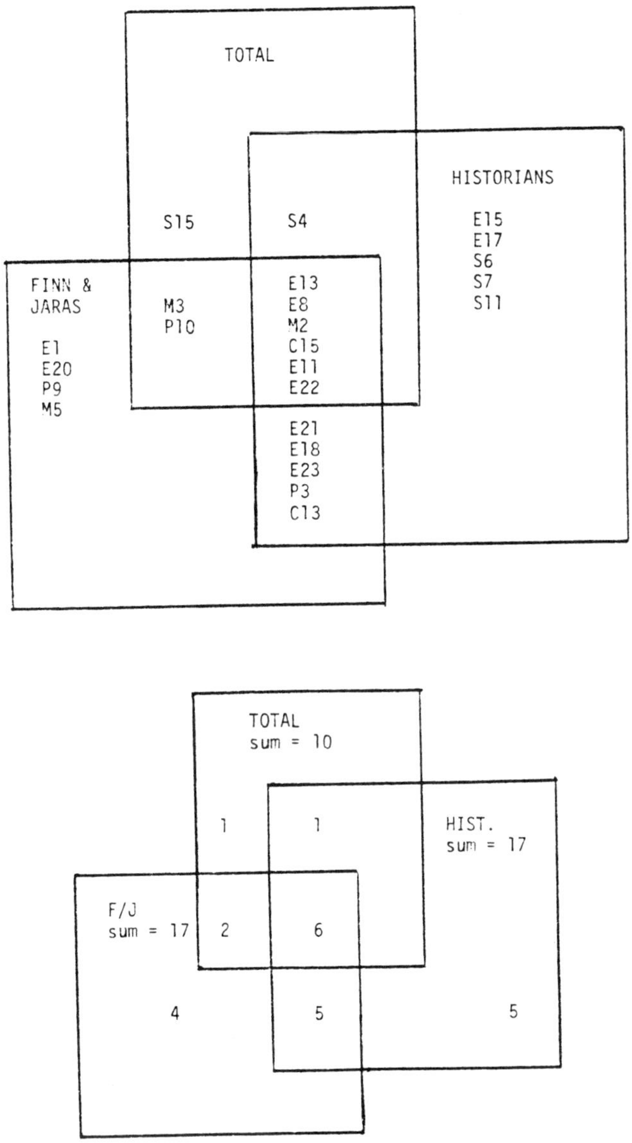

FIG. III–7.—Impacts considered most likely.

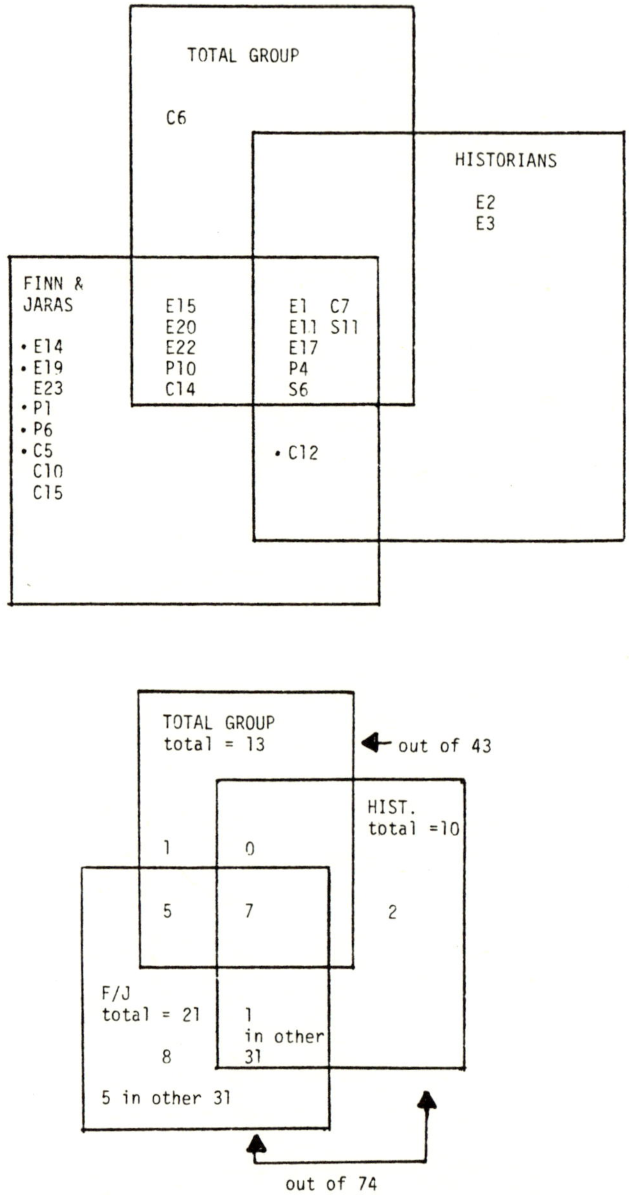

FIG.III–8.—Impacts considered very important.

Notes and References

Chapter 1

1. Samuel Carter, *Cyrus Field: A Man of Two Worlds* (New York: Putnam, 1968), 99.
2. *Charles Bright Scrapbooks* (London: Institution of Electrical Engineers, 12 vols.), I–19.
3. Robert Luther Thompson, *Wiring a Continent* (Princeton: Princeton University Press, 1947), 299.
4. Carter, op. cit., 96.
5. *Ibid.*, 96–97.
6. George Saward, *The Transatlantic Submarine Telegraph* (London: Printed for private circulation, 1878), 3.
7. Sir William Howard Russell, *The Atlantic Telegraph* (London: Dawson, 1866), 22.
8. Carter, *op. cit.*, 97.
9. *Ibid.*, 99.
10. Saward, *op. cit.*, 4.
11. George Saward, *Deep Sea Telegraphs: Their Past History and Future Progress* (London: Mechanics Magazine, 1861), 15.
12. Carter, *op. cit.*, 104.
13. James Dugan, *The Great Iron Ship* (New York: Harper, 1953), 166.
14. Saward, *Transatlantic Telegraph*, 8.
15. R. J. Mann, *The Atlantic Telegraph: A History of Preliminary Experimental Proceedings and Descriptive Account of the Present State and Prospectus of the Undertaking* (London: Harrold, 1857), 34.
16. Charles Bright, *The Story of the Atlantic Cable* (New York: Appleton, 1903), 41.
17. *Ibid.*, 13.
18. Thomas Jaras, "Promoters and public servants: The role of the British government in the expansion of submarine telegraphy" (Ph.D. Dissertation, Georgetown University, 1975), 58, fn.
19. Carter, *op. cit.*, 127.
20. Henry M. Field, *The Story of the Atlantic Telegraph* (New York: Scribners, 1892), 69.
21. Saward, *Transatlantic Telegraph*, 12.
22. Jaras, *op. cit.*, 48.
23. Field, *op cit.*, 81.
24. Saward, *Transatlantic Telegraph*, 6.
25. Carter, *op. cit.*, 131.
26. Jaras, op. cit., 57.
27. *Ibid.*
28. Isabella Field Judson, *Cyrus W. Field: His Life and Work* (New York: Harper, 1896), 76.
29. Carter, *op. cit.*, 132.
30. Fred Perkins, "Atlantic telegraphy," *Galaxy* 2 (September 1866), 92.
31. Carter, *op. cit.*, 135.
33. Saward, *Transatlantic Telegraph*, 10.
34. Carter, *op. cit.*, 133.
35. Saward, *Transatlantic Telegraph*, 14.
36. Carter, *op. cit.*, 145.
37. Saward, *Transatlantic Telegraph*, 30.
38. *Ibid.*, 33–34.
39. *Ibid.*, 33.
40. Carter, *op. cit.*, 179.
41. Saward, *Transatlantic Telegraph*, 33.
42. *Ibid.*, 42.
43. *Ibid.*

44. *Ibid.*, 43.
45. *Ibid.*
46. Jaras, *op. cit.*, 207.
47. *Bright Scrapbooks*, II–24.
48. Saward, *Transatlantic Telegraph*, 46.
49. Carter, *op. cit.*, 181.
50. *Ibid.*, 187.
51. *Ibid.*, 191.
52. *Ibid.*, 192.
53. *Ibid.*
54. *Ibid.*, 193.
55. *Ibid.*, 194.
56. *Bright Scrapbooks*, II–24.
57. Carter, *op. cit.*, 201.
58. C. T. McClenachan, *Detailed Report of the Proceedings Held in Commemoration of the Successful Laying of the Atlantic Cable, 1858* (New York: Edmund Jones and Co., 1863), 14.
59. Carter, *op. cit.*, 205.
60. *Ibid.*, 201.
61. Saward, *Transatlantic Telegraph*, 49.
62. Carter, *op. cit.*, 213.
63. Atlantic Telegraph Company, *Report and Minutes of Proceedings at the Seventh Ordinary Annual Meeting on Wednesday 10 March 1864 and Minutes of Proceedings at the Extraordinary General Meeting on Thursday 31 March 1864* (London: William Brown & Co., 1864), 5.
64. Carter, *op. cit.*, 224.
65. Saward, *Transatlantic Telegraph*, 63.
66. Phillip B. McDonald, *A Saga of the Seas* (New York: Wilson-Erickson, 1937), 170–171.

Chapter 2

1. Sir William Thomson, *Mathematical and Physical Papers* (Cambridge: University Press, 1884), II, 61–76.
2. *Ibid.*, 133–134.
3. *Ibid.*, 77–91.
4. *Ibid.*, 133–134.
5. J. E. Brittain, "The introduction of the loading coil: George A. Campbell and Michael J. Pupin," *Technology and Culture* 11:36–57, 1970.
6. Charles Bright, *Submarine Telegraphs* (London: Crosby Lockwood, 1898), 52–53.
7. R. L. Thompson, *Wiring a Continent* (Princeton: Princeton University Press, 1947), 437.
8. Bright, *op. cit.*, 35.
9. *Report to the Lords of the Committee of Privy Council for Trade*, 1861, 8.
10. *Charles Bright Scrapbooks* (London: Institution of Electrical Engineers, 12 vols.), IV–45.
11. John Canington, "The talking drums of Africa," *Scientific American* 225 (No. 6): 90–94, 1971.
12. *Report, op. cit.*
13. *Ibid.*, 9.
14. *Ibid.*
15. *Ibid.*, 36.
16. Bright, *op. cit.*, 11.
17. *Ibid*, footnote, 11.
18. *Ibid.*
19. *Report, op. cit.*, 10.
20. *Ibid.*, 32.
21. *Ibid.*, 25–26.
22. J. D. Bernal, *The Social Function of Science* (Cambridge, Mass.: MIT Press, 1939), 129.
23. Thompson, *op. cit.*, 242.
24. Ibid., 310.
25. Ibid., 311.
26. P. B. McDonald, *A Saga of the Seas* (New York: Wilson-Erickson, 1937), 113.

27. Thompson, *op. cit.*, 371–372.

28. *Ibid.*, 427–428.

29. *Ibid.*, 428.

30. *Ibid.*, 437.

31. Bright, *op. cit.*, 33, fn.

32. George Saward, *The Transatlantic Submarine Telegraph* (London: Printed for private circulation, 1878), 64.

33. *Ibid.*, 68.

34. *Ibid.*, 73.

35. *Ibid.*, 73–78.

36. Samuel Carter III, *Cyrus Field: Man of Two Worlds* (New York: Putnam, 1968), 283.

37. Thompson, *op. cit.*, 413.

38. Thompson, *op. cit.*, Ch. 28.

39. Thompson, *op. cit.*, 424.

40. *Ibid.*, 439.

41. *Ibid.*

42. Thomas Jaras, "Promoters and Public Servants: The Role of the British Government in the Expansion of Submarine Telegraphy (1860–1870)" (Ph. D. Dissertation, Georgetown University, Washinton, D. C., 1975), 41.

43. E. O. Wilson and W. H. Bossert, *A Primer of Population Biology* (Stamford, Conn.: Sinauer Associates, 1971), 157.

44. Allen Pred, *The Spatial Dynamics of U.S. Urban-industrial Growth, 1800–1914: Theoretical and Interpretive Essays* (Cambridge, Mass.: MIT Press, 1966).

Chapter 3

1. C. E. Fayle, *A Short History of the World's Shipping Industry* (London: Allen & Unwin, 1933), 263.

2. From B. R. Mitchell, *Abstract of British Historical Statistics* (London: Cambridge University Press, 1962), and B. M. Deakin, *Shipping Conferences: A Study of Their Origins, Development and Economic Practices* (London: Cambridge University Press, 1973), 19.

3. Frederick Martin, *The History of Lloyd's and of Marine Insurance in Great Britain* (New York: Burt Franklin, 1971; first published 1876), 365-373.

4. Charles Wright and E. C. Fayle, *A History of Lloyd's* (London: Macmillan, 1928), 351.

5. Martin, *op. cit.*

6. Wright, *op. cit.*, 384–387.

7. J. R. Smith, *The Organization of Ocean Commerce* (Philadelphia: University of Pennsylvania, 1905), 13–14.

8. D. E. W. Gibb, *Lloyd's of London: A Study in Individualism* (London: Macmillan, 1957), 155–157, and H. M. Grey, *Lloyd's Yesterday and To-day* (London: John Haddon, 1893), 54–55.

9. A. W. Kirkaldy, *British Shipping, Its History, Organization and Importance* (New York: Augustus M. Kelley, 1970; first published 1914), 157–160.

10. G. H. Knap, *A Century of Shipping: The History of the Royal Netherlands Steamship Company 1856–1956* (Amsterdam: De Bussy, 1957), 19.

11. *Report of the Royal Commission on Shipping Rings* (*Parl. Papers*, 1909; Cd. 4668). Testimony of Sir James L. MacKay, Director of the British India Steam Navigation Company, QQ. 19569–19838. The Commission fully endorsed this view in its report, 38.

12. Deakin, *op. cit.*, 15.

13. H. C. Emery, *Speculation on the Stock and Produce Exchanges of the United States* (New York: 1896), 96.

14. *Ibid.*, 97, quotes C. W. Smith, *Commercial Gambling* (London: 1893) and J. F. Hume, *Art of Investing* (New York: 1888). Also see Stanley Drumbell, "The origins of cotton futures," *Economic History* 1, No. 2 (May 1927) and J. H. Clapham, *Free Trade and Steel*, in *An Economic History of Modern Britain* (London: Cambridge University Press, 1932), 2:320.

15. They include: Emery, *op. cit.* (1896); J. E. Boyle, *Cotton and the New Orleans Cotton Exchange: A Century of Commercial Evolution* (Garden City, N.Y.: Country Life Press, 1934); A. B. Shepperson, *Cotton Futures* (New York: Alfred B. Shepperson, 1911); G. W. Hoffman, *Future Trading upon Organized Commodity Markets in the United States* (Philadelphia: University of Pennsylvania Press, 1932); and a number of articles on commodity exchanges in *Annals Am. Acad. Political and Social Science* (September 1911).

16. Boyle, *op. cit.*, 82.

17. Stanley Drumbell, *op. cit.*, 259–262.

18. *Report of the Commissioner of Corporations on Cotton Exchanges*, Washington, D. C., 1909, 1: 39–40.

19. *Ibid.*, 40.

20. Hoffman, *op. cit.*, 40.

21. *Ibid.*, 43.

22. Thomas Ellison, *The Cotton Trade of Great Britain: New Impression* (London: Frank Carr, 1968, first published 1886), 269; Arthur Marsh, "Cotton exchanges and their economic functions," *Annals Am. Acad. Political and Social Science* 38:274–275, 1911; *Report of Commissioner of Corporations, supra,* 40–42.

23. S. S. Huebner, "The functions of produce exchanges," *Annals Am. Acad. Political and Social Science* 38:342–343, 1911.

24. *Ibid.*, 332–333, 347–349.

25. J. E. Boyle, *Chicago Wheat Prices for Eighty-one Years: Daily, Monthly and Yearly Fluctuations and Their Causes* (Ithaca, N.Y.: Cornell University Press, 1922); Hoffman, *op. cit.,* 428–429.

26. Hoffman, *op. cit.,* 44.

27. *New York Cotton Exchange Annual Reports*, 1874–1898.

28. Hoffman, *op. cit.,* 33–34, 45–47.

29. J. H. Clapham, *op. cit.,* 381.

30. Boyle, *New Orleans Cotton Exchange*, 86–91.

31. See evidence of Robert Palmer Harding, Assistant Secretary to the Board of Trade, and of George Lord, Director of the Manchester Chamber of Commerce, in *Report of the Royal Commission on Depression of Trade and Industry (Parl. Papers*, 1886, vol. 21: C 4621), Q. 49 and enclosure Appendix A, 105.

32. Alan Barnard, *The Australian Wool Market 1840–1900* (London and New York: Cambridge University Press, 1958); based on importer's commission of 2½% from 1856 to 1870, 2% from 1871 to 1872, and thereafter 1½%. (Table XXIV, 227.)

33. Franklin Escher, *Modern Foreign Exchange* (New York: Macmillan, 1932), 126.

34. S. S. Pratt, *The Work of Wall Street* (New York: Appleton, 1903), 115.

35. *Ibid.*, 141–142.

36. *Report of the Commissioners of the London Stock Exchange Commission (Parl. Papers* 1878, 29: C2157), QQ 8772–8831.

37. E. C. Stedman, Ed., *The New York Stock Exchange*, first published 1905 (New York: Greenwood Press, reprint, 1969), 195; N.S.B. Gras and H. M. Larson, "The development of the New York Stock Exchange," in *Casebook in American Business History* (New York: F. S. Crofts, 1939), 337–338.

38. *Report of the Royal Commission on the Depression of Trade and Industry Parl. Papers*, 1886, 21: C 4621, QQ 888–891).

39. J. C. Brown, *A Hundred Years of Merchant Banking* (New York, privately printed, 1909), 287.

40. M. G. Myers, *The New York Money Market,* in *Origins and Development* (New York: Columbia University Press, 1931), 1:348.

41. A. H. Cole, "Seasonal variation in sterling exchange," *Journal of Economic and Business History*, 1:204, 1929–30.

42. *Ibid.*, 213.

43. *Report on the Maritime Canal Connecting the Mediterranean at Port Said with the Red Sea at Suez*, by Capt. Richards, R.N., and Lt. Col. Clark, C.B., Rt. (*Parl. Papers*, 1870, 44: C 92). They stressed the advantages of shortening the route to China by 36 days coupled with the new cable; 2.

44. N. A. Pelcovits, *Old China Hands and the Foreign Office* (New York: King's Crown Press, 1948), 102–103.

45. Report of the Royal Commission on Depression of Trade and Industry (Parl. Papers, 1886, 23: C 4794), par. 464, 472.

46. Pratt, *op. cit.*, 139.

47. *Ibid.*, 142–145.

48. *The History of The Times*, *The Traditional Establishment* (London: The Times, 1939), 2:87–88; Graham Storey, *Reuter's Century 1851–1951* (London: Max Parrish, 1951), 6.

49. H. R. F. Bourne, *English Newspapers: Chapters in the History of Journalism* (London: Chatto and Windus, 1887), 2:255–257.

50. *The History of The Times*, 83.

51. See R. L. Thompson, in *Wiring A Continent: The History of the Telegraph Industry in the United States 1832–1866* (Princeton, N. J.: Princeton University Press, 1947), for a detailed account of domestic U. S. telegraph expansion.

52. F. L. Mott, *American Journalism: A History of Newspapers in the United States through 250 years, 1690–1940* (New York: Macmillan, 1941), 387; and A. M. Lee, *The Daily Newspaper in America: The Evolution of a Social Instrument* (New York: Macmillan, 1937), 508.

53. *History of The Times*, 90–91.

54. D. G. O. Ayerst, *The Manchester Guardian: Biography of a Newspaper* (Ithaca, N. Y.: Cornell University Press, 1971), 142–143.

55. *Ibid.*, 144. No record exists of Pender's reply.

56. Lee, *op. cit.*, 503–504; and Oliver Grambling, *AP: The Story of News* (New York: Farrar & Rinehart, 1940), 20–26.

57. Storey, *op. cit.*, 33.

58. *History of The Times*, 271.

59. Grambling, *op. cit.*, 34; and *History of The Times*, 272.

60. P. R. Knights, *The Press Association War of 1866–1867* (Austin, Tex.: Association for Education in Journalism, 1967), 48.

61. H. W. Baehr Jr., *The New York Tribune Since the Civil War* (New York: Dodd, Mead, 1936), 32–34.

62. Grambling, *op. cit.*, 88.

63. Baehr, *op. cit.*, 39, 76; and Mott, *op. cit.*, 380–381.

64. C. S. Diehl, *The Staff Correspondent* (San Antonio, Tex.: The Clegy Company, 1931), 271–274; The Associated Press, *"MES" His Book*, 160.

65. Lee, *op. cit.*, 528; and Grambling, *op. cit.*, 287.

66. D. L. Shaw, "News bias and the telegraph: A study of historical change," *Journalism Quarterly* 44: 5–6, Spring 1967.

67. *Ibid.*, 3–5.

68. *History of The Times*, 187–189.

69. Sir Edward Cook, *The Press in War-time* (London: Macmillan, 1924), 21–22.

70. Grambling, *op. cit.*, 142.

71. Storey, *op. cit.*, 112–115.

72. H. S. Foster Jr., "Studies in America's News of the European War" (Ph.D. Dissertation, University of Chicago, June 1932). The author documents the importance of the Entente Powers' advantage in having almost exclusive access to the U. S. public through control of the bulk of transatlantic communications. See Walter Millis, *Road to War: America 1914–1917* (Boston: Houghton Mifflin, 1935), for a more general approach to the American position.

73. Kent Cooper, *The Story of the News Agency Epoch: Barriers Down*, first published 1942 (Port Washington, N. Y.: Kennikat Press, 1969), 45–47. The author, a leading figure in postwar AP expansion, relates the AP confrontation with Reuters. Sir Roderick Jones, the head of Reuters, provides the British view in *A Life in Reuters*.

74. This observation was made in 1913 by AP president Frank B. Noyes, in *The Associated Press: An Article Relating to the Methods of Operation, Organization and Collection and Distribution of News Matter by the Associated Press* (Washington, D. C.: GPO, 1913), 5.

75. This is one of several definitions of diplomacy discussed by Lord William Strang in *The Foreign Office* (Washington, D. C.), 115.

76. G. H. Stuart, *American Diplomacy and Consular Practice*, 2d ed. (New York: Appleton-Century-Crofts, 1952, first published 1936), 167.

77. Strang, *op. cit.*, 115.

78. S. S. Cox, *The Folly and Cost of Diplomacy* (Washington, D. C.: 1874), 4. See W. F. Ilchman, *Professional Diplomacy in the United States 1779–1939: A Study in Administrative History* (Chicago: University of Chicago Press, 1961), 20–21, for other attacks on diplomatic representation abroad.

79. Agatha Ramm, Ed., *The Political Correspondence of Mr. Gladstone and Lord Granville*, I:240.

80. Schenck to Department of State, 7 May 1872.

81. U. S. Department of State, *Instructions to the Diplomatic Officers of the United States* (Washington, D. C.: Department of State, 1897), 37.

82. Another $5484.72 was accounted for as telegram expenditures but according to the figures through 1880 a maximum of $641 per year was expended on telegrams as opposed to cablegrams. It is likely most of this

$5484.72 was also spent on cablegrams in 1867. Figures are from ''Statement of the expenditures for the contingency fund for the expenses of foreign intercourse, including the contingent expenses of all missions abroad,'' provided annually in *House Documents* from 1866 to 1887 (excluding 1885) and from 1907 to 1917.

83. U. S. Department of State, *Outline of the Organization and Work of the Department of State* (Washington, D. C.: 1911), 151–154.

84. Stuart, *op. cit.*, 51.

85. Lord Newton, *Lord Lyons: A Record of British Diplomacy* (London: Edward Arnold, 1913), I: 77–78.

86. R. J. Cain, ''Telegraph Cables in the British Empire, 1850—1900'' (Ph.D. Dissertation, Duke University, 1971), 189.

87. *Ibid.*, 188–189. Also see B. L. Blakely, *The Colonial Office 1868–1892* (Durham, N. C.: Duke University Press, 1972), 116–117.

88. *First Report from the Select Committee on Diplomatic and Consular Services* (238) (*Parl. Papers*, 1871, 7:12–13).

89. J. W. Cell, *British Colonial Administration in the Mid-19th Century: The Policy-Making Process* (New Haven, Conn.: Yale University Press, 1970), 41.

90. *Ibid.*, 41–43.

91. *Ibid.*, 41.

92. *Ibid.*, 36–37.

93. R. V. Kubicek, *The Administration of Imperialism: Joseph Chamberlain at the Colonial Office* (Durham, N. C.: Duke University Press, 1969), 30.

94. Blakely, *op. cit.*, 65.

95. Cain, *op. cit.*, 195.

96. Kubicek, *op. cit.*, 33, 197.

97. Stuart, *op. cit.*, 26–27, 174.

98. Department of State, *The Department of State, 1930–1955* (Washington, D. C.: GPO, 1960), 30–31.

99. Strang, *op. cit.*, 41.

100. *Ibid.*, 40–41.

101. *Ibid.*

102. G. N. Clark, *Unifying the World* (London: Swarthmore Press, 1920), 98–99.

103. G. A. Schreiner, *Cables and Wireless* (Boston: Stratford Co., 1924), Introduction, 5.

104. Naval War College, *Submarine Telegraph Cables in Their International Relations*, lectures by G. W. Wilson (Washington, D.C.: GPO, 1901), 14.

105. *Ibid.*, 11.

106. John Colombos, *The International Law of the Sea*, 6th ed., (New York: Longmans, Green, 1967), 539.

107. Naval War College, *op. cit.*, 32.

108. *Ibid.*, 12.

109. Colombos, *op. cit.*, 538.

110. Naval War College, *op. cit.*, 32.

111. Colombos, *op. cit.*, 539.

112. Schreiner, *op. cit.*, 29; and Colombos, *op. cit.*, 536.

113. Michael Howard, *Studies in War and Peace* (New York: Viking Press, 1971).

114. *Ibid.*, 185.

115. *Ibid.*

116. *Ibid.*, 107.

117. B. H. Liddell Hart, *Strategy* (New York: Praeger, 1954; 2d ed., 1967).

118. *Ibid.*, 189.

119. W. L. McElwell, *The Art of War: Waterloo to Mons* (Bloomington, Ind.: Indiana University Press, 1974).

120. *Ibid.*

121. J. F. C. Fuller, *War and Western Civilization, 1832–1932* (Freeport, N.Y.: Books for Libraries Press, 1932).

122. Howard, *op. cit.*, 109.

123. Fuller, *op. cit.*, 72.

124. G. S. Graham, *The Politics of Naval Supremacy: Studies in British Maritime Ascendancy* (London:

Cambridge University Press, 1965), 14.

125. *Ibid.*, 28–29.

126. J. W. Brett, *The Origin and Progress of Brett's Submarine Oceanic and Subterranean Electric Telegraph* (London, 1858), 171.

127. *Report of the Inter-departmental Committee on Cable Communications* (*Parl. Papers*, 1902, Cd 1056), Appendix D, 54.

128. Charles Bright, *Submarine Telegraphs: Their History, Construction and Working* (London: Crosby Lockwood and Son, 1898), 153–154.

129. *Minutes of Evidence Taken Before the Inter-departmental Committee on Cable Communications* (*Parl. Papers*, 1902, Cd 1118), Q 1551.

130. *Report of the Select Committee on the Measures Resorted to or Which Were Available, and as to the Laws of Communication Adopted for Reenforcing Our Army, During the Pending Revolt in India (382) (Parl. Papers*, 1857–58, 10), vii.

131. *Ibid.*

132. *Ibid.*, xvi, xxvi.

133. P. H. Colomb, *Essays on Naval Defense* (London: Allen and Co., 1896), 14–16.

134. *Ibid.*

135. *Report from the Select Committee on Colonial Military Expenditures (423)(Parl. Papers*, 1861, vol. 13 Q 3154).

136. D. C. Gordon, *The Dominion Partnership in Imperial Defense 1870–1914* (Baltimore, Md.: Johns Hopkins Press, 1965), 61. Quoted from Sir Henry Lucy, *A Diary of Two Parliaments* (London: 1885–86), 1:429.

137. *Distribution and Mobilization of the Fleet (Parl. Papers*, 1905, vol. 28, Cd 2335), 1.

138. *Ibid.*, 2–4; and A. J. Marder, *From the Dreadnought to Scapa Flow: The Royal Navy in the Fisher Era, 1904–1919*, in *The Road to War 1904–1914* (London: Oxford University Press, 1961), 1:40–43.

139. P. K. Kemp, Ed., *The Papers of Admiral Sir John Fisher* (London: Navy Records Society, 1960), 1, 161.

140. *Proceedings of Conference Between Secretary of State for the Colonies and the Premiers of the Self-governing Colonies (Parl. Papers*, 1897, vol. 59), 16–17.

141. *Report of S.C. on Colonial Military Expenditures (423) (Parl. Papers*, 1861, vol. 13), Q 3085.

142. *Proceedings of the Colonial Conference 1887 (Parl. Papers*, 1887, vol. 56, C 5091), 1:9.

143. *Telegram from the Adjutant-General to the Forces to the General Commanding in Egypt, Relative to a Force to be Collected at Suakein (Parl. Papers*, 1884, vol. 48, C 3889), 3–4.

144. The information on AEF communications is taken from *Cablegrams Exchanged Between General Headquarters, American Expeditionary Forces and the War Department, 1917–1919*, National Archives Microfilm Publications Pamphlet Describing M930 (Washington: National Archives and Records Service, 1973), 1–8.

145. Bronsart von Schellendorff, *The Duties of the General Staff* (London: HMSO, 1908, 3rd ed.; first published 1857), 2.

146. J. D. Hittle, *The Military Staff: Its History and Development* (Harrisburg, Pa.: The Stackpole Co., 1961, 3rd ed., first published 1944), 78.

147. Schellendorff, *op. cit.*, 5–6.

148. *Ibid.*

149. Laurence Lafore, *The Long Fuse: An Interpretaion of the Origins of World War I* (Philadelphia, New York and Toronto: Lippincott, 1971, 2nd ed.).

150. *Conference with Representatives of the Self-governing Dominions on the Naval and Military Defence of the Empire (Parl. Papers*, 1909, vol. 59), 32–33.

151. See ONI's series "Information from Abroad No. III," on "Report of British Naval and Military Operations in Egypt, 1882," published in 1885, held by the U.S. Navy Department Library.

152. Michael Howard, *The Franco-Prussian War* (London: Collins, 1967).

153. Tim Nenninger of the National Archives says that the Navy used the cable more than did the Military Intelligence Department before 1918, and the bulk of the MID files were destroyed in the 1930s (*National Archives*, Record Group 38, Entry 101); Schroeder, Chief Intelligence Officer to Naval Attaché, St. Petersburg, 1 March 1905, 265.

154. Kahn, *op. cit.*, 252.

155. R. G. 38, Entry 101. Schroeder to N/A, St. Petersburg, 17 March 1904, ff.

156. *Ibid.*, Schroeder to N/A, St. Petersburg, 1 March 1905, 264.

157. *Ibid.*, Schroeder to N/A, St. Petersburg, 22 April 1905, 315; N/A, Tokyo, 31 May 1905, 363; N/A, St. Petersburg, 2 June 1905, 366. Information was readily passing from Tokyo to Washington. Schroeder wrote Marble in Tokyo on 18 July 1905 that there had been confusion, but not delays, on the cables incoming from Tokyo.

158. *Ibid.*, Schroeder to N/A, Vienna, 18 March 1904, 14.

159. *Ibid.*, Schroeder to N/A, Paris, 14 February 1905.

160. James Gerard, *My Four Years in Germany* (New York: 1917), 58.

161. *Ibid.*, 56.

162. Kahn, *op. cit.*, 262.

163. Admiral Sir William James, *The Code-breakers of Room 40* (New York: 1956), 206.

164. Kahn, *op. cit.*, 284.

165. Barbara Tuchman, *The Zimmermann Telegram* (New York: 1967), 8–14. Kahn, *op. cit.*, 285–293.

166. Tuchman, *op. cit.*, 151–152.

167. *Ibid.*, 161–169.

168. Sir George Cockerill, *What Fools We Were* (London: 1944), 16–21.

169. Quoted in: Arthur Marwick, *The Deluge* (Boston: 1965), 36–37.

170. Cockerill, *op. cit.*, 22–25.

171. *Public Records Office*, London. Cab 38/87, ''Forecast of the first phase of a war between Russia and Japan;'' Cab 38/6/100, ''Control of the press in time of war or threat of war,'' 12 October 1904, W.O., 6 pp; Cab 38/8/12, ''Control of naval and military information;'' Cab 38/87.

172. Cab 38/3/72, ''Use by private persons of wireless telegraph stations in war-time.'' 18 November 1903.

173. Cab 38/11/1, ''Ceylon cable.'' H.H.A., 9 January 1906.

174. Cab 38/12/47; Cab 38/12/51; Cab 38/12/55.

175. Cab 16/8, ''The Report of the Sub-committee to consider the question of Foreign Espionage in the United Kindom,'' July 1909.

176. Cab 38/15/14, 16.

177. Cab 4/3/2/128B, ''Establishments of a chain of wireless stations throughout the Empire,'' June 1911.

178. Cab 38/24/26; Tuchman, *op. cit.*, 15.

179. V F. Lenzen, *Benjamin Peirce and the U.S. Coast Survey* (San Francisco: San Francisco Press, 1968).

180. Susan Schlee, *The Edge of an Unfamiliar World: A History of Oeanography* (New York: Dutton, 1973), 23*ff*.

181. M. F. Maury, *The Physical Geography of the Sea* (New York: Harper, 1855), 210.

182. American Society of Civil Engineers, *Standard Time for the United States of America, Canada and Mexico* (New York: 1882), 8–9. Report by Sandford Fleming, 17 January 1882.

183. *Ibid.*, 12.

184. *Ibid.*, 23.

185. *Ibid.*, 24.

186. *Proceedings of the International Meridian Conference* (House of Representatives, 48th Cong., 2d sess., Executive Document No. 14), 57.

187. G. M. Beard, ''Causes of American Nervousness,'' from *American Nervousness, Its Causes and Consequences* (New York, 1881), 96–99, quoted in H. N. Smith, Ed., *Popular Culture and Industrialism 1865–1890* (Garden City, N.Y.: Doubleday, 1967), Document 7.

188. Thoreau, *op. cit.*

Chapter 4

1. J. B. Bury, *The Idea of Progress* (New York: Macmillan, 1932), Chap. 18.

2. A. Ekirch Jr., *The Idea of Progress in America, 1815–1860* (New York: Peter Smith, 1951), Chap. 4.

3. Leo Marx, *The Machine in the Garden: Technology and the Pastoral Ideal in America* (New York: Oxford University Press, 1964).

4. Ekirch, *op. cit.*

5. *Ibid.*, 111.

6. *Ibid.*, Chap. 3.

7. *Ibid.*

8. Asa Briggs, Ed., *The Nineteenth Century: Contradictions of Progress* (London: Thames and Hudson, 1970), 332–336 and 340 *ff*.

9. Marx, *op. cit*.

10. Ekrich, *op. cit*., Chap 6; and Briggs, *op. cit*., 320 *ff*.

11. Briggs, *op. cit*., 28.

12. J. A. Froude, *Oceania or England and Her Colonies* (New York: Scribner and Sons, 1866).

13. H. B. Franklin, *Future Perfect: American Science Fiction of the Nineteenth Century* (New York: Oxford University Press, 1966).

14. Ekrich, *op. cit*., Chap. 3.

15. Henry Field, *The Story of the Atlantic Telegraph* (New York: Scribners, 1892), 108.

16. *Ibid*., 110–111.

17. *Ibid*., 100–101.

18. *Congressional Globe*, 34th Congress, 3d Session, vol. 26, 1857, 421.

19. Samuel Carter III, *Cyrus Field: A Man of Two Worlds* (New York: Putnam, 1968), 193.

20. *Congressional Globe*, 398.

21. George Saward, "Deep sea telegraphs: Their past history and future progress," *Mechanics Magazine*, 46–48, 1861.

22. *Congressional Globe*, 425.

23. *Ibid*., 397.

24. W. H. Russell, *The Atlantic Telegraph* (London: Dawson, 1866), 45.

25. Cyrus Field, "Prospects of the Atlantic cable," paper read before the American Geographical and Statistical Society, 1862, 3–4.

26. *Congressional Globe*, 398.

27. *Ibid*., 755.

28. *Ibid*., 416.

29. *Ibid*., 398.

30. I. F. Judson, *Cyrus W. Field: His Life and Work* (New York: Harper, 1896), 102.

31. Carter, *op. cit*., 166.

32. "The oceanic telegraph," *North American Review* 87: 539–540, 1858.

33. *Commercial and Financial Chronicle* 1: 195, 12 August 1865.

34. C. T. McClenachan, *Detailed Report of the Proceedings Held in Commemoration of the Successful Laying of the Atlantic Cable, 1858* (New York: Edmund Jones, 1863), 7.

35. *Ibid*., 271.

36. *Ibid*., 77.

37. "Criticism of H. M. Field's history of the Atlantic cable," *Southern Review*, 49–50, July 1867.

38. R. J. Mann, *The Atlantic Telegraph : A History of Preliminary Experimental Proceedings and Descriptive Account of the Present State and Prospectus of the Undertaking* (London: Jarrold and Sons, 1857), 6.

39. *Ibid*., 7.

40. Field, *op. cit*., 103.

41. *Congressional Globe*, 396.

42. Carter, *op. cit*., 132.

43. *Congressional Globe*, 416.

44. Field, *op. cit*., 104–105.

45. *Ibid*., 107.

46. McClenachan, *op. cit*., 77.

47. *Ibid*., 75–76.

48. *Ibid*., 7.

49. *Congressional Globe*, 397.

50. *Ibid*., 423.

51. *Ibid*., 244–245.

52. McClenachan, *op. cit*., 246–247.

53. *Ibid*., ii.

54. *Ibid*., 204.

55. J. W. Brett, *On the Origin and Progress of the Oceanic Electric Telegraph* (London: Nassau Steam Press, 1858), 104.

56. "Criticism of H. M. Field's history," *op cit.,* 45.

57. *Ibid.*

58. McClenachan, *op. cit.*, 199–200.

59. *Ibid.*, 273–274.

60. *Congressional Globe*, 415.

61. W. C. Anderson, *The Ocean Telegraph: A Discourse Delivered in the First Presbyterian Church, San Francisco* (San Francisco: Towne and Bacon, 1858), 7.

62. McClenachan, *op. cit.*, 72.

63. *Ibid.*

64. Judson, *op. cit.*, 102.

65. Field, *op. cit.*, 373.

66. H. D. Thoreau, *Walden* (New York: Cromwell, 1961), 67.

67. McClenachan, *op. cit.*, 134.

68. *Ibid.*, 72.

69. *Ibid.*, 160.

70. *Ibid.*, 33.

71. *Ibid.*, 212.

72. *Ibid.*, 208–209.

73. *Ibid.*, 232.

74. "The oceanic telegraph."

75. McClenachan, *op cit.*, 244–245.

76. *Ibid.*, 72.

77. *Ibid.*

78. *Ibid.*, 160

79. *Ibid.*, 72.

80. Anderson, *op. cit.*, 7.

81. Carter, *op. cit.*, 106.

82. McClenachan, *op. cit.*, 128.

83. *Ibid.*, 121–122.

84. J. B. Bury, *The Idea of Progress* (New York: Macmillan, 1932).

85. K. F. Schuessler, "Prediction," *International Encyclopedia of Social Sciences*, vol. 12, 418–425.

86. Victor Zarnowitz, "Prediction and forcasting," *International Encyclopedia of Social Sciences*, vol. 12, 425-437.

87. D. J. Boorstin, *The Americans, The Democratic Experience* (New York: Random House, 1973, vol. 3, 167.

88. *Ibid.*

89. André Marchal, "Juglar, Clément," *International Encyclopedia of Social Sciences*, vol. 8, 324–326.

90. S. N. Eisenstadt, "Development of sociological thought," *International Encylcopedia of Social Sciences*, vol. 15, 23–35.

91. Bernard Lécuyer and A. R. Oberschall, "Early history of social research," *International Encyclopedia of Social Sciences*, vol. 15, 36–51.

92. Schuessler, *op. cit.*, 418.

93. Boorstin, *op. cit.*, 200.

Chapter 5

1. Compiled from *Lists of Cables Forming the World Submarine Network* published by the International Telecommunications Union (and its predecessors) in Geneva at intervals 1877-1972.

2. *Ibid.*

3. Charles Bright, *Submarine Telegraphs* (London: Crosby Lockwood & Son, 1898), 592.

4. *Ibid.*, 105, 540.

5. *Ibid.*, 118, 604.

6. *Ibid.*, 123.

7. *Ibid.*, 552, 662.

8. *Ibid.*, 141.

9. J. W. Millnor, "Submarine cable telegraphy," *J. AIEE* 41:118–136, 1922; H. H. Haglund, "Atlantic telegraph cable centennial," *Western Union Tech. Rev.* 88:81–102, 1958.

10. H. H. Haglund, "Ocean cable crossroads," *Western Union Tech. Rev.* 83:106–110, 1953. I. Coggeshall, "Submarine telegraphy," *J. Am. IEE* 49: 217–220, 1930.

11. H. Kingsbury and R. A. Goodman, "Methods and equipment in cable telegraphy," *J. IEE* 70:477–521, 489, 1932.

12. C. H. Cramer, "Submerged repeaters for long submarine cables," *Western Union Tech. Rev.* 81:90, 1951; "Submerged telegraph repeaters come of age," *ibid.* 86:176–184, 1956.

13. J. D. Scott, *Siemens Brothers, 1858–1958* (London: Weidenfeld and Nicholson, 1958), 41.

14. "Transatlantic submarine telephone cable system," *Bell System Tech. J.* 36:1-326, 1957.

15. *Lists*, note 1.

Chapter 6

1. Talcott Parsons, "The present status of 'structural-functional' theory in sociology," in L. A. Coser, Ed., *The Idea of Social Structure* (New York: Harcourt-Brace-Jovanovich, 1975), 74–79.

2. *Ibid.*

Appendix I

1. L. F. Richardson, "Mathematical psychology of war," *Nature* 135: 830–831, 1025, 1935; reprinted in J. R. Newman, Ed., *The World of Mathematics* (New York: Simon and Schuster, 1956).

2. Walter Christaller, *Die zentralen Orte in Süddeutschland: Eine ökonomisch-geographische Untersuchung über die Gesetzmässigkeit der Verbreitung und Entwicklung der Siedlungen mit städtischen Funktionen* (Jena: Gustav Fischer Verlag, 1933); and August Losch, *The Economics of Location* (New Haven, Conn.: Yale University Press, 1954).

3. B. J. L. Berry and F. E. Horton, *Geographic Perspectives on Urban Systems* (Englewood Cliffs, N.J.: Prentice Hall, 1970), 170–175.

4. B. J. L. Berry, *Geography of Market Centers and Retail Distribution* (Englewood Cliffs, N.J.: Prentice Hall, 1976), 26–40.

5. Allen Pred, *Urban Growth and the Circulation of Information: The United States System of Cities, 1790–1840* (Cambridge, Mass.: Harvard University Press, 1973).

6. D. S. Wilson, "Evolution of the level of communities," *Science* 192: 1358–1360, 1976.

7. John von Neumann and Oskar Morgenstern, *The Theory of Games and Economic Behavior* (Princeton, N.J.: Princeton University Press), 1944.

8. J. K. Galbraith, *The Age of Uncertainty* (Boston: Houghton Mifflin, 1977), 32.

9. *Ibid.*, p. 28; T. R. Malthus, "An essay on the principle of population as it affects the future improvement of society" (1798), reprinted in part in Newman (note 1), vol. 2, pp. 1189–1199.

10. *Ibid.*, p. 32; and Galbraith's comments on the BBC series "The Age of Uncertainty."

11. *Ibid.*, pp. 32–33.

12. *Ibid.*, pp. 33 and 40.

13. Augustin Cournot, *Recherches sur les principles mathématiques de la théorie des richesses* (1838); and W. S. Jevons, *The Theory of Political Economy and the Principles of Science* (1871), see Newman, *op. cit.*, pp. 1201–1237; and Vilfredo Pareto, *Course d'économique politique* (Lausanne and Paris, 1897).

14. K. J. Arrow, "Economic welfare and the allocation of resources for invention," in *The Rate and Direction of Inventive Activity* (Princeton, N.J.: Princeton University Press, 1962), 609–625.

15. K. J. Arrow, *Essays in the Theory of Risk Bearing*, Amsterdam: North-Holland, 1971.

16. The "competitive exclusion principle" of systems ecology has been stated by E. O. Wilson and W. H. Bossert in *A Primer of Population Biology* (Sinauer Associates, 1971) as "No two species that are ecologically identical can long coexist"; in other words, "competitive exclusion occurs if one species produces enough individuals to prevent the population of the other from increasing."

17. Richardson, *op. cit.*, note 1.

18. R. H. May, *Stability and Complexity in Model Ecosystems* (Princeton, N.J.: Princeton University Press, 1973), 5, 26–30, 102–107.

19. *Ibid.*, pp. 18, 62–68; Richard Levins, "Evolution in communities near equilibrium," in M. L. Cody and J. M. Diamond, Eds., *Evolution of Communities* (Cambridge, Mass.: Harvard University Press, 1975), 16–50.

20. May, note 18, pp. 62–69.

21. May, note 18—a theme throughout this book, summarized on pp. 172–174. Also see the quantitative representation by J. W. MacArthur, "Environmental fluctuations and species diversity," in Cody and Diamond, note 19, pp. 74–80. Also see M. R. Gardner and W. R. Ashby, "Connectance of large dynamic

(cybernetic) systems: Critical values for stability,'' *Nature* 228: 784, 1970.

22. Marvin Zelen, Ed., *Statistical Theory of Reliability* (Madison: University of Wisconsin Press), 1963; see esp. Chap. 1 by G. H. Weiss, and Chap. 2 by Frank Proscham. These principles have been observed in ecological systems in the form of ''predator switching'' in the food web according to prey abundances.

23. E. P. Odum, ''The emergence of ecology as a new integrative discipline,'' *Science* 195: 1289–1293, 1977. See also D. J. Rapport and J. E. Turner, ''Economic models in ecology,'' *Science* 195: 367–373, 1977.

24. François Jacob, ''Evolution and tinkering,'' *Science* 196: 1161–1166, 1977.

Typeset by Friedrich Typography
Santa Barbara, California